Learning Materials in Biosciences

Learning Materials in Biosciences textbooks compactly and concisely discuss a specific biological, biomedical, biochemical, bioengineering or cell biologic topic. The textbooks in this series are based on lectures for upper-level undergraduates, master's and graduate students, presented and written by authoritative figures in the field at leading universities around the globe.

The titles are organized to guide the reader to a deeper understanding of the concepts covered.

Each textbook provides readers with fundamental insights into the subject and prepares them to independently pursue further thinking and research on the topic. Colored figures, step-by-step protocols and take-home messages offer an accessible approach to learning and understanding.

In addition to being designed to benefit students, Learning Materials textbooks represent a valuable tool for lecturers and teachers, helping them to prepare their own respective coursework.

Francisco J. Sáez • Iker Badiola

Essential Oral Histology

Springer

Francisco J. Sáez
School of Medicine and Nursing
University of the Basque Country UPV/EHU
Leioa, Spain

Iker Badiola
School of Medicine and Nursing
University of the Basque Country UPV/EHU
Leioa, Spain

ISSN 2509-6125 ISSN 2509-6133 (electronic)
Learning Materials in Biosciences
ISBN 978-3-032-03164-8 ISBN 978-3-032-03165-5 (eBook)
https://doi.org/10.1007/978-3-032-03165-5

This Springer imprint is published by the registered company Springer Nature Switzerland AG
The registered company address is: Gewerbestrasse 11, 6330 Cham, Switzerland

Preface

Health science studies at university level are designed to provide training in basic disciplines that deal with the organisation and functioning of the human body. One of these disciplines is Histology, the science that studies the structure of tissues and how they are organised to form the organs of our body. These basic disciplines provide the knowledge of the organisation of the human body necessary to be able to deal with the different specialities of the health sciences. In this context, dentistry students and dentists need a basic knowledge of Histology in order to be able to develop their profession.

This book is aimed at practising and future dentists, and attempts to present in a clear and concise manner the basic information on the subject. Furthermore, rather than presenting the information as complex texts, the book has a concise presentation in outline format to facilitate the student initially to gain an overview of each topic and later to review them. This schematic format is important nowadays because health science curricula have greatly reduced the time devoted to the study of basic disciplines.

This manual focuses on the histology of the basic tissues of the human body (General Histology) and the tissues and structures of the oral cavity (Oral Histology). The treatment of all organs and systems of the human body (Special Histology or Microscopic Organography) has been omitted, as they are not normally included in dental curricula. However, in the topic dedicated to nervous tissue, the histological organisation of nerves and nerve endings is described, as reference is made to innervation when studying oral histology. For the same reason, a topic dedicated to the circulatory apparatus has been included in which the blood vessels that can be found in the oral cavity are analysed in some detail.

Human Histology is an eminently visual discipline. The basic tool for studying histology is the microscope. The first histological investigations completed their descriptions with detailed drawings. Later, these drawings were replaced by photographs taken with the microscope (micrographs). The description of any histological structure is much easier to explain and understand if it is accompanied by explanatory images, whether micrographs or illustrations, and in this book, an effort has been made to include explanatory illustrations of what is described in the text.

Many histology textbooks and numerous articles published in specialised journals have been consulted in the preparation of this book. Some of the books used as sources of information are listed here. This list, which is not exhaustive, includes the most recent editions of some treatises, but also some older editions that have been considered interesting for their presentation of the concepts:

- Berkovitz BKB, Holland GR, Moxham BJ (2018) Oral anatomy, histology & embryology, 5th edn. Elsevier
- Bloom W, Fawcett DW (1983) Tratado de Histología, 9th edn. Labor
- Brüel A, Christensen EI, Tranum-Jensen J, Qvortrup K, Geneser F (2015) Geneser – Histología, 4ª edn. Panamericana
- Carrascal E (2001) Histología Humana. Librería Cervantes
- Chiego DJ (2014) Principios de Histología y Embriología Bucal con orientación clínica, 5th edn. Elsevier
- Eynard AR, Valentich MA, Rovasio RA (2008) Histología y Embriología del ser Humano. Bases celulares y moleculares, 4th edn. Panamericana
- Gartner LP, Hiatt JL (2008) Texto Atlas de Histología, 3rd edn. McGraw-Hill Interamericana
- Gómez de Ferraris ME, Campos Muñoz A (2009) Histología, Embriología e Ingeniería Tisular Bucodental, 3rd edn. Panamericana
- Ham AW, Cormack DH (1985) Tratado de Histología, 8th edn. Interamericana
- Hand AR, Frank ME (2014) Fundamentals of oral histology and physiology. Wiley Blackwell
- Junqueira LC, Carneiro J (2015) Histología Básica. Texto y Atlas, 12th edn. Panamericana
- Kierszenbaum AL, Tres LL (2016) Histología y Biología Celular. Introducción a la anatomía patológica, 4th edn. Elsevier
- Krstić RV (1985) General histology of the mammal. Springer
- Krstić RV (1991) Human microscopic anatomy. Springer
- Nanci A (2013) Ten Cate's oral histology, 8th edn. Elsevier
- Pawlina W (2024) Histology. A text and atlas, with correlated cell and molecular biology. Wolters Kluwer
- Villaro AC (2021) Histología para estudiantes. Panamericana
- Weis L (ed) (1988) Cell and tissue biology. A textbook of histology, 6th edn. Urban & Schwarzenberg
- Welsch (2014) Sobotta Histología, 3rd edn. Panamericana
- Young B, O'Dowd G, Woodford P (2014) Wheater – Histología Funcional. Texto y Atlas en color, 6th edn. Elsevier

As this manual is intended as an introduction to Histology, and not as a treatise on the speciality, references to sources are not included in the text. However, a brief list of further reading has been included for each chapter.

The etymological origin of scientific terms has been given a great deal of importance, as it can facilitate the understanding of their meaning. To meet this objective, two main works have been consulted:

- Diccionario de la lengua española. Real Academia Española. https://www.rae.es/
- Diccionario médico-biológico, histórico y etimológico. Universidad de Salamanca. https://dicciomed.usal.es

The authors (Francisco J. Sáez and Iker Badiola) teach Human Histology in the degree of Odontology at the School of Medicine and Nursing of the University of the Basque Country UPV/EHU. Both have wide experience in university teaching in several degrees (Biology, Biochemistry, Medicine, Dentistry) and an extensive curriculum in Histology research.

Leioa, Spain Francisco J. Sáez
Leioa, Spain Iker Badiola

Acknowledgements

We would like to thank our respective families for their patience and continued support during the preparation of this book. We also thank our colleagues for their continued collaboration. In addition, we are indebted to all the books and articles we have consulted for this book, without which it would not have been possible. The successes of this handbook are mainly due to the sources consulted; any errors are our own responsibility. Finally, we would like to thank Springer Nature for their interest in this project.

Contents

Histology and the Study of Human Tissues

1

What You Will Learn in This Chapter

In this chapter you will learn what a biological tissue is and what Histology is. You will learn some of the most commonly used vocabulary in the study of tissues and you will see how to interpret the two-dimensional histological sections that are studied with microscopes. You will also review some of the basic concepts of cell organisation, as cells are the fundamental components of tissues.

1.1 What Is Histology and What Is Tissue?

A. The Encyclopaedia Britannica states that the fundamental aim of *histology*[1] is to determine how tissues are organised at all structural levels, from cells and intercellular substances to organs.

B. The term histology is formed by the prefix *histo-*, which comes from the classical Greek ἱστός (histós), which means *tissue*, and the suffix *-logy*, from the Greek λογία (logía), which means *treatise, study* or *science*.

C. Histology is an evolution of microscopic anatomy that developed mainly from the nineteenth century onwards through the conjunction of several milestones:

1. Emergence of the concept of tissue. The origin of the term tissue applied to animal organisms is due to Marie François Xavier Bichat (1771–1802).
   a. He distinguished what he called simple tissues that combined in the organism form the organs.
   b. It should be noted that Bichat never used a microscope.
2. Evolution of microscopes and microscopic technique. New microscopes with higher resolution and fewer optical aberrations were developed, improving microscopic studies.

[1] Some authors have recently introduced the term 'tissular biology'. If we understand them to refer to the science that studies tissues, this would be a new and unnecessary way of referring to Histology.

F. J. Sáez, I. Badiola, *Essential Oral Histology*, Learning Materials in Biosciences,
https://doi.org/10.1007/978-3-032-03165-5_1

3. Statement of the *cell theory* (1838–1839), which led to a breakthrough in the study of the organisation of living organisms. This theory proposed that all living things, animals, plants and micro-organisms, are made up of *cells* (from Latin cella 'cell, small room').
4. Statement of the *theory of evolution* (Darwin 1868) and the *theory of heredity* (Mendel 1865), which together with cell theory revolutionised the biological sciences.

D. Cells are living entities capable of very complex chemical reactions (*metabolism*) that allow them to produce, maintain and renew their components.
E. Most cells are capable of dividing by *mitosis* to produce two new cells.
F. The Encyclopaedia Britannica defines tissue as 'a level of organisation in multicellular organisms; it consists of a group of structurally and functionally similar cells and their intercellular material'.
G. A tissue is made up of:
   1. *Cells*:
      a. They can be all identical—of the same *cell type*—or they can be different—of several cell types—each specialising in one job, but always working in a more or less coordinated way.
      b. All cell types of a tissue may have a common *embryonic origin*, or there may be cells of different origin.
   2. *Extracellular matrix*: The medium in which cells are immersed.
      a. Mostly it is usually produced by one of the cell types of the tissue.
      b. It is a very hydrated medium, such as the cells themselves.

1.2 Some Basic Vocabulary in Histology

A. All the cells of a human being come from a single cell, the *zygote*, formed by the fusion of the paternal *sperm* and the maternal *egg*.
B. The zygote divides into two cells by *mitosis*. Through successive mitoses, the resulting cells produce many cells.
C. The new cells that are produced become specialised to do different functions. This process of specialisation is called *differentiation*.
D. In each tissue, groups of cells can be found that have differentiated in different ways. Each of these groups of cells constitutes a *cell type*. One cell type can be present in different tissues.
E. Cells of a cell type that are in the same tissue, in the same environment, constitute a *cell population*.
F. Each cell type is given a name that identifies it. Usually, this name is a word made up of:
   1. A *root* or *lexeme*. Lexemes are usually referred to:

a. To the tissue in which it is. Examples are mio- (from Greek μῦς, μυός myos 'muscle'), osteo- (ὀστεο osteo 'bone'), condro- (χόνδρος chondros 'cartilage').
b. To its function, to its role in the tissue or to some characteristic of the cell, e.g. erythro-, from the Greek ἐρυθρο, meaning red.

2. A *suffix*. Suffixes are usually two:
   a. *Cyte*: The suffix -cyte comes from the Greek κύτος kýtos, meaning vessel, receptacle, and is the equivalent of the Latin cella, meaning cell, from which the word cell comes.
   b. *Blast*: The suffix -blast comes from the Greek βλαστός blastós, meaning germ, which forms or originates something. This suffix is often used for:
      (1) To designate an *immature cell*, i.e. a cell that has not completed its differentiation process and is about to transform into a well-differentiated cell. Thus, an immature cartilage cell is called a chondroblast, while a mature, well-differentiated cell is called a chondrocyte.
      (2) To denote a *cell that produces* or *originates some structure*. There are three examples:
         (a) The fibroblast, the mature connective tissue cell, which is thus called a 'fibre-forming' cell.
         (b) The odontoblast, which forms the dentine of the tooth.
         (c) The ameloblast, which forms tooth enamel.
3. In some cases, the word begins with a *prefix*. Some prefixes indicate:
   a. A previous stage of cell maturation, such as:
      (1) Meta- (from Ancient Greek μετα-, 'next to, after, between').
      (2) Pre- (from Latin prae-, 'previous').
      (3) Pro- (from Latin pro-, 'before, in front of').
   b. A characteristic of the cell, e.g. mega- (from Greek μεγα- mega-, 'large').

G. The following table shows the components of the names of some cell types.

Cell type	Prefix	Root or lexeme	Suffix
Odontoblast		Odonto	Blast
Preodontoblast	Pre	Odonto	Blast
Ameloblast		Amelo	Blast
Preameloblast	Pre	Amelo	Blast
Chondrocyte		Chondro	Cyte
Chondroblast		Chondro	Blast
Erythrocyte		Erythro	Cyte
Proerythroblast	Pro	Erythro	Blast
Metamyelocyte	Meta	Myelo	Cyte
Megakaryocyte	Mega	Kario	Cyte

H. *General histology* is the part of histology that studies the basic tissues, their characteristics and embryonic origin.
I. *Special histology* studies the organisation and arrangement of tissues in each organ. This part of histology is also called *microscopic organography*.
J. *Organs* are grouped into *apparatuses* and *systems*. Although these concepts have different meanings in different countries, an apparatus or system can be considered to be a set of organs that collaborate in a specific function of the organism.
K. In organs, two inseparable components are usually distinguished:
   1. The organ tissues that perform their main function form what is called *parenchyma* (from the Ancient Greek παρέγχυμα parénchyma 'substance of the organs').
   2. The tissues that have a secondary or supporting function constitute the *stroma* (from the Greek στρῶμα strôma 'tapestry').
L. Tissues are studied under the *light microscope* in sections 4–10 μm thick.
   1. In these sections, the tissues are transparent under the microscope, colourless.
   2. In order to distinguish the structure of the tissues, histological sections are treated with a wide variety of *staining substances*.
   3. The most common *staining* technique combines two dyes:
      a. *Haematoxylin* is an alkaline substance that stains the cell in purple colours where acidic components predominate.
         (1) It stains intensely the nucleus, which has a high deoxyribonucleic acid (DNA) content.
         (2) Cell regions that stain with haematoxylin are called *basophilic* (from the Greek βάσις básis 'base' and φιλία philía 'friendship, sympathy'), because of their affinity for alkaline dyes.
      b. *Eosin* is an acidic substance that stains cellular regions with abundant alkaline components. Cellular components that stain with eosin are called *acidophils* or *eosinophils*.

1.3 Interpretation of Histological Sections

A. The most common microscopic observation of tissues is performed in *bright-field microscopes*. These microscopes use visible light passing through the sample and glass lenses that magnify the image.
B. To study a sample of an organ under a bright-field microscope requires a multi-step process:
   1. *Fixation* with chemicals, which prevents degradation of some tissue components and destruction of the structures.

2. *Embedding* in a solid medium, such as *paraffins* or *resins*. This process requires *dehydration of* the sample with alcohol and xylene.
3. *Cutting* or *microtomy* (from classical Greek μικρο- mikro- 'small' and -τομία -tomía, derived from τέμνειν témnein 'to cut') with a blade to obtain very thin sections of tissue (4–10 μm thick) that can be placed on a thin glass *slide*.
4. *Staining of the* section with *dyes*, which requires *rehydration of* the sample with solvents and water.
5. *Mounting*, which consists of covering the histological section with a very thin glass slide (*coverslip*), which is adhered to the sample with a resin or *mounting medium* that solidifies and preserves the sample. This requires *re-dehydration* of the section with solvents.

C. The preparation process can lead to sample alterations. Alterations due to sample processing are called *artefacts*. Some possible alterations are:
1. Shrinkage due to chemicals used or heat during paraffin embedding. Shrinkage may cause:
   a. Decrease in size.
   b. Appearance of spaces between cells and other tissue components.
2. Void spaces appear due to the loss of substances, either because they have not been properly fixed or because they are soluble in the chemicals used.
3. Alterations in shape due to breakage of the tissue during procurement.
4. Tissue degeneration due to insufficient fixation.
5. Appearance of folds.
6. Appearance of lines due to blade imperfections.
7. Precipitates of dyes or dirt.
8. Bubbles in the mounting medium.

D. The Histology student must take into account:
1. The characteristics of a tissue described in the books have been known by analysing many samples with different microscopic or biochemical techniques. Therefore, you should not expect to be able to observe all the characteristics of a tissue in the preparations you study in a single sample.
2. Under the microscope, a *two-dimensional* specimen (length and width) is observed, but tissues have a *three-dimensional* organisation (length, width and height).
   a. The same three-dimensional structure can give rise to very different apparent two-dimensional structures (Fig. 1.1).
   b. It is necessary to interpret what is observed and to imagine the possible real structure of what is observed.

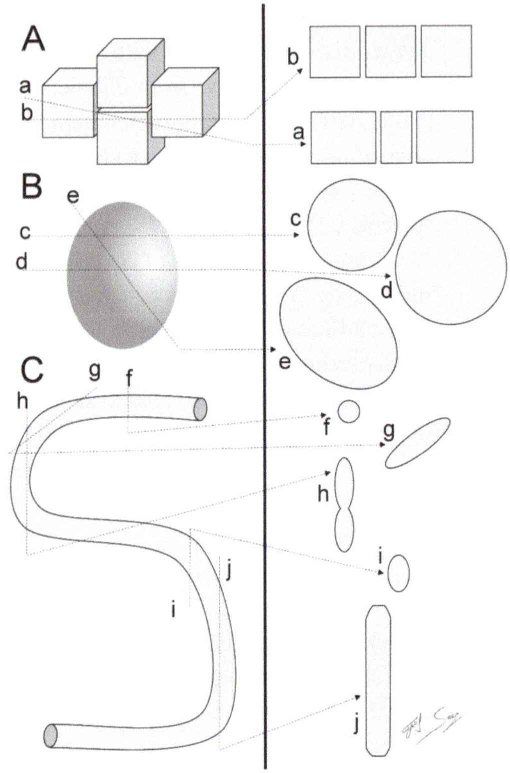

Fig. 1.1 Representation of the different two-dimensional images that can be observed if sections of three-dimensional objects are made in different planes. Sections of polyhedral shapes, e.g. a regular hexahedron (**A**), can give rise to different polygonal shapes (a, b). Depending on the complexity of the polyhedra sectioned, the diversity of polygonal shapes that can be observed will vary. Sections of rounded or ovoid structures (**B**) can produce circular (c, d) or oval (e) images. Tubular structures (**C**) can be cut in very different planes, resulting in circular (f), oval (g, i), irregular (h) and elongated (j) shapes

1.4 Cellular Organisation

A. Before beginning the study of histology, it is useful to recall some basic concepts about the cell.
B. Human cells, such as the cells of other organisms, consist primarily of water and organic molecules of various kinds, which constitute the so-called *protoplasm* (from the Greek οπρωτο, prōto, 'first' and πλάσμα, plásma, 'substance').
C. All cellular contents have a boundary, the *plasma membrane (cell membrane)*, a semi-permeable physical barrier that encloses the rest of the protoplasm.
D. Within the cell, there are other membranes that divide the cell protoplasm, creating compartments called *organelles*, each one specialised in different functions.
E. The cell is divided into two main compartments (Fig. 1.2):
   1. The *nucleus* is a compartments only visible when the cell is not dividing (*interphase*).
      a. When the cell divides (*mitosis*), the nuclear material condenses to form *chromosomes*, which are linear DNA molecules containing the genetic information in their nucleotide sequence.
      b. In the interphase nucleus you can see:

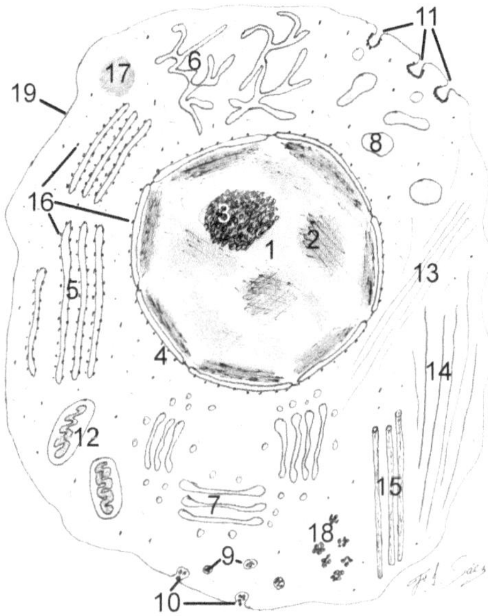

Fig. 1.2 Simplified representation of the structure of a hypothetical cell with its main components: (1) nucleus; (2) heterochromatin; (3) nucleolus; (4) nuclear envelope with ribosomes; (5) rough endoplasmic reticulum; (6) smooth endoplasmic reticulum; 7, Golgi apparatus; (8) lysosomes and endosomes; (9) secretory vesicles; (10) exocytosis of vesicles; (11) formation of endocytosis vesicles; (12) mitochondria; (13) actin microfilaments; (14) intermediate microfilaments; (15) microtubules; (16) ribosomes; (17) lipid droplets; (18) glycogen granules; (19) plasma membrane

(1) *Nuclear envelope*, which has two membranes that form the boundary of the nucleus. *Ribosomes* are attached to the outside of the nuclear envelope.
(2) *Chromatin*, consisting of partially decondensed chromosomes. Darker regions (*heterochromatin*) and lighter areas (*euchromatin*) can be observed.
(3) *Nucleoli*, one or more, which are the regions where ribosome subunits are produced.

2. The *cytoplasm* is the remainder of the protoplasm that together with the nucleus forms the cell. The cytoplasm consists of membrane-bounded compartments surrounded by the cytosol.
   a. Compartments surrounded by membranes, the cavities of which are called *lumen* (from Latin lumen, 'light, opening'). The most important are:
   (1) *Endoplasmic reticulum* (from Latin reticŭlum, 'net', from Greek ἐνδο- endo-, 'inside', and the diminutive suffix -ico): made up of membrane-bounded cavities that are distributed throughout the cytoplasm forming a network. Many different functions are performed in the endoplasmic reticulum, such as the import of newly synthesised proteins by ribosomes for modification and transport to their destination in the cell or outside the cell.

(a) Some regions of this organelle may have particles (*ribosomes*) attached to the outer surface of its membrane. These regions are called the *rough endoplasmic reticulum.*

(b) Other areas lack ribosomes and form the *smooth endoplasmic reticulum.*

(2) *Golgi apparatus*: a set of stacked cisternae that receives materials from the endoplasmic reticulum, modifies them and transports them to their final destination in the cell or outside the cell.

(3) *Lysosomes* and *endosomes*: vesicles (membrane-bounded cavities) filled with enzymes that are responsible for degrading other substances of endogenous (the cell itself) or exogenous (the extracellular environment) origin. These vesicles contain *enzymes* that come from the Golgi apparatus.

(4) *Transport vesicles*: vesicles that are produced by budding of the membrane of one organelle and fuse with the membrane of another organelle. In this way, components of the membrane or lumen are transported between the two organelles.

(a) Some vesicles (*secretory vesicles*) move materials from the Golgi apparatus to the cell surface (plasma membrane), and their contents are secreted to the extracellular space by a process called *exocytosis.*

(b) Vesicles are produced in the plasma membrane into the cell (*endocytosis vesicles*) that can fuse with other organelles, such as endosomes, thereby incorporating extracellular materials into the cell.

(5) *Mitochondria* (from Ancient Greek μίτος mítos 'thread' and κόνδρος kóndros 'granule'): organelles bounded by two membranes. Inside take place the *Krebs cycle* and *oxidative phosphorylation*, i.e. the production of *ATP* (adenosine triphosphate).

(a) ATP is produced mainly from the energy obtained from the oxidation of organic molecules in the Krebs cycle.

(b) ATP is then distributed throughout the cell and the hydrolysis of this molecule is used to provide the energy needed to carry out many biochemical reactions in the cell.

b. *Cytosol* (from the Greek κύτος kýtos 'cell' and Latin solu- 'dissolve') or *hyaloplasm* (from the Greek ὕαλος hýalos 'glass' and ὑάλινος hyálinos 'translucent'): The medium surrounding the cytoplasmic compartments:

(1) The cytosol itself is another compartment between the plasma membrane and the membranes of the nucleus and cytoplasmic organelles.

(2) In the cytosol we can find structures that are visible under the microscope but do not have a membrane, for example:

(a) Protein filaments that make up the *cytoskeleton*. They can be of various types:

1. *Microfilaments* or *actin filaments*. They are the thinnest (about 6 nm).
2. *Intermediate filaments*, of intermediate thickness, as their name suggests (between 8 and 10 nm). There are various types, depending on

the protein that forms them: *vimentin filaments*, *desmin filaments*, *tonofilaments* (*keratin*), *neurofilaments*, *gliofilaments*, etc.

3. *Thick filaments* or *myosin filaments*. They are thicker (25 nm). They are only distinguishable in *muscle cells*. They are associated with microfilaments and glide over them, resulting in movement.
4. *Microtubules*. These are hollow filaments (tubes) formed by the protein *tubulin*.

(b) *Ribosomes* (from ribose, the pentose that forms ribonucleic acid, and from the Greek σῶμα sôma 'body'):

1. They are ribonucleoprotein macromolecular complexes, i.e. made up of proteins and *ribonucleic acids (RNA)*.
2. Its function is to produce proteins by reading messenger RNAs from the nucleus to translate their nucleotide sequence into the amino acid sequences of each protein.

(c) *Lipid droplets*: These are lipid inclusions, i.e. accumulations of fats or oils which, not being soluble in water, form a non-aqueous phase isolated from the rest of the cytosol.

(d) *Glycogen granules*: Glycogen is a polymer of glucose, which is produced to store glucose in the cell. As it can be a large molecule, it can be seen as small inclusions in the cytosol under the transmission electron microscope.

F. Many small molecules (water, ions, amino acids, sugars, nucleotides, etc.) must pass through cell membranes for cells to function properly.

1. Normally this transport is done in favour of the *chemical gradient* or *electrochemical gradient* of the transported substance, i.e. by *diffusion*.
2. However, there are several different mechanisms (Fig. 1.3):
   a. *Simple* or *passive diffusion*: If the substance is fat-soluble, it can pass through the lipid bilayer of the membrane.
   b. *Facilitated diffusion*: If the molecule is not fat-soluble, it needs the help of *transmembrane proteins*. In this case we can find variants:
      (1) Facilitated diffusion of ions is normally carried out by *channel proteins*.
      (2) Substances with no electrical charge usually pass through membranes via *transporters*.
         (a) Some transporters (*uniporters*) carry a single substance in favour of its gradient.
         (b) Other transporters (*coupled transporters*) transport two different molecules simultaneously, usually one of them down and the other against their respective gradients.
   c. *Active transport*: Certain proteins, called *pumps*, transport substances, usually ions, against their gradients due to the energy provided by ATP hydrolysis.

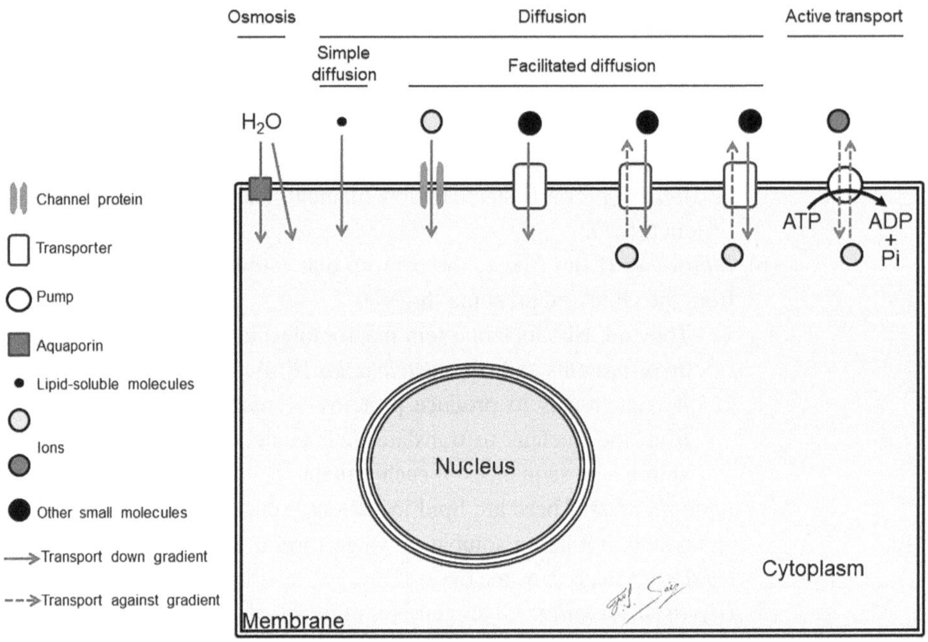

Fig. 1.3 Mechanisms of transport of small molecules across membranes. Water crosses membranes by osmotic pressure, mostly through aquaporins, but to a lesser extent through the lipid bilayer. Lipid-soluble molecules pass through the lipid bilayer in favour of their chemical gradient by simple diffusion. Other molecules cross the membrane in favour of their electrochemical or chemical gradient by facilitated diffusion, ions by channel proteins and non-electrically charged molecules by transporters. Some transporters transport two different substances (coupled transporters), one of them down and the other against their respective gradients. There are also pumps that transport substances against their gradient by means of active transport, which occurs due to energy provided from the hydrolysis of ATP

d. *Osmosis*: Water passes through the membranes under *osmotic pressure*.[2] It can occur in two ways:
   (1) Across the lipid bilayer. This transport is slow.
   (2) Via transmembrane proteins called *aquaporins*. As the passage of water through the lipid bilayer is slow, transmembrane water transport depends primarily on aquaporins.

[2] Osmotic pressure is the pressure that drives the flow of water across a membrane to equalise the concentrations of dissolved substances on both sides of the membrane, when these substances cannot pass through the membrane.

Take-Home Message

Histology is the science that studies biological tissues, which are aggregates of cells and intercellular material organised in one particular place in an animal organism to perform some function. Histology uses a specialised vocabulary that is essential to know. One of the basic tools used to study tissues is the microscope, and as it uses two-dimensional sections of tissues, it is necessary to know how to interpret microscopic images to understand the three-dimensional organisation of tissues. In addition, as cells are the fundamental components of tissues, it is necessary to have an in-depth knowledge of the organisation of cells.

Further Reading

Gartner LP (2017a) Cytoplasm. In: Textbook of histology, 4th edn. Elsevier, Philadelphia, pp 13–54

Gartner LP (2017b) Introduction to histology and basic histological techniques. In: Textbook of histology, 4th edn. Elsevier, Philadelphia, pp 1–12

Gartner LP (2017c) Nucleus. In: Textbook of histology, 4th edn. Elsevier, Philadelphia, pp 55–80

Gómez de Ferraris ME, Campos Muñoz A (2009a) Histología, Embriología e Ingeniería Tisular Bucodental. In: Histología, Embriología e Ingeniería Tisular Bucodental, 3rd edn. Médica Panamericana, pp 1–12

Gómez de Ferraris ME, Campos Muñoz A (2009b) Métodos y técnicas de estudio en histología e ingeniería tisular bucodental. In: Histología, Embriología e Ingeniería Tisular Bucodental, 3rd edn. Médica Panamericana, pp 13–26

Krstić RV (1985) Introduction. In: General histology of the mammal. An atlas for students of medicine and biology. Springer Verlag, Berlin, pp 1–13

Krstić RV (1991) Introduction. In: Human microscopic anatomy. An atlas for students of medicine and biology. Springer Verlag, Berlin, pp 1–3

Lowe JS, Anderson PG (2015a) Histology. In: Stevens & Lowe's human histology, 4th edn. Elsevier, Philadelphia, pp 1–10

Lowe JS, Anderson PG (2015b) The cell. In: Stevens & Lowe's human histology, 4th edn. Elsevier, Philadelphia, pp 11–36

O'Dowd G, Bell S, Wright S (2023) Cell structure and function. In: Wheater's functional histology. Elsevier, Amsterdam, pp 2–32

Padykula HA (1988) Histochemistry and cytochemistry. In: Weiss L (ed) Cell and tissue biology – a textbook of histology, 6th edn. Urban & Schwarzenberg, Baltimore, pp 93–112

Pawlina W (2024a) Methods. In: Histology a text and atlas with correlated cell and molecular biology, 9th edn. Wolters Kluwer, Philadelphia, pp 1–27

Pawlina W (2024b) Tissues: concept and classification. In: Histology a text and atlas with correlated cell and molecular biology, 9th edn. Wolters Kluwer, Philadelphia, pp 112–120

2 General Embryology

What You Will Learn in This Chapter

The development of a human being can be divided into several stages from fertilisation to birth and from birth to old age. Development depends on genetic and epigenetic factors, and is carried out by a combination of cellular and molecular mechanisms that result in the formation of tissues (histogenesis) and shape organs and the human being as a whole (morphogenesis). During embryonic development, a structure with two cell layers (bilaminar embryo) is produced. Cells from the dorsal layer (epiblast) of this embryo migrate and a trilaminar embryo is produced with the three germinative layers: ectoderm, mesoderm and endoderm. In addition, the ectoderm forms the neural tube and neural crests, whose cells migrate through the mesoderm and form many cells and tissues of the organism. Finally, you will see that each germinative leaf can generate different human tissues and organs.

2.1 Stages of Human Development

A. The development of an individual begins with *fertilisation* and the formation of the *zygote*.
B. The development of an individual progresses through these stages:
   1. *Prenatal stage*, which includes:
      a. *Embryonic period*: This covers the first 8 weeks during which the *embryo* and its *extra-embryonic membranes* develop. The latter will form the *placenta*.
      b. *Foetal period*: from the ninth week until birth. In this period, the embryo is called *foetus*.
   2. *Postnatal stage*, which is divided into several periods:
      a. *Neonatal period*: the first 4 weeks of life.
      b. *Breastfeeding (lactation) period*: from the first month to 12 or 14 months. From the sixth month onwards, the first primary teeth begin to appear.
      c. *Childhood*: Divided into two periods:

F. J. Sáez, I. Badiola, *Essential Oral Histology*, Learning Materials in Biosciences,
https://doi.org/10.1007/978-3-032-03165-5_2

(1) *Early childhood*, from 12 months to 6 years of age. In the first 3 years, the hatching of the primary teeth continues.
(2) *Second childhood*, from 7 to 13 years of age, a period in which the permanent dentition progressively replaces the primary dentition.

d. *Puberty*: from 12 to 14 years of age in boys and from 11 to 14 years of age in girls. During this period, the development of the sexual organs and secondary sexual characteristics takes place.
e. *Adolescence*: maturity is reached during this period and the third permanent molar usually appears.
f. *Adult period*: from 20 years of age onwards. Development is complete and eventually gives way to *maturity* and *senility*.

C. *Embryology* is the science that studies the development of a living being (in our case, the human being) from fertilisation to birth. A distinction is usually made:
   1. *General embryology*, which studies early development up to the appearance of *germinative leaves* in the embryo and the formation of organ outlines.
   2. *Special embryology*, which studies *organogenesis*, i.e. the formation of organs, apparatus and systems in the foetus.

D. This chapter will describe the first weeks of development, up to the appearance of the germinative leaves and the neural tube.

2.2 Factors Regulating Development

A. The entire development of a human being is regulated by many factors that can be grouped into two categories:
   1. *Genetic factors*: each organism has genetic information in the zygote that conditions the developmental programme.
   2. *Epigenetic factors*: these are the set of factors that regulate how the genetic information contained in chromosomes is expressed.
      a. They usually depend on the *environment* in which development takes place.
      b. As an example of the influence of these epigenetic factors, we can give *homozygous twins* possessing exactly the same genetic information but often displaying some different characteristic.

2.3 Mechanisms Regulating Development

A. *Cell proliferation*:
   1. The zygote divides by *mitosis*, and the resulting cells divide by mitosis, and so on to give rise to all the cells of the organism.
   2. Mitosis allows from one cell to form two cells whose genetic material (*genotype*) is identical to that of the initial cell.

B. *Differentiation*:
  1. During development, certain genes are activated in each cell and others are silenced.
     a. As a result, each cell that forms acquires different characteristics.
     b. This process is called *differentiation*.
  2. The set of characteristics of an organism or cell is called a *phenotype*.
     a. Two individuals with the same genotype (e.g. homozygous twins) can have different characteristics (different phenotypes).
     b. Two cells of an organism, with the same genotype, can have very different phenotypes.
  3. During development, groups of cells differentiate and specialise in different functions, i.e. they acquire different phenotypes. Cells with the same phenotype constitute a *cell type*.
  4. Specialisation results in the cell being incapable of transforming into other cell types. This process is progressive and we can distinguish cells that have different differentiation potentials:
     a. *Totipotent cell*: the zygote is a totipotent cell, as it can give rise to all the cells and tissues of the individual and the whole individual.
     b. *Pluripotent cell*: they can produce any cell type in the organism, but are not capable of producing a complete individual.
     c. *Multipotent cells*: can give rise to several cell types of one lineage, but not of a different lineage. An example is the haemopoietic stem cell, which can give rise to any blood cell.
     d. *Monopotent cells*: they produce only one cell type. An example is the erythroblast, which can only give rise to blood erythrocytes.

C. *Migration*: Some cells have the ability to move, so that their differentiation ends at a different place in the embryo from the original one.

D. *Cell death*:
  1. Some cells receive signals to activate internal cell death mechanisms.
  2. There are many forms of cell death through mechanisms that are programmed into the cell in response to a particular situation:
  3. *Apoptosis* is a widely used and best known mechanism. It is very frequent during development.
  4. Other mechanisms of cell death, such as *autophagy*, *entosis*, *paraptosis*, *mitoptosis*, *pyroptosis*, *NETosis* and *necroptosis*, are less well understood and less frequent or situation-specific.
  5. When cell death affects all or an extensive part of a tissue, this structure is said to undergo *involution* or *regression*.
  6. Involution allows some structures that are no longer needed to disappear and the space can be occupied by new tissues.

E. *Cellular communication*:
  1. Development is regulated by communication mechanisms between cells.
  2. Cells can use many mechanisms to communicate with each other, but we can reduce all of them to three basic strategies (Fig. 2.1):

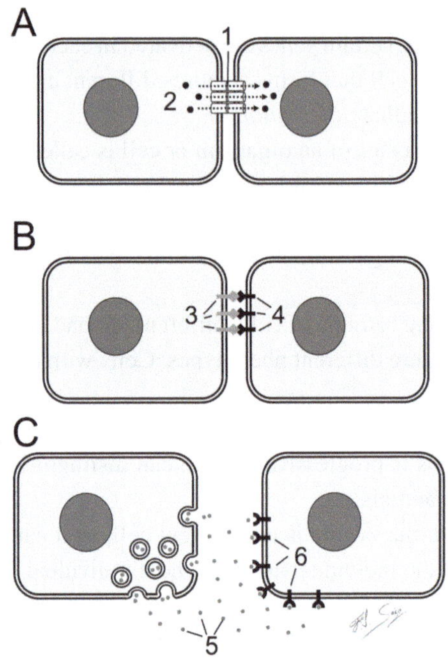

Fig. 2.1 The three basic mechanisms of communication between cells. In communicating junctions (**A**), cells are joined by transmembrane complexes (1) through which small molecules (2) can pass from one cell to the other. In direct contact junctions (**B**), the contact is between a molecule on the membrane surface of one cell, which we call a ligand (3) and another molecule on the membrane surface of the other cell, which we call a receptor (4). In communication by soluble molecules (**C**), one of the cells releases, usually by exocytosis, a soluble signal molecule (5) which is taken up by specific receptors on the surface of another cell (6)

a. *Communicating junctions*: In *gap junctions*, which we will study in Chap. 3, small molecules are transmitted from the inside of one cell to the inside of the other.
b. *Direct contact*: The membrane of one cell has a molecule on its surface (*ligand*) that binds to another molecule (*receptor*) on the surface of the membrane of the other cell. The receptor bound to the ligand stimulates other molecules inside the cell.
c. *Communication by soluble molecules*: One cell releases a *soluble signal molecule* that diffuses to the surface of another cell, where it binds to a *receptor*. Diffusion of the signal molecule can occur:
   (1) In a very small microenvironment, as in the nerve *synapse*.
   (2) Within a tissue or between neighbouring tissues. Signal molecules are called *local chemical mediators*.

(3) Travelling through the blood to distant organs. The signal molecules are called *hormones*.

3. Some cells develop according to the place they occupy, which is called their *cell niche*. Cells in different niches develop differently because they receive different signals in each niche.
4. At each point in space, the cell receives *positional information*. This information is mediated by molecules called *morphogens* (Fig. 2.2).
   a. Morphogens are produced in cells and spread by diffusion, whereby a *concentration gradient of the morphogen* is formed:
      (1) Near the point of origin, the concentration is higher.
      (2) As we move away, concentration decreases.
   b. Cells differentiate differently depending on their distance from the morphogen source because they respond differently depending on the concentration of morphogen in their environment. This developmental system is called the *French flag model*.

F. *Induction*:
1. An *inductor tissue* causes another tissue capable of receiving a signal (*competent tissue*) to differentiate into another tissue type (Fig. 2.3).
2. Induction processes are frequent in the embryo and are very important in tooth development.

G. *Morphogenesis*:
1. Morphogenesis involves changes in the shape of the embryo to generate tissues and organs.
2. Molecular, cellular and tissue mechanisms are involved in these processes, among those mentioned above.
3. Many mechanisms of morphogenesis can be distinguished. Some of the most important ones are (Fig. 2.4):
   a. *Invagination*: a concave-shaped depression is produced in a flat sheet of cells.
   b. *Evagination*: in contrast to the previous case, there is a convex elevation.
   c. *Emboly*: when a depression occurs in a spherical structure at a point that extends into the interior of the sphere.
   d. *Epiboly*: a group of cells on top of others proliferate and spread to cover those below them.
   e. *Canalisation* or *folding*: a fold is formed in a cell sheet that forms a channel. If the channel closes, a tube is formed.
   f. *Mamelonation*: in a group of cells, the cells grow at one point, forming a bud that protrudes from the initial surface.
   g. *Delamination*: a lamina consisting of one cell layer is divided, producing two cell layers.

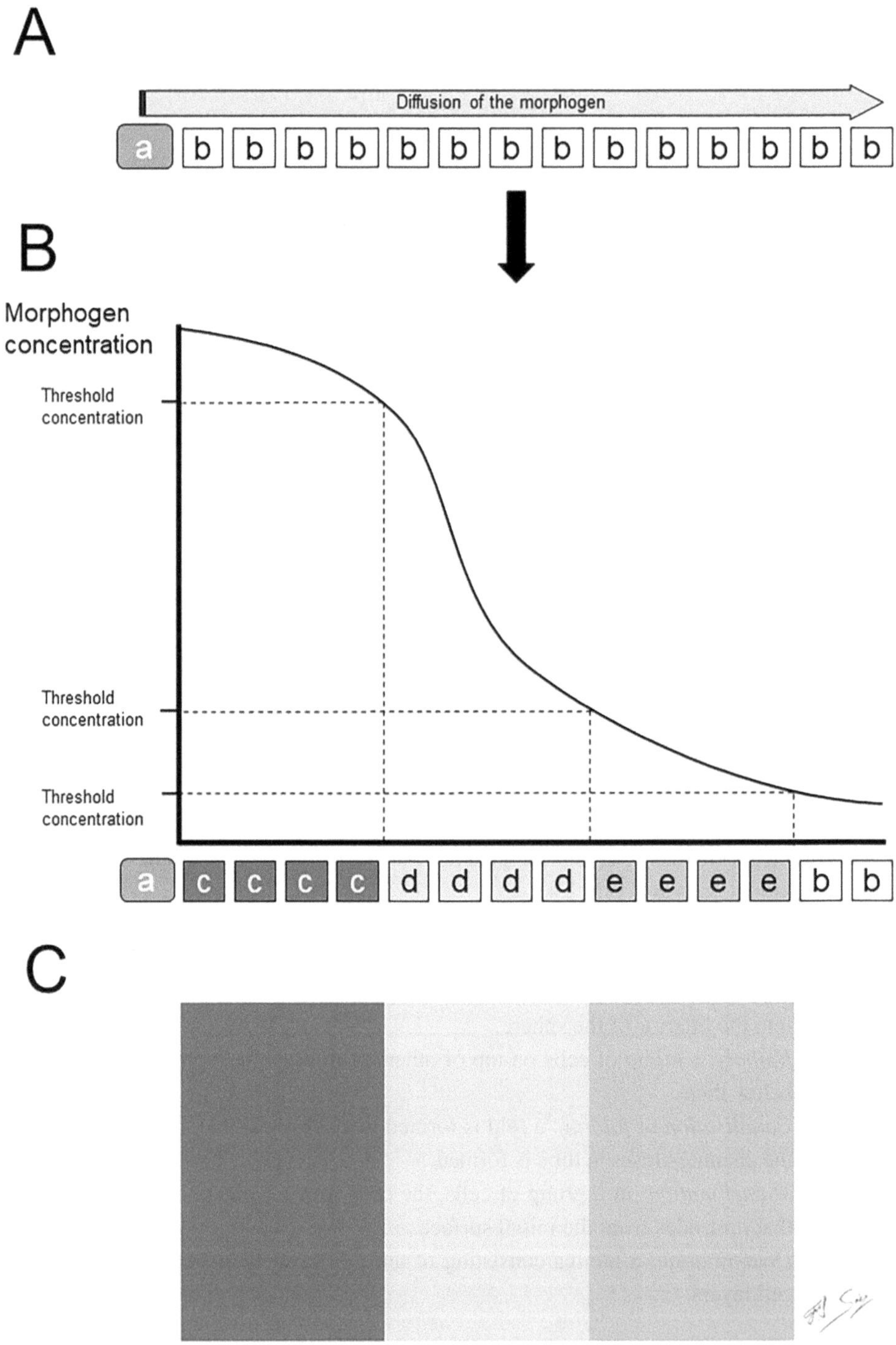

Fig. 2.2 Schematic representation of the action of morphogens. For simplicity, cells are depicted arranged in one dimension, but the process can occur in two and three dimensions. The small

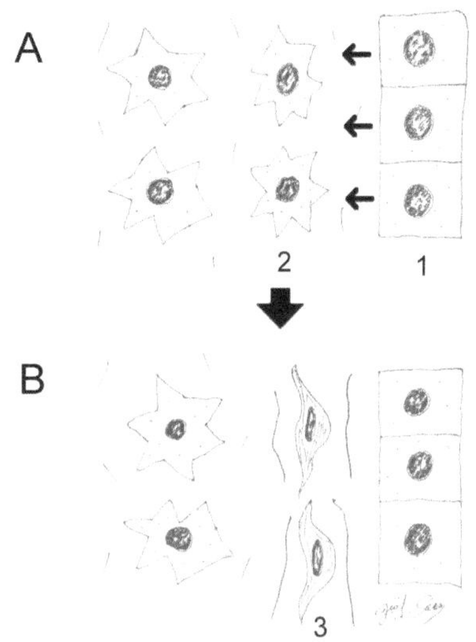

Fig. 2.3 Representation of an induction process before (**A**) and after (**B**) the process. (1) Inducer tissue; (2) competent tissue; (3) differentiated tissue

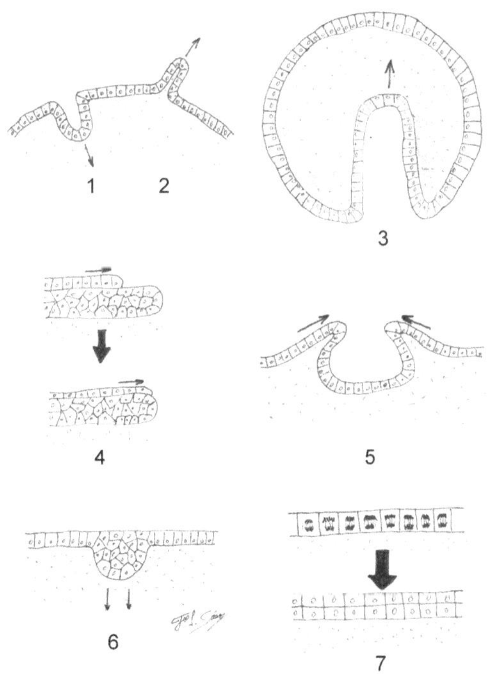

Fig. 2.4 Some mechanisms of morphogenesis. (1) Invagination; (2) evagination; (3) emboly; (4) epiboly; (5) canalisation or folding; (6) mamelonation; (7) delamination

Fig. 2.2 (continued) quadrilaterals symbolise cells. A cell (a) produces a substance (morphogen) that diffuses into neighbouring undifferentiated cells (b) in the direction of the arrow (**A**). The concentration of morphogen is lower the greater the distance from the morphogen-producing cell, resulting in a gradient of morphogen concentration (**B**). There are threshold concentrations of morphogen, above and below which undifferentiated cells respond differently and develop into different cell types (c–e). This morphogenesis phenomenon has been called the French flag model (**C**)

2.4 Early Human Development

A. *Fertilisation*:
   1. Fusion of *sperm* and *egg* in the woman's oviducts (fallopian tubes).
   2. As a result of this fusion, the *zygote* is produced, in which the paternal and maternal genetic material is mixed.

B. *Segmentation*:
   1. The zygote divides by mitosis to form two cells, which undergo successive mitoses. The periods between two divisions (interphase) are very short.
   2. The cells are called *blastomeres* (from the Greek βλαστός blastós, 'germ' and μερής merḗs, 'parts or portion').
   3. Segmentation has two phases:
      a. The blastomeres, all identical, organise themselves into a solid ball called *morula*.
      b. At the end, *compaction* takes place:
         (1) The outer blastomeres are flattened and tightly bound to each other (*outer cell mass*).
         (2) Inner blastomeres are polyhedral and not tightly bound (*inner cell mass*).

C. *Cavitation*:
   1. When the embryo reaches the uterus (Fig. 2.5), several changes occur in the morula, which becomes known as a *blastocyst* (from the Greek κύστις kýstis 'bladder, pouch').
   2. A large cavity appears inside, the *blastocoel* (from the ancient Greek κοίλē koílē, 'hollow, cavity').

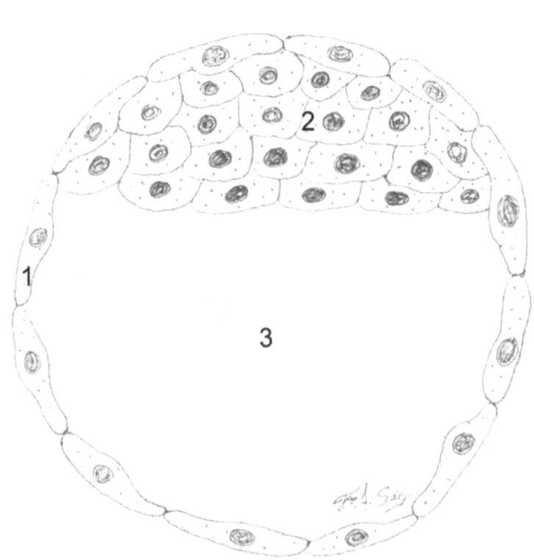

Fig. 2.5 Representation of a cut of a blastocyst. (1) Trophoblast; (2) embryoblast (inner cell mass); (3) blastocoel

3. The inner cell mass is displaced to one pole. This pole will be the dorsal pole of the embryo and the opposite pole will be the ventral pole. The inner cell mass is called the *embryoblast*.
4. The outer cell mass becomes known as the *trophoblast* (from the Greek τροφος tropos, 'feeding').

D. *Hatching*: In the uterus, 5 or 6 days after fertilisation, the blastocyst is released from its envelope (*zona pellucida*).

E. *Implantation*:
   1. The embryo contacts the *endometrium* of the maternal uterus and attaches to its surface.
   2. Subsequently, trophoblast cells invade the endometrium and the trophoblast enters the endometrium.
   3. The trophoblast begins to produce the *placenta*.
   4. The trophoblast then divides into two structures (Fig. 2.6):
      a. The *syncytiotrophoblast*, on the outside, a multinucleated structure formed by the fusion of cells.
      b. The *cytotrophoblast*, inside, which forms a sheet of epithelial cells.

F. *Formation of the bilaminar disc* (Fig. 2.6): During the second week of development, several processes take place in the embryoblast:
   1. Two cell layers form a discoidal structure, the *bilaminar embryonic disc*:
      a. The dorsal layer is the *epiblast* or *primitive ectoderm*.

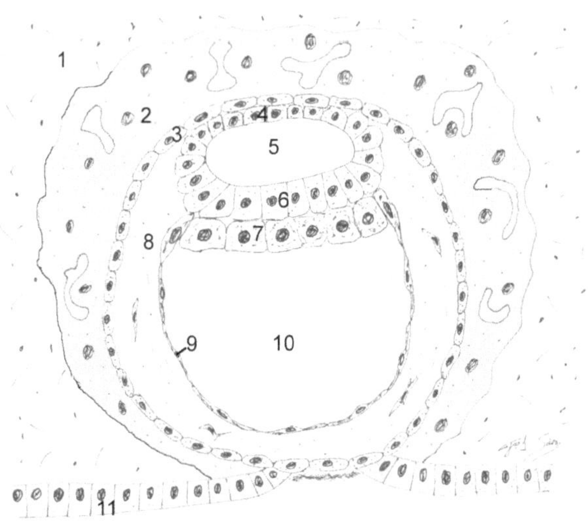

Fig. 2.6 Representation of a section of a bilaminar embryo. (1) Uterine endometrium; (2) syncytiotrophoblast; (3) cytotrophoblast; (4) amniotic membrane; (5) amniotic cavity; (6) epiblast or primitive ectoderm; (7) hypoblast or primitive endoderm; (8) extra-embryonic mesenchyme; (9) exocellomic membrane or Heuser's membrane; (10) exocellomic cavity or primitive yolk sac; (11) uterine epithelium

   b. The ventral layer is the *hypoblast* or *primitive endoderm.*
2. Another cavity appears in the embryo, dorsal to the epiblast, called the *amniotic cavity* (from the Greek ἀμνίον amnion, 'foetal membrane').
   a. This cavity is located between a roof formed by a layer of flattened cells, the *amniotic membrane* formed by the *amnioblasts* and a floor formed by the *epiblast.*
   b. The amniotic cavity will continue to grow and later, from the eighth week onwards, it will completely surround the embryo.
3. *Yolk sac* formation (Fig. 2.6):
   a. As the amniotic membrane forms, hypoblast cells migrate ventrally over the cytotrophoblast cells. Two structures are formed:
      (1) The flattened cells form the *exocellomic membrane* or *Heuser's membrane* that surrounds the entire blastocoel.
      (2) The newly formed cavity is called an *exocellomic cavity* or *primitive yolk sac* (from Latin vitellum 'egg yolk').
   b. Between Heuser's membrane and the cytotrophoblast is the *extra-embryonic reticulum*, which gradually incorporates some cells of different origins that form the *extra-embryonic mesenchyme* (from the Greek μεσο- meso- 'middle or intermediate' and ἔγχυμα énchyma 'infusion, injection').
4. Formation of the *chorionic cavity* (Fig. 2.7):
   a. In the extra-embryonic mesenchyme, several cavities are produced that merge into a single cavity, called the *extra-embryonic coelom* or *chorionic cavity.*

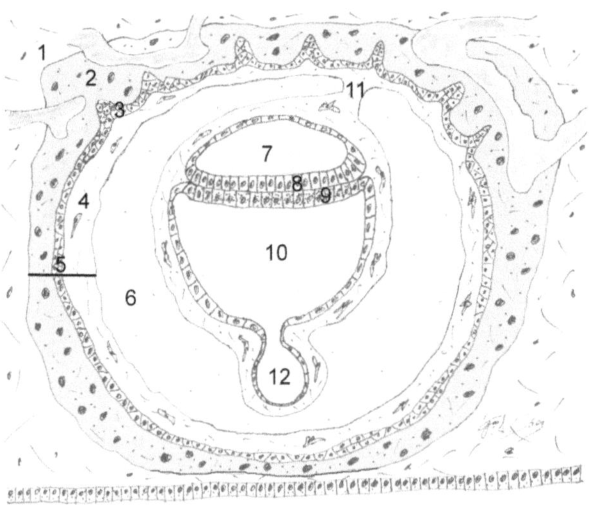

Fig. 2.7 Representation of chorion formation in a bilaminar embryo. (1) Uterine endometrium; (2) syncytiotrophoblast; (3) cytotrophoblast; (4) extra-embryonic mesenchyme; (5) chorion; (6) extra-embryonic coelom or chorionic cavity; (7) amniotic cavity; (8) epiblast; (9) hypoblast; (10) secondary yolk sac; (11) connecting stalk or attachment pedicle; (12) remains of the primitive yolk sac

b. This cavity expands and surrounds the entire embryo (bilaminar disc, amniotic cavity and yolk sac), which is separated from the trophoblast, except for a *connecting stalk* or *attachment pedicle*, which will later form the *umbilical cord.*
c. The outer wall of the chorionic cavity, consisting of extra-embryonic mesenchyme, cytotrophoblast and syncytiotrophoblast, is called the *chorion* (from Latin corĭum 'skin, leather' and from Greek χόριον chórion 'skin, hide, membrane').

5. A new migration of hypoblast cells displaces Heuser's membrane cells, which move outwards, and a new *secondary yolk sac* is formed.

G. *Formation of the leaves* or *germinative layers* (Fig. 2.8):

1. In the third week of development, *gastrulation* or formation of the *gastrula* (from the scientific Latin gastrula, from the Ancient Greek γαστήρ, γαστρός gastḗr, gastrós, 'stomach' and from the Latin diminutive suffix -ŭla) takes place.
2. A line called the *primitive line* appears in the epiblast and marks the cephalocaudal axis of the embryo. The primitive line runs towards the cephalic region and at its end the *primitive node* or *node of Hensen* is formed (Fig. 2.8).
3. Cells of the epiblast at the primitive line are introduced into the bilaminar disc.
   a. Some are located between epiblast and hypoblast, forming the *intraembryonic mesoderm* (Greek μεσο- meso-, 'middle' or 'intermediate' and δέρμα derma, 'skin').
   b. Others displace the hypoblast cells, forming a new layer called *endoderm* (from the Greek ἐνδο- endo-, 'within', 'inside').
   c. The cells of the epiblast that remain in the surface layer of the embryonic disc form the *ectoderm* (from the Greek ἐκτο- ekto-, 'outside', 'on the outside').
4. Ectoderm, mesoderm and endoderm are the three *germinative layers* of the embryo, which is now a *trilaminar embryonic disc*.

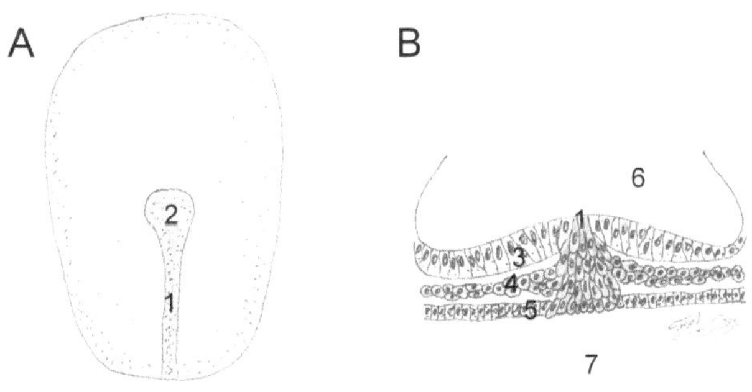

Fig. 2.8 Dorsal view (**A**) and cross-section (**B**) of an embryo in the third week of development, during gastrulation. (1) Primitive line; (2) primitive node or node of Hensen; (3) ectoderm; (4) mesoderm; (5) endoderm; (6) amniotic cavity; (7) secondary yolk sac

5. The three germinative leaves originate from the epiblast.

H. *Formation of the notochord*:
   1. Some of the cells that penetrate the mesoderm form a tube-like structure, the *notochordal canal.*
   2. The notochordal canal loses its cavity and flattens, forming the *notochordal plate.*
   3. Finally, a solid cylindrical cord of cells called *notochord* (from the Greek νῶτον nôton 'dorsum' and χορδή chordḗ 'chord') is formed.

I. *Neurulation* (Fig. 2.9):
   1. Neurulation (from the Greek νευρο- neuro-, 'nerve') occurs at the end of the third week of development. This process continues into the fourth week of development and the embryo at this stage is called the *neurula.*
   2. In the ectoderm there is a thickening over the notochord, the *neural plate.*
      a. The neural plate is produced by inducing signals from the notochord.
      b. The plate grows so that it is wider in the cephalic region and narrower in the caudal region.
   3. The lateral edges of the plate fold (*neural folds*), and the central part collapses (*neural groove*), in a process of canalisation or folding.
   4. The neural folds on both sides approach each other, merge and form a hollow cylinder, the *neural tube*, which invaginates and lies below the ectoderm and above the notochord.

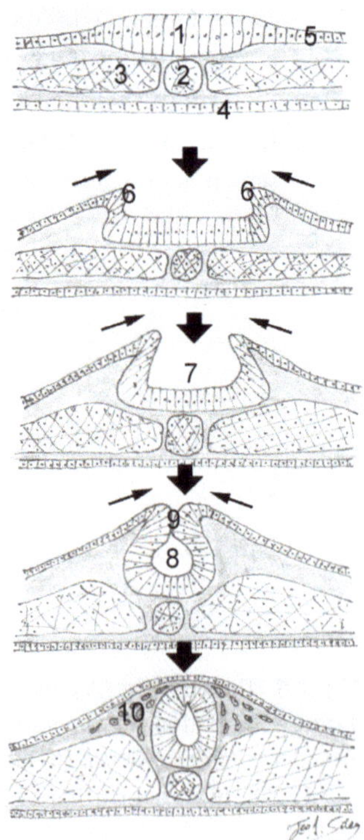

Fig. 2.9 Cross-sections of an embryo during different stages of the neurulation process. The thick arrows indicate the chronological progress of the process, and the thin arrows indicate the direction of movement of the neural folds. (1) Neural plate; (2) notochord; (3) mesoderm; (4) endoderm; (5) ectoderm; (6) neural folds; (7) neural groove; (8) neural tube; (9) neural crests; (10) neural crest cells

5. As the neural folds fuse, some cells from the crests of these folds detach and migrate to lateral and ventral areas of the neural tube.
   a. These *neural crest cells are* directed to very diverse regions of the embryo to participate in the formation of many tissues.
   b. Neural crest cells are involved in the formation of tooth outlines.

2.5 Differentiation of Germinative Leaves

A. After gastrulation, the embryo, so far a disc, folds like a book towards the ventral side, rotating on the axis formed by the notochord and neural tube.
B. Figure 2.10 shows an idealised and simplified cross-section of the trunk of a vertebrate embryo during the folding and closure of the digestive tract.

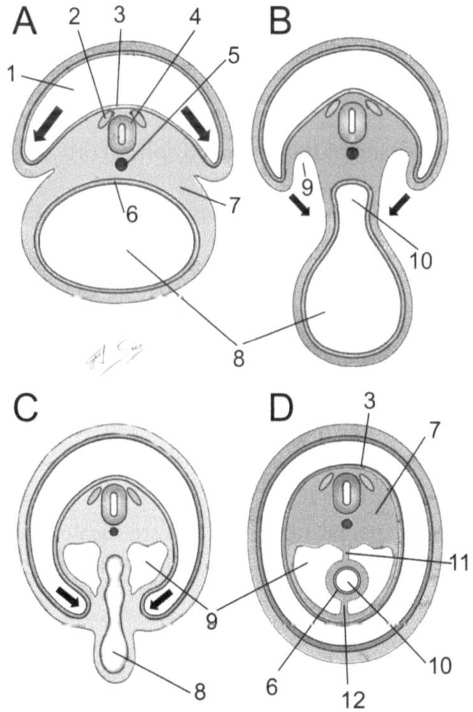

Fig. 2.10 Simplified and idealised cross-sections of embryos to show the process of folding and formation of the digestive tract. Many extra-embryonic structures surrounding the embryo have been removed from the diagram. The process begins at stage **A** and progresses through stages **B–D**, in which the fully closed digestive tract is already visible. The arrows indicate the direction of folding. (1) Amniotic cavity; (2) neural crests; (3) surface ectoderm; (4) neural tube; (5) notochord; (6) endoderm; (7) mesoderm; (8) yolk sac; (9) intraembryonic coelom; (10) digestive tube or intestine; (11) dorsal mesentery; (12) ventral mesentery

1. In this folding, the lateral ends meet in the middle zone and the endoderm, which was once a lamina, forms a tube, the *primitive gut*.
2. In the fourth week, there is a depression in the cephalic area, the *stomodeum* or primitive mouth, a cavity that later fuses and communicates with the anterior part of the primitive intestine.
3. From this point on, development is a very complex process that we will not explore in detail.

C. At the end of development, each germinative leaf produces the following tissues and organs:
1. *Ectoderm*: From the ectoderm originate:
   a. *Superficial ectoderm*:
      (1) Epithelium of the skin (epidermis) and the epithelium of the glands whose excretory ducts lead to the surface of the body
      (2) Adenohypophysis
      (3) Dental enamel
      (4) Inner ear
      (5) Epithelium of the cornea and lens
   b. *Neuroectoderm*:
      (1) *Neural tube*:
         (a) Nervous tissue of the brain and spinal cord
         (b) Retina
         (c) Pineal gland
         (d) Neurohypophysis
      (2) *Neural crests*:
         (a) Sensory and cranial nerves and ganglia.
         (b) Schwann cells.
         (c) In the cephalic region, neural crest produce mesenchyme (*ectomesenchyme*), which will give rise to connective and supportive tissues of the head and tooth.
         (d) Cartilages of the pharyngeal arches.
         (e) Cells with special functions that are located in various tissues, including endocrine and paracrine cells and pigment cells of the skin and epithelium of the mouth.
         (f) Adrenal medulla.
2. *Mesoderm*: Apart from the notochord, which gradually disappears, the cells of the mesoderm differentiate very early into two lineages:
   a. *Mesenchymal cells*: They produce the *mesenchyme* of almost the entire body—except in some cephalic regions. The mesenchyme gives rise to:
      (1) Connective tissues, including those of organ stroma, skin (dermis and hypodermis) and the connective tissues of nerves and muscles.

(2) Specialised connective or supportive tissues (cartilage and bone).
(3) Muscle.
(4) Epithelial-type tissues lining the serous cavities of the body (cavities derived from the coelom of the embryo).
(5) Epithelial tissues lining the lumen of blood vessels.
(6) Epithelia of the excretory system (kidney and urinary tract) and reproductive system.

b. *Haemopoietic stem cells*: They produce all the cells that circulate in the blood.

3. *Endoderm*: The endoderm produces:
   a. The epithelium of the digestive tract.
   b. The epithelium (parenchyma) of the glands that discharge their secretions into the lumen of the digestive tract (liver, pancreas, etc.).
   c. Epithelium of the bladder.
   d. Epithelium of thyroid, tympanic cavity, auditory tube, tonsils and parathyroid glands.

D. Because of their interest in the formation of the oral cavity, we point out here those derived from the cephalic area:

1. *Derivatives of the ectoderm*:
   a. Derivatives of the superficial ectoderm:
      (1) Epithelium of the oral mucosa.
      (2) Epithelia of *salivary glands* (at least parotid glands and other minor glands).
      (3) Odontogenetic epithelium (*enamel organ*).
   b. Derivatives of the neuroectoderm:
      (1) Neural tube derivatives (already mentioned above): nervous tissue of the brain, retina, pineal gland and neurohypophysis.
      (2) Derived from neural crests:
         (a) Connective tissues and bones of the face and skull.
         (b) *Dental papilla* (which forms dental pulp and dentine).
         (c) Dental sac or follicle (which gives rise to the periodontal ligament, alveolar bone and cementum of the dental root).
2. *Derived from mesoderm*: some connective tissues and bones of the head, although most of the cephalic mesenchyme is derived from the neuroectoderm (ectomesenchyme).
3. *Derived from the endoderm*:
   a. Epithelia of the root of the tongue, pharynx and larynx.
   b. Submaxillary, sublingual and lingual *salivary glands* of von Ebner (some authors attribute all these glands to be of ectodermal origin).
   c. Epithelium of the *tonsils*.

Take-Home Message

Humans undergo several stages of development, some of which include the development of teeth and other associated structures. Development occurs through cellular (proliferation, differentiation, migration, cell death) and molecular (cell communication, induction) mechanisms that shape the embryo (morphogenesis) and produce tissues (histogenesis). In early stages of development, the three germinative leaves of the embryo originate: ectoderm (external), mesoderm (intermediate) and endoderm (internal), from which all tissues of the organism will be formed. It is important to remember that through neurulation, part of the ectoderm enters the mesoderm and forms the neural tube (which will form the central nervous system) and the neural crests. The cells of the neural crests migrate through the embryonic mesoderm and give rise to cells and tissues throughout the human body. It is important to note that most of the dental tissues and periodontium originate from the surface ectoderm and neural crests. It is also important to know which tissues and organs of the human body originate from each germinative leaf.

Further Reading

Artus J, Chazaud C (2014) A close look at the mammalian blastocyst: epiblast and primitive endoderm formation. Cell Mol Life Sci 71:3327–3338

Carleton AE, Duncan MC, Taniguchi K (2022) Human epiblast lumenogenesis: from a cell aggregate to luminal cyst. Semin Cell Dev Biol 131:117–123

Carlson BM (2014) Human embryology & developmental biology, 5th edn. Elsevier, Philadelphia

Chiego DJ (2023) Development and structure of cells and tissues. In: Essentials of oral histology and embryology, 6th edn. Elsevier, St. Louis, pp 1–17

Dudek RW (2011) Embryology, 6th edn. Wolters Kluwer, Barcelona

Fehrenbach MJ, Popowics T (2020) Prenatal development. In: Illustrated dental embryology, histology, and anatomy, 5th edn. Elsevier, St. Louis, pp 19–32

Ghimire S, Mantziou V, Moris N, Arias AM (2021) Human gastrulation: the embryo and its models. Dev Biol 474:100–108

Gómez de Ferraris ME, Campos Muñoz A (2009) Embriología General Humana. In: Histología, Embriología e Ingeniería Tisular Bucodental, 3rd edn. Médica Panamericana, pp 27–56

Keller R (2005) Cell migration during gastrulation. Curr Opin Cell Biol 17:533–541

Martik ML, Bronner ME (2021) Riding the crest to get a head: neural crest evolution in vertebrates. Nat Rev Neurosci 22:616–626

McClay DR (1991) Gastrulation. Curr Opin Genet Dev 1:191–195

Mitchell B, Sharma R (2009) Embryology. Churchill Livingstone, Edinburgh

Moon LD, Xiong FZ (2022) Mechanics of neural tube morphogenesis. Semin Cell Dev Biol 130:56–59

Nanci A (2013) General embryology. In: Ten Cate's oral histology, 8th edn. Elsevier, London, pp 1–13

Rodrigues da Silva MA, da Silveira GD, Taufer CR, Calloni GW (2022) The mesenchymal potential of trunk neural crest cells. Int J Dev Biol 66:317–331

Roth DM, Bayona F, Baddam P, Graf D (2021) Craniofacial development: neural crest in molecular embryology. Head Neck Pathol 15:1–15

Schauer A, Heisenberg CP (2021) Reassembling gastrulation. Dev Biol 474:71–81

Sheng GJ (2015) Epiblast morphogenesis before gastrulation. Dev Biol 401:17–24

Solnica-Krezel L, Sepich DS (2012) Gastrulation: making and shaping germ layers. Annu Rev Cell Dev Biol 28:687–717

Van der Spuy M, Wang JX, Kociszewska D, White MD (2023) The cellular dynamics of neural tube formation. Biochem Soc Trans 51:343–352

Epithelial Tissues

3

What You Will Learn in This Chapter

In this chapter you will learn what epithelia are, what their main characteristics are, how they are classified, what their embryonic origin is and what their functions are. You will learn how cells organise themselves to form epithelia, and that epithelia and their cells are polarised structures. You will also learn that epithelial cells are constantly renewed. Finally, you will learn that some epithelia (glandular epithelia) are specialised in the production of substances that they secrete onto surfaces or free spaces or into other neighbouring tissues. You will see what types of glandular epithelia exist, how their secretion mechanisms are and how these epithelia originate in the embryo.

3.1 Epithelium or Epithelial Tissue

A. The human body displays a wide variety of tissues that we consider *epithelial tissues* or *epithelia* (from the Ancient Greek ἐπι epi 'on, over' and θηλή thēlḗ 'nipple'), which show certain common characteristics.

B. Epithelial tissues are characterised by:
   1. They are made up of tightly packed cells with little space between them and little extracellular matrix.
   2. They form *sheets* that cover or upholster other tissues, usually connective tissues, which we call *underlying tissues*.
   3. They separate underlying tissues from non-tissue occupied spaces, which may be:
      a. *Internal ducts* or *cavities* of the organism that are isolated from the environment surrounding the human body. For example:
         (1) The *serous cavities* of the body.
         (2) The ducts of the blood vessels.

F. J. Sáez, I. Badiola, *Essential Oral Histology*, Learning Materials in Biosciences,
https://doi.org/10.1007/978-3-032-03165-5_3

   b. The outside of our organism. This is what happens in the epithelium of the skin.
   c. *Ducts* in our body that communicate with the outside world. For example, the ducts of the digestive and respiratory systems.
4. They are *polarised*, as the region of the tissue facing the underlying tissue is different from the region facing the space or cavity. This is also reflected in the organisation of the cells of the epithelium. Therefore, they can be distinguished:
   a. *Tissue polarisation*:
      (1) A *basal pole* or *region*, in contact with the underlying connective tissue.
      (2) An *apical pole* or *region*, oriented towards the free space on the other side.
   b. *Cell polarisation*:
      (1) It is visible mainly in the cells of simple epithelia (see below).
      (2) In multi-layered epithelia this can be seen only in one of the layers, for example in the basal cells.
      (3) They can be distinguished:
         (a) Cell surfaces:
            1. *Apical surface*: The opposite of the underlying tissue, in contact with the free surface (in a simple epithelium).
            2. *Lateral surface*: It serves to join with the rest of the epithelial cells.
            3. *Basal surface*: In contact with the underlying tissue.
         (b) Cellular domains: In a simple epithelium, the membranes of its cells have two domains, i.e. two different areas of organisation:
            1. *Apical domain*: closest to the free surface.
            2. *Basolateral domain*: formed by the lateral and basal regions. The separation of the two domains is achieved by the *tight junctions* (see below).
5. They have a single cell type or several cell types.
6. They are organised by cells arranged in single or multiple *layers* or *strata*.
7. They are *avascular* tissues, i.e. they lack blood vessels.

C. Not all epithelia form sheets; some form masses of cells surrounded by other tissues. In this case, we usually call them *epithelioid tissues*. This is common in endocrine glands.

3.2 Embryonic Origin

A. All *germinative leaves* of the embryo can produce some form of epithelium (Fig. 3.1).
B. The *surface ectoderm* originates:
   1. The epithelium of the skin (epidermal epithelium or epidermis).
   2. The epithelia of the glands that shed their secretion on the skin surface (sweat, sebaceous and mammary glands).
   3. Part of the epithelium of the digestive tract (e.g. oral epithelium).

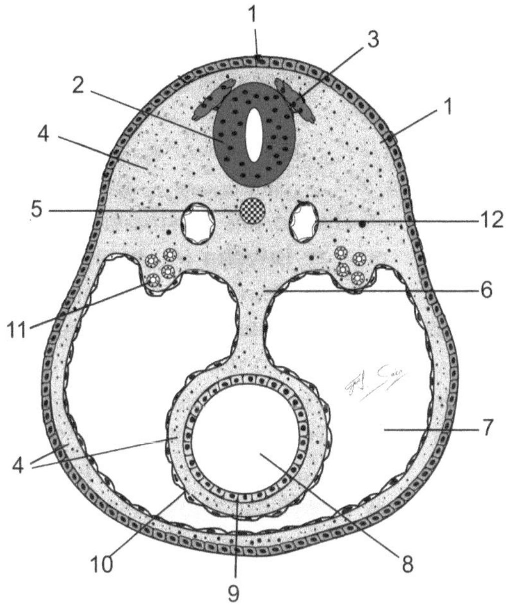

Fig. 3.1 Simplified representation of a cross-section of an early embryo to indicate the different origins of epithelial tissues. The surface ectoderm and endoderm are already epithelial tissues in origin and retain this condition in the structures they form in adults. The mesoderm is not an epithelial tissue, but during development it produces epithelial tissues. (1) Surface ectoderm; (2) neural tube; (3) neural crests; (4) mesoderm; (5) notochord; (6) mesentery; (7) coelom (gives rise to the serous cavities of the body); (8) lumen of the digestive tract; (9) endoderm; (10) mesothelium (epithelium lining the walls of the serous cavities); (11) epithelia of the excretory and reproductive apparatus; (12) endothelium lining the blood vessels

C. The *neuroectoderm* give rise to the epithelia of the adrenal medulla, which is produced from the *neural crests*.

D. The *endoderm* produces:
   1. Most of the epithelia lining the digestive and respiratory tracts.
   2. The epithelia lining the glands attached to these apparatuses (pancreas, liver, etc.).
   3. The epithelia of some endocrine glands (thyroid and parathyroid).

E. Some epithelia (formerly called *false epithelia*) are also formed from the *mesoderm*:
   1. The epithelia of the excretory system, the reproductive system and the adrenal cortex.
   2. The *mesothelium* or epithelium that lines the *serous cavities*[1] of the body (pericardial cavity, pleural cavity and peritoneal cavity).
   3. The *endothelium* or epithelium that lines the insides of blood vessels (which carry blood) and lymphatic vessels (which carry lymph).

[1]The serous cavities originate from the intraembryonic coelom.

3.3 Functions of Epithelia

A. Each epithelium may be specialised in performing one or more of the following functions:

B. *Protection*:
   1. Physical protection against mechanical aggressions.
   2. Protection from loss of underlying tissue components.

C. *Secretion*:
   1. Cells of the epithelium may specialise in producing substances (*glandular cells*).
   2. Sometimes all the cells of an epithelium are glandular, and that *glandular epithelium* is often called a *gland*.

D. *Absorption*:
   1. To allow the passage of necessary substances from the apical environment to the underlying connective tissue.
   2. These substances can be nutrients, metabolites, gases, solvents.

E. *Transport*:
   1. When the passage of substances through an epithelium requires energy input from the epithelial cells.
   2. This process can occur in processes of:
      a. Excretion
      b. Absorption
      c. Secretion

F. *Sensory receptivity*:
   1. Through nerve endings that pick up stimuli (e.g. painful stimuli).
   2. By epithelial cells specialised in stimulus uptake.

3.4 Classification of Epithelia

A. They are usually distinguished:
   1. *Lining epithelia*, epithelia whose main function is to cover a surface.
   2. *Glandular epithelia*, consisting mainly of secretory cells.
   3. *Transport epithelia*, specialised in transporting substances from the basal to the apical zone or vice versa.

B. This classification is artificial, as many epithelia mix features of lining, glandular or transport epithelia, but if taken with caution it can serve to better understand epithelial tissues.

3.5 Cytological and Histological Organisation of Epithelia

A. All variants of *specialised cell junctions* described in cell biology text books can be found among epithelial cells (Fig. 3.2). These types of junctions are:

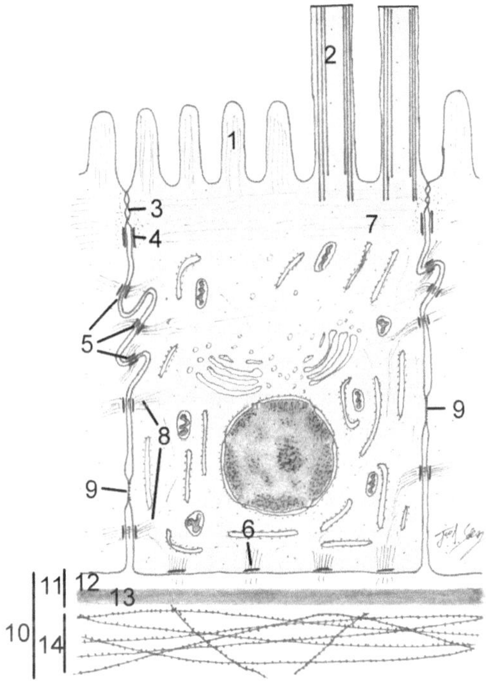

Fig. 3.2 Representation of a hypothetical epithelial cell showing the main features of epithelial cells. (1) Microvilli; (2) cilia; (3) tight junction (zonula occludens); (4) adherens junction (zonula adherens); (5) desmosome (macula adherens); (6) hemidesmosome; (7) terminal web of actin filaments; (8) intermediate filaments (tonofilaments); (9) gap junction; (10) basal membrane; (11) basal lamina; (12) lamina rara or lamina lucida; (13) lamina densa; (14) lamina reticularis

1. *Tight junction* or *zonula occludens*:
   a. The membranes of two cells are in intimate contact.
   b. It forms a belt (*zonula*) around each cell.
   c. It is on the lateral membrane near the apical membrane.
   d. It serves to seal the space between cells, preventing or hindering the passage of substances from one side of the tissue to the other (*paracellular transport*).
   e. It divides the plasma membrane of the cell into an apical and a basolateral domain.
2. *Adherens junction*, *banded desmosome* or *zonula adherens*:
   a. Cells adhere to each other.
   b. It has the shape of a belt surrounding the cell (*zonula*).
   c. It is located very close to the tight junction, in a basal position with respect to it.
   d. On its inner side it is associated with *actin microfilaments* that form an internal meshwork called the *terminal web*.
3. *Desmosome* or *macula adherens*:
   a. Cells are anchored to each other.
   b. It is a focal junction (*macula*).
   c. It is found on all cell surfaces that contact another cell.
   d. *Intermediate keratin filaments* are anchored inside the cell.

4. *Gap junction*:
   a. They are communicating junctions, i.e. they allow cells to send signals to each other and to couple their functioning.
   b. There are small channels that connect the cytosol of both cells and allow the passage of small substances (ions, cAMP).
5. *Hemidesmosome*:
   a. It is an anchoring junction of the cell with the *basal lamina*, which is explained below.
   b. It is a focal junction.
   c. Intermediate keratin filaments are anchored inside.

B. The epithelia are separated from the underlying tissue by a *basal membrane*.
   1. The basal membrane is visible under the light microscope.
   2. When this basal membrane is observed under the transmission electron microscope, it can be seen to be made up of:
      a. *Basal lamina*. This is a special structure attached to the epithelium:
         (1) It is a double sheet:
            (a) Clear region (*lamina rara* or *lamina lucida*) close to the epithelium.
            (b) Darker region (*lamina densa*) close to the underlying connective tissue.
         (2) It is synthesised by the epithelium.
         (3) It consists of non-fibrillar *collagen* molecules (*type IV* and *type VII*) and *laminin*.
      b. *Lamina reticularis*.
         (1) It is synthesised by the underlying connective tissue.
         (2) It consists of *type III collagen* fibres (see Chap. 4).

C. Epithelia are *avascular* tissues:
   1. They lack blood vessels.
   2. Nutrients from the blood vessels in the underlying tissue must diffuse through the basal lamina.
   3. Exceptionally, very fine capillary loops with a small connective tissue sheath penetrating between the epithelial cells and forming *intraepithelial capillaries* are observed in some epithelium.

D. In some epithelia, the *sensory nerve endings* pass through the basal lamina and are located in the interstices between the epithelial cells.

3.6 Renewal and Maintenance of Epithelia

A. Epithelia have a variable *turnover rate*, from a few days (small intestine epithelium) to a month (epidermis). In some cases the rate of replacement is very slow (epithelium of the renal tubules).

B. Older cells are replaced by younger cells.

C. Several compartments can be seen in the epithelium:

1. *Mitotic compartment*: Contains cells that can divide by mitosis. It consists of two zones:
   a. *Restricted zone* or *stem cell niche*: Stem cells are characterised by the following:
      (1) They are morphologically indistinguishable.
      (2) When they divide, they produce two cells: the cell that remains in the niche is still the stem cell, and the one that is displaced from the niche is the one that differentiates and renews the epithelium.
   b. *Area without stem cells*: Cells displaced from the niche that continue to divide.
2. *Post-mitotic* or *differentiation compartment*: Its cells no longer divide and complete their differentiation.

D. The renewal procedure involves:
1. *Cell division* (in the mitotic compartment).
2. *Migration*: New cells move from the niches through the epithelium.
3. *Differentiation*: In some epithelia, cells change their characteristics as they move.
4. *Senescence*: At the end of the process, cells age and lose vitality.
5. *Removal* of older cells from certain areas of the epithelium (Fig. 3.3):
   a. In simple epithelia:
      (1) There are points of removal of old cells in the single cell layer at points away from the mitotic compartments.
      (2) The cells move within the stratum from the stem cell niches to a point of elimination.
   b. In stratified epithelia:
      (1) The niches are usually in the basal stratum.
      (2) The cells move upwards towards the upper strata.
      (3) Cells that reach the apical surface of the epithelium are removed.

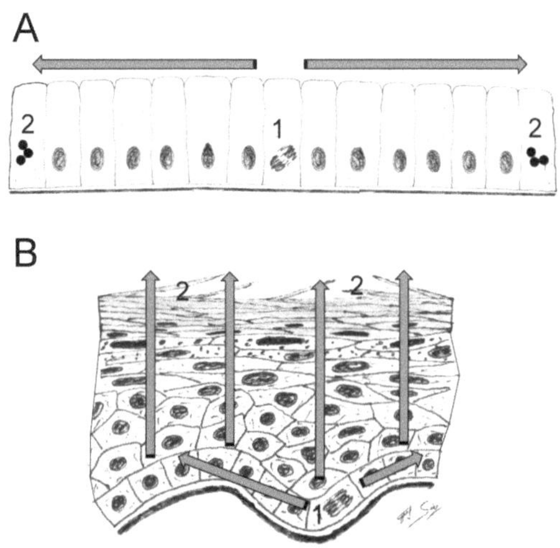

Fig. 3.3 Cell renewal in a simple (**A**) and a stratified (**B**) epithelium. Stem cells (1) divide and produce new epithelial cells that move in the direction indicated by the arrows. At certain points, the cells die (2) and are destroyed or detached from the epithelium

3.7 Classification of the Lining Epithelia

A. Different types of epithelium can be seen in Figs. 3.4, 3.5 and 3.6.
B. Lining epithelia are usually classified according to several morphological criteria:
   1. *Shape* of the cells when viewed in histological sections made perpendicular to the apical and basal planes. There are three basic types (although we can find epithelia whose cells have an intermediate shape):
      a. *Squamous epithelium*, if the cells are flattened.
      b. *Cuboidal epithelium*, if the appearance of the cell is square or trapezoidal.
      c. *Cylindrical*, *columnar* or *prismatic epithelium*, if the cell looks elongated, rectangular.
   2. *Number of layers* or *strata* of cells. Three types are distinguished:
      a. *Simple epithelium*: a single layer of cells (Fig. 3.3).
      b. *Stratified epithelium*: two or more cell layers (Fig. 3.4).
      c. *Pseudostratified epithelium*: all cells in contact with the basal lamina (one layer) but there are low cells and high cells, giving the impression of stratified epithelium (Fig. 3.5). Only the upper cells reach the free surface of the epithelium.

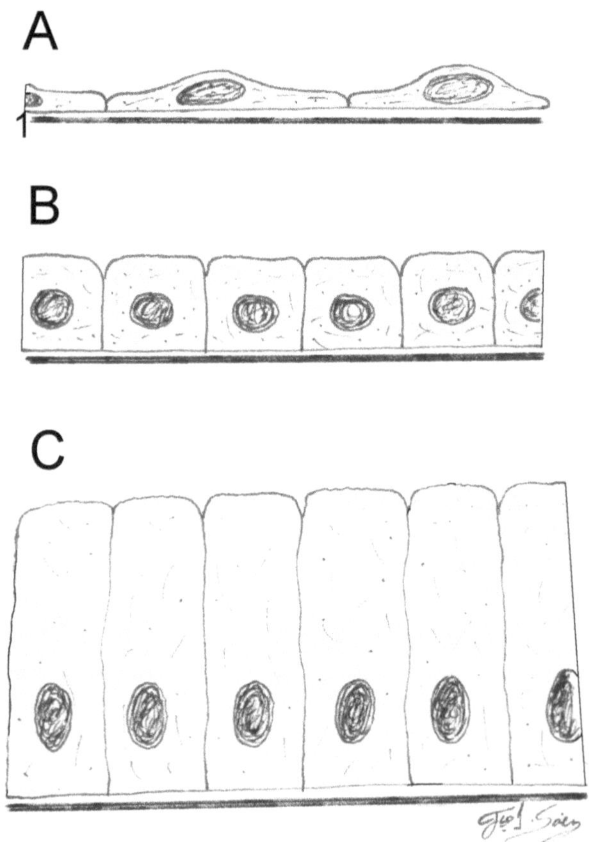

Fig. 3.4 Simple epithelia depicted in cross-sections. (**A**) Simple squamous epithelium; (**B**) simple cuboidal epithelium; (**C**) simple cylindrical or columnar epithelium; (1) basal lamina

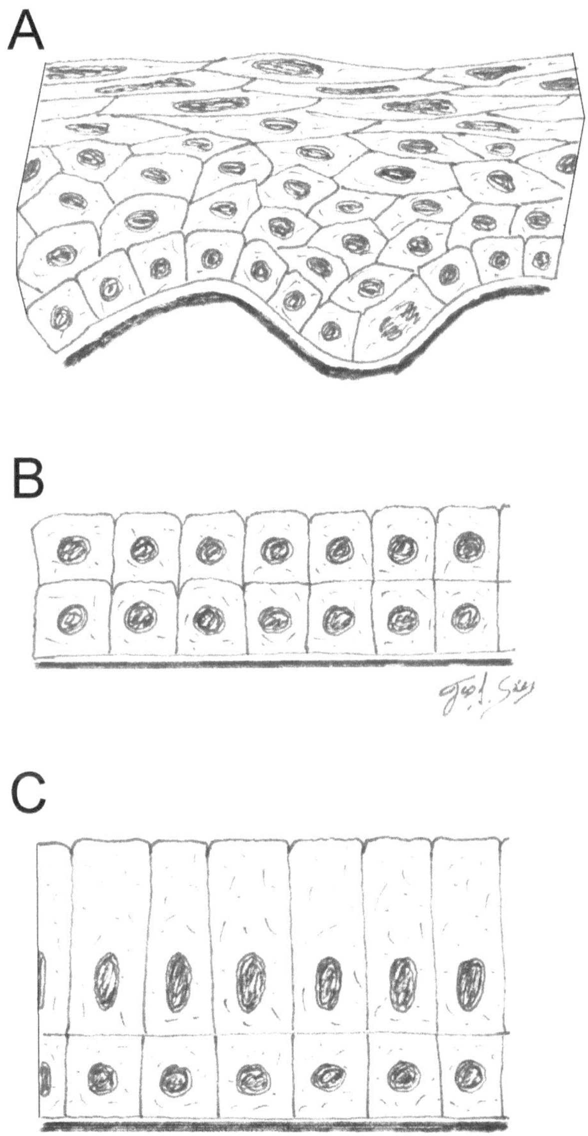

Fig. 3.5 Stratified epithelia depicted in cross-sections. (**A**) Stratified squamous epithelium; (**B**) stratified cuboidal epithelium; (**C**) stratified columnar epithelium

3. *Surface differentiations* on the apical side of epithelial cells. Several types of epithelia are distinguished when the cells show some specialised structure on the free surface:
   a. *Ciliated epithelium*, which has cells with *cilia* on the free epithelial surface.
   b. *Epithelium with striated border* or *brush border*, possessing many large *microvilli*.
   c. *Epithelium with stereocilia*, showing *stereocilia*, which are long and immotile microvilli.

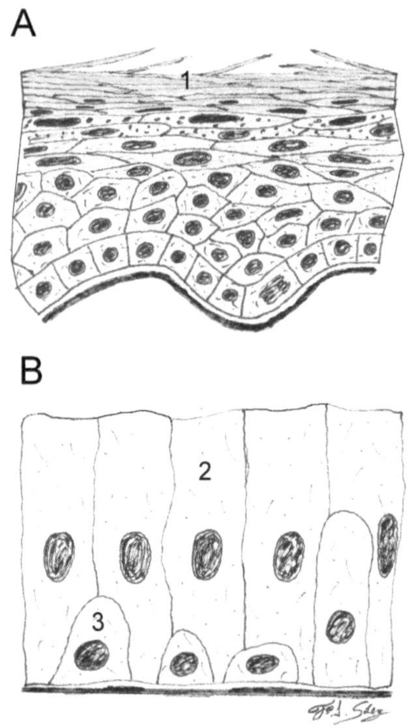

Fig. 3.6 Some epithelia with special features depicted in cross-sections. (**A**) Keratinised or cornified stratified flat epithelium; (**B**) columnar pseudostratified epithelium; (1) cornified cell layer; (2) high cells; (3) low cells

   d. *Keratinised* or *cornified epithelium*, with the flat, dead, *keratin-filled* apical cells forming a superficial layer that is shed (Fig. 3.5).

C. When an epithelium is stratified, i.e. has several cell layers, the shape of the cells in the apical layer is always taken into account.

D. These classification criteria apply to lining and transport epithelia and, in many cases, to glandular epithelia.

E. Some epithelia are difficult to classify:
   1. *Urinary epithelium* or *transitional epithelium (urothelium)* of the urinary bladder. The number of layers depends on whether the organ is empty (many layers) or full (few layers).
   2. *Seminiferous epithelium* of the testis.

F. Some epithelia are given special names:
   1. *Mesothelium*, which lines the serous cavities.
   2. *Endothelium*, which lines the lumen of blood and lymphatic vessels.

3.8 Glandular Epithelia: Glands

A. Glandular epithelia are epithelia whose cells, or most of them, are specialised in the production of one or more substances that are released into the extracellular environment.
B. Glandular epithelia can be organised in various ways:
   1. Areas of lining epithelia whose cells are secretory.
   2. Epithelia specialised in secretion. These epithelia are called *glands*.
   3. The term gland is used to denote:
      a. A glandular epithelium found within an organ. For example, minor salivary glands scattered on the tongue.
      b. An organ whose main constituent is glandular epithelia. Examples are the major salivary glands (parotid, submaxillary and sublingual). In these glands:
         (1) The glandular tissue, the epithelium, is called *parenchyma*.
         (2) The rest of the organ tissues surrounding the parenchyma (connective tissues, blood vessels and nerves) form the *stroma*.[2]
C. Considering the destination of the secretion, glands can be classified into the following types (Fig. 3.7):
   1. *Exocrine glands*:
      a. Secretion occurs from the apical pole of the epithelium.
      b. They discharge their secretion directly or through an epithelial lining duct:
         (1) To the outside of the organism.
         (2) In body cavities communicating with the outside, such as those of the digestive or respiratory tract.
   2. *Endocrine glands*:
      a. They discharge their secretion from the basal pole.
      b. If secreted substances are transported in the blood and lymph, they carry signals to other cells or tissues in a distant part of the body. This process is called *endocrine communication* or *regulation*. In this case, the molecules released are called *hormones*.
      c. If it is transported by diffusion into nearby tissues to send signals to those tissues, it is called a *paracrine secretion*.
      d. A variant of this paracrine secretion is the so-called *autocrine secretion*,[3] in which the signal produced by one cell type regulates the activity of the same cell type.

---

[2] The concepts of organ parenchyma and stroma have been explained in Chap. 1.

[3] Paracrine and autocrine communication takes place between neighbouring cells and can be performed by non-epithelial cells.

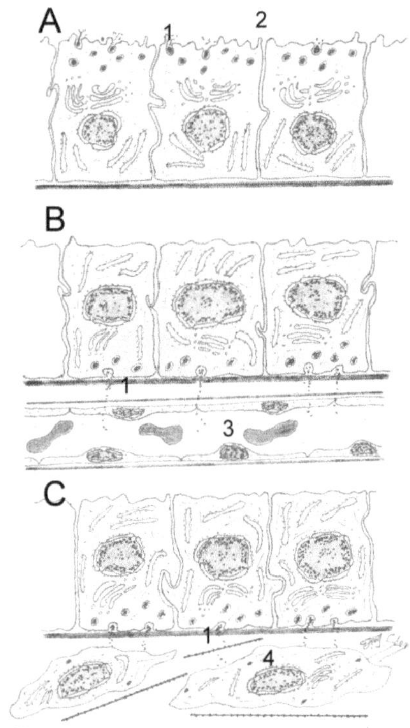

Fig. 3.7 Representation of exocrine (**A**), endocrine (**B**) and paracrine (**C**) glands. (1) Release of secretion; (2) free surface; (3) blood capillary; (4) cells close to the epithelium

3. *Amphicrine glands*: when a single cell type performs a double secretion, one substance by exocrine secretion and the other by endocrine secretion. An example is the liver.
4. *Mixed glands*: when a gland has at least two different types of secretory cells, one of which performs exocrine secretions and the other endocrine secretions. An example is the pancreas.

3.9 Exocrine Glands

A. Several classifications of exocrine glands can be made according to different criteria.
B. Depending on the mechanism of secretion release, three types of exocrine glands are considered (Fig. 3.8):
   1. *Eccrine* or *merocrine*:
      a. When the secretion product is released by *exocytosis*.
      b. This mechanism is the most common.
   2. *Apocrine*:
      a. When the apical end of the cell is cleaved, so that the entire cytoplasmic contents of this region are released.

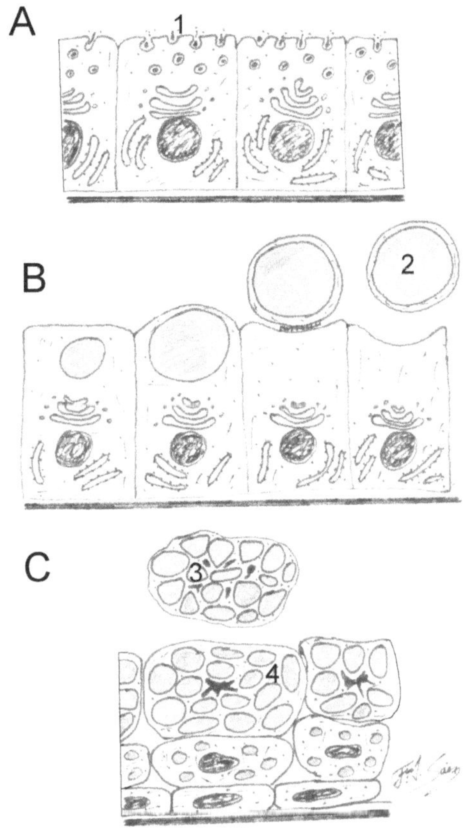

Fig. 3.8 Representation of eccrine or merocrine (**A**), apocrine (**B**) and holocrine (**C**) secretion mechanisms. (1) Secretion by exocytosis; (2) cleaved apical cytoplasm; (3) released dead cell; (4) cells undergoing cell death

   b. The basal region of the cell, with the nucleus and most of the organelles, remains intact and continues to produce materials for secretion.
   c. It is not very common. It occurs in the mammary gland.
3. *Holocrine*:
   a. When it is the whole cell that dies, it ruptures, and all its contents are released. An example is the sebaceous gland in the skin.
   b. Holocrine secretion is also considered to be the release of the living cell that separates from the epithelium. An example would be the release of spermatozoa.

C. According to the organisation of the gland (*morphological classification*), exocrine glands can be:
1. *Intraepithelial glands* (Fig. 3.9): A lining epithelium that has some secretory cells. They can be of two types:
   a. *Unicellular intraepithelial gland*, if the secretory cells are isolated between non-secretory epithelial cells.
   b. *Multicellular intraepithelial gland*, if there are clusters of secretory cells in the lining epithelium.

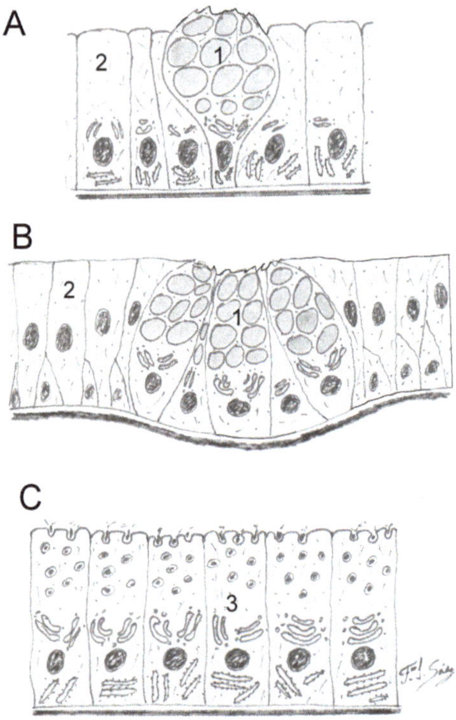

Fig. 3.9 Unicellular intraepithelial gland (**A**), multicellular intraepithelial gland (**B**) and secretory sheet (**C**). (1) Secretory cells; (2) lining, non-secretory epithelial cells; (3) secretory lining cells

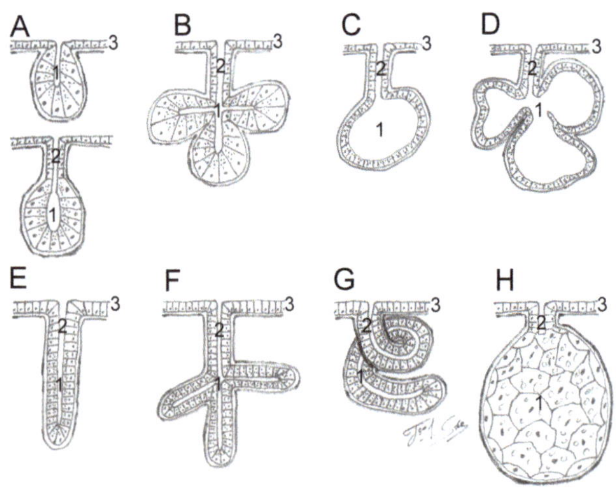

Fig. 3.10 Representation of simple acinar (**A**), simple branched acinar (**B**), simple alveolar (**C**), simple branched alveolar (**D**), simple tubular (**E**), simple branched tubular (**F**), glomerular (**G**) and simple saccular (**H**) glands. (1) Adenomeres; (2) excretory ducts; (3) surface epithelia

2. *Secretory sheet* (Fig. 3.9): These are lining epithelia whose cells are all secretory.
3. *Extra-epithelial glands* (Figs. 3.10 and 3.11): The secretory epithelium is not in the lining epithelium that covers a surface:
   a. They form an invagination that is located in the underlying connective tissue.
   b. The epithelium envelops a space (*lumen*) that is connected to the free surface of the lining epithelium.

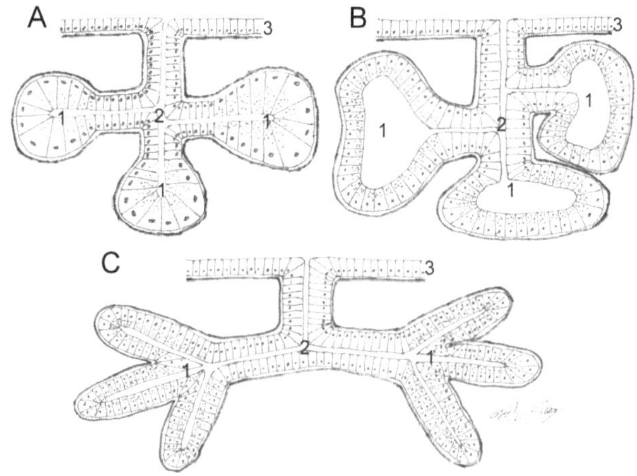

Fig. 3.11 Compound acinar gland (**A**), compound alveolar gland (**B**) and branched compound tubular gland (**C**). (1) Adenomeres; (2) excretory ducts; (3) surface epithelia

c. Two zones can be distinguished in the invagination:
   (1) The secretory portion is called the *adenomere.*
   (2) There is continuity between the lining epithelium and the adenomere via a non-secretory tubular epithelial duct, the *excretory duct.*
d. Extra-epithelial exocrine glands can have a very varied organisation and are classified according to several criteria.
   (1) Depending on the shape of the adenomere, the glands can be:
      (a) *Acinar*, formed by *acini* (singular *acinus*, from the Latin acinus and this from the Greek ακινος akinos 'grape seed'), which are not very large spherical structures with a very small lumen.
      (b) *Alveolar*, consisting of relatively large spherical structures with a wide span (alveoli, from Latin alveŏlus, diminutive of alveus 'cavity').
      (c) *Tubular*, consisting of structures in the form of a hollow tube or cylinder. If this tube is rolled into a *glomerulus* (from the Latin glomus 'ball'), the gland is called a *glomerular gland.*
      (d) *Saccular*, dilated structures filled with epithelial cells. This gland has no internal lumen.
      (e) Intermediate structures:
         1. Tubular adenomere ending in an acinus (*tubuloacinar*).
         2. Tubular adenomere terminating in an alveolus (*tubuloalveolar*).
      (f) If the adenomeres are divided, or if there are several adenomeres at the end of an excretory duct, it is called a *branched gland* (branched acinar, branched alveolar or branched tubular).
   (2) According to the organisation of the excretory duct, the glands can be:
      (a) *Simple gland*: has no excretory duct or a single duct without branches (Fig. 3.10).
      (b) *Compound gland*: has a branched excretory duct (Fig. 3.11).

   e. Extra-epithelial exocrine glands are described by combining the above criteria: simple acinar, compound acinar, simple tubular, compound tubular, simple branched acinar, simple branched tubular glands, etc.

D. According to the substance it produces and releases, two types of cells can be distinguished (Fig. 3.12):
   1. *Mucous cells* (from Latin muccōsus, 'mucus or mucus-like'):
      a. They produce *mucus*, a viscous material made up of glycoproteins called *mucins*.
      b. The secretion accumulates at the apical pole as voluminous and numerous *mucinogen granules* that occupy a large part of the cell cytoplasm.
      c. The rest of the organelles and the nucleus are compressed at the basal pole of the cell.
      d. In standard histological preparations, the mucinogen granules appear pale and poorly stained and give the impression that the cell is empty.
      e. Two types of mucous cells have been distinguished:
         (1) *Closed-pole mucous cells*, the apical surface of which appears to be intact.
         (2) *Open-pole mucous cells*, the apical pole of which appears to be open, which is an artefact. These open-pole mucous cells are usually cup-shaped or calyx-shaped. They are called *goblet cells*.
   2. *Serous cells* (from Latin serum, 'serum'):

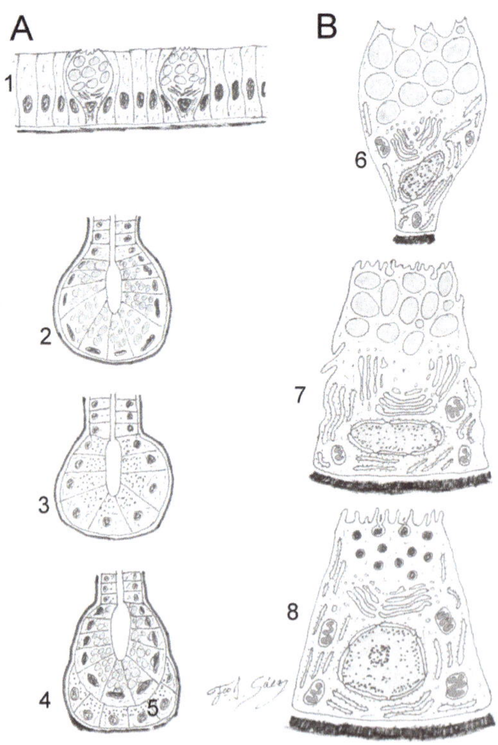

Fig. 3.12 Different types of exocrine glands (**A**) and glandular cells (**B**) according to the substance they release. (1) Unicellular mucous glands; (2) mucous gland; (3) serous gland; (4) mixed gland; (5) serous demilune (crescent of Giannuzzi or von Ebner); (6) open-pole mucous cell (globet cell); (7) closed-pole mucous cell; (8) serous cell

a. Serous cells produce a watery, proteinaceous, rather fluid secretion that usually contains *enzymes*.
b. The secretory granules (*zymogen granules*) are small and do not compress the other cellular components towards the basal pole.
c. The granules are stained with the usual histological stains and the apical cytoplasm is well stained and granular in appearance.

3. The combination of these cells in the glandular epithelia can produce three types of glands (Fig. 3.12):
   a. *Mucous glands*, consisting of mucous cells.
   b. *Serous glands*, consisting of serous cells.
   c. *Mixed glands*, consisting of mucous and serous cells. In these cases, the adenomere consists of:
      (1) A mucous acinus or tubuloacinus.
      (2) A cap of serous cells called *serous Demilune* (or *crescent of Giannuzzi*, or *crescent of von Ebner's*, or *crescent of Heindenhain*) lining the end of the mucous tubuloacinus.

3.10 Endocrine Glands

A. *Morphological classification*: There is a great variety of endocrine glands, but they can be classified as follows:
   1. *Unicellular glands*: isolated cells in lining epithelia or exocrine glandular epithelia.
   2. *Islet glands*: more or less spherical groups of cells surrounded by connective tissue.
   3. *Cord-shaped glands*: elongated groups of cells lined with connective tissue.
   4. *Follicular glands*:
      a. The epithelium forms hollow spheres (follicles, from Latin follicŭlus 'pouch').
      b. The epithelium is usually simple cubic.
      c. The follicles are surrounded by connective tissue.

B. A classification *according to secretion* is almost impossible because the substances produced by the endocrine glands are so varied. A simple classification is:
   1. Glands that produce *proteins* or *polypeptides*.
   2. *Steroid-producing* glands.
   3. Glands that produce simpler substances, such as *modified amino acids*.

3.11 Embryonic Origin of the Glands

A. Exocrine and endocrine glands develop from a lining epithelium (Fig. 3.13).
B. *Mamelonation* or thickening forms due to an inductive action of the underlying embryonic mesenchyme (connective tissue of the embryo).

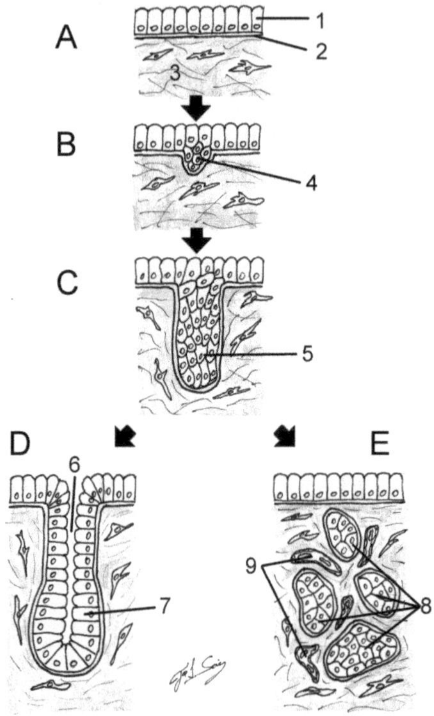

Fig. 3.13 Process of formation of the extra-epithelial exocrine (**A–D**) and endocrine (**A–C** and **E**) glands. Various time points have been depicted. The arrows indicate the temporal progression. In the final phase of the formation of the extra-epithelial exocrine gland (**D**), an excretory duct is produced. In the final stage of endocrine gland formation (**E**), the connection between the lining epithelium and the glandular epithelium is lost. (1) Lining epithelium; (2) basal membrane; (3) connective tissue; (4) mamelonation of epithelial cells; (5) growth of an epithelial bud; (6) formation of the excretory duct; (7) extra-epithelial exocrine gland; (8) endocrine gland; (9) blood capillaries

C. The mamelonate gradually grows and invades the mesenchyme to form a solid *bud* or excrescence.
D. Bud growth occurs, which varies according to the type of gland:
   1. In the exocrine glands:
      a. A cavity is formed, which will be the *lumen* of the gland.
      b. The gland always remains attached to the lining epithelium.
   2. In the endocrine glands:
      a. In these glands, no inner lumen is formed (except in follicular glands).
      b. The growing bud separates from the lining epithelium and is completely surrounded by mesenchyme.

Take-Home Message

Epithelial tissues or epithelia are characterised by tightly joined cells that form sheets that cover a connective tissue (underlying tissue) and separate it from a space or cavity. Epithelia and their cells are polarised structures showing a basal pole (in contact with the underlying tissue) and an apical pole (opposite to the basal pole). Epithelia originating from the three germinative sheets are found in the human body and can perform many functions, although a simple classification distinguishes between lining epithelia, glandular epithelia and transport epithelia. Epithelial cells show specialised cytological features, many of which are responsible for attaching the cells to each other and the epithelium to the underlying tissue. Between the underlying tissue and the epithelium there is a basal membrane, which contains a basal lamina intimately attached to the epithelium. The cells of the epithelia are constantly renewed by stem cells that produce new cells. Lining epithelia are classified according to the number of cell layers and the shape of the cells. Glandular epithelia (glands) are classified according to the destination of their secretion (exocrine, endocrine and paracrine). Exocrine glands are classified according to the cellular mechanism of their secretion (eccrine, apocrine and holocrine), according to tissue organisation (intraepithelial, secretory sheets and extra-epithelial), according to the morphology of the secretory portion (acinar, tubular, alveolar, saccular) and according to the shape of their non-secretory duct (simple and compound). They can also be classified according to the product they secrete (mucous and serous). Endocrine glands are difficult to classify because of their diversity. Finally, remember that glands are formed from lining epithelia.

Further Reading

Basbaum CB, Jany B, Finkbeiner WE (1990) The serous cell. Annu Rev Physiol 52:97–113

Botstein Griepp E, Robbins ES (1988) Epithelium. In: Weiss L (ed) Cell and tissue biology – a textbook of histology, 6th edn. Urban & Schwarzenberg, Baltimore, pp 113–154

Gartner LP (2017) Epithelium and glands. In: Textbook of histology, 4th edn. Elsevier, Philadelphia, pp 99–126

Gómez de Ferraris ME, Campos Muñoz A (2009) Histología general humana. In: Histología, Embriología e Ingeniería Tisular Bucodental, 3rd edn. Médica Panamericana, pp 57–78

Higashi T, Saito AC, Chiba H (2024) Damage control of epithelial barrier function in dynamic environments. Eur J Cell Biol 103:151410

Imai M, Shibata T, Moriguchi K, Sakai J (1986) Mucous cell types. Okajimas Folia Anat Jpn 63:1–5

Khan S, Fitch S, Knox S, Arora R (2022) Exocrine gland structure-function relationships. Development 149:dev197657

Kierszenbaum AL, Tres LL (2019a) Epithelial glands – cell biology. In: Histology and cell biology – an introduction to pathology, 5th edn. Elsevier, Philadelphia, pp 67–98

Kierszenbaum AL, Tres LL (2019b) Epithelium – cell biology. In: Histology and cell biology – an introduction to pathology, 5th edn. Elsevier, Philadelphia, pp 1–66

Krstić RV (1985) Epithelial tissue. In: General histology of the mammal. An atlas for students of medicine and biology. Springer Verlag, Berlin, pp 15–97

Lowe JS, Anderson PG (2015) Epithelial cells. In: Stevens & Lowe's human histology, 4th edn. Elsevier, Philadelphia, pp 37–54

O'Dowd G, Bell S, Wright S (2023) Epithelial tissues. In: Wheater's functional histology. Elsevier, Amsterdam, pp 82–100

Pawlina W (2024) Epithelial tissue. In: Histology a text and atlas with correlated cell and molecular biology, 9th edn. Wolters Kluwer, Philadelphia, pp 122–175

Polito MP, Romaldini A, Rinaldo S, Enzo E (2024) Coordinating energy metabolism and signaling pathways in epithelial self-renewal and differentiation. Biol Direct 19:63

Zhu GL, Hu HL, Xi RW (2021) The cellular niche for intestinal stem cells: a team effort. Cell Regen 10:1

Connective Tissues

4

What You Will Learn in This Chapter
In this chapter you will learn the general characteristics of connective tissues, their embryonic origin and the various types of cells that can be found in them. You will see that some cell types of connective tissues come from blood. You will see that one of the fundamental components of connective tissue is the extracellular matrix, formed by protein fibres embedded in a ground substance hydrated by the tissue fluid. Finally, you will learn that there are many varieties of connective tissues, some of which are specialised in forming skeletal structures of the human body.

4.1 General Characteristics

A. *Connective tissue* is so called because it provides the support and connective framework for the rest of the body's tissues.
B. All the body's epithelia, nerves and muscles are supported or covered by connective tissue.
C. Connective tissue is characterised by the abundance and importance of the *extracellular matrix* (Fig. 4.1).
   1. In this respect, connective tissue differs from the epithelia studied in the previous chapter:
      a. The cells are relatively few in number.
      b. A large part of the tissue is made up of the extracellular matrix that they synthesise.
   2. The extracellular matrix consists of *protein fibres* that are embedded in a gel-like *ground substance*, a ground substance that is hydrated by *tissue fluid*.
D. There are many varieties of connective tissue, so it would be more appropriate to speak of connective tissues.
   1. The varieties of connective tissue are often classified according to:
      a. The abundance of fibres.

F. J. Sáez, I. Badiola, *Essential Oral Histology*, Learning Materials in Biosciences,
https://doi.org/10.1007/978-3-032-03165-5_4

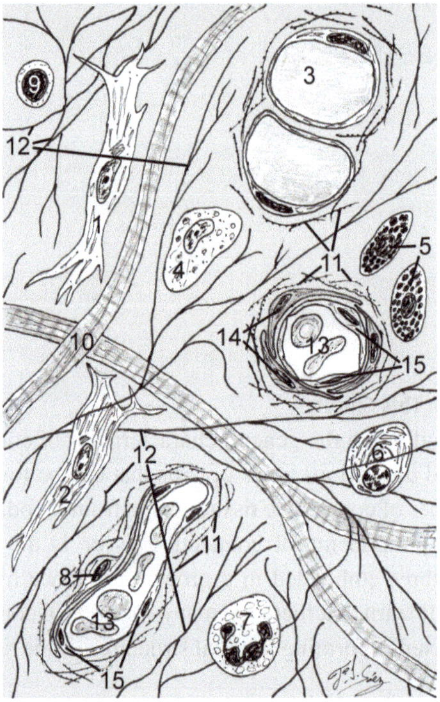

Fig. 4.1 Idealised representation of connective tissue using a loose connective tissue as an example. The main cell types and fibres that can be found in the tissue are shown. The grey background represents the ground substance hydrated by the tissue fluid. Two small blood vessels are also shown. No nerve fibres or nerves are depicted. The different components are not drawn to scale. (1) Fibroblast; (2) myofibroblast; (3) adipocyte; (4) macrophage; (5) mast cell; (6) plasma cell; (7) neutrophil; (8) pericyte; (9) lymphocyte; (10) collagen fibres; (11) reticular fibres; (12) elastic fibres; (13) blood vessels; (14) smooth muscle cells in the wall of a blood vessel; (15) endothelial cells of the blood vessel wall

   b. The predominant fibre type.
   c. If the fibres are loosely intertwined or tightly packed.
   d. Whether the fibres are ordered or randomly arranged.
2. Some varieties of connective tissue are classified depending on the predominant cell type in the tissue.
3. A simple classification of connective tissues would be as follows:
   a. *Connective tissue proper*. The varieties will be discussed at the end of this chapter.
   b. *Specialised connective tissues*:
      (1) Adipose tissue
      (2) Cartilaginous tissues
      (3) Bone tissue
      (4) Blood
      (5) Haemopoietic tissue
4. Separate chapters will be devoted to cartilage, bone, blood and haemopoietic tissue.

E. Connective tissues can perform different functions depending on the variety of connective tissue. For example:
   a. The cells of the *immune system*, whose function is to defend the body against external pathogens, are abundant in the loose connective tissue.
   b. The regular dense connective tissue present in tendons has a high density of fibres in its matrix whose function is to provide strength.

4.2 Embryonic Origin

A. Connective tissues are derived from *embryonic mesenchyme* (from the Greek μεσο meso 'medium' or 'intermediate' and ἔγχυμα énchyma 'infusion', 'injection').
   1. Mesenchyme is a simple variant of connective tissue, consisting of few undifferentiated cells, the mesenchymal cells, and an extracellular matrix with thin, sparse fibres.
   2. The embryonic mesenchyme lies between the ectoderm and endoderm.

B. Although most mesenchyme is derived from *mesoderm*, in the cephalic region of the embryo, mesenchyme is also produced from cells originating from the *neural crest* (Fig. 4.2). This cephalic mesenchyme of neuroectodermal origin is called *ectomesenchyme*.

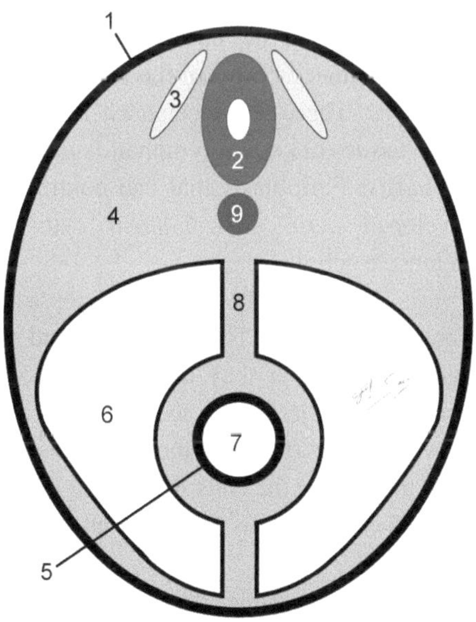

Fig. 4.2 Simplified schematic representation of a cross-section in the trunk region of an early embryo, once the digestive tract has been folded and closed. The connective tissues are mainly derived from mesenchyme (4) produced by the mesoderm. In the cephalic region, not shown here, the connective tissues are also derived from the ectomesenchyme that forms from the neural crests (3). (1) Surface ectoderm; (2) neural tube; (5) endoderm; (6) coelom; (7) digestive tract lumen; (8) mesentery; (9) notochord

C. Several cell types may exist in connective tissues, the origin of which may be different:
   1. Some cell types originate from embryonic mesenchyme (*cells of mesenchymal origin*).
   2. Other cell types are derived from blood (*cells of haemopoietic origin*).
   3. In any case, both are of mesodermal origin (with the aforementioned exception of cephalic ectomesenchyme).
   4. Traditionally, cells have been organised into two populations not necessarily related to origin, *resident cells* and *transient cells*, which can lead to some confusion.

4.3 Connective Tissue Cells

A. Some of the connective tissue cells are shown in Fig. 4.1.
B. Connective tissue cells have traditionally been classified into two categories:
   1. *Resident cells*:
      a. They are cells that move little in the tissue and are considered permanent residents of the tissue.
      b. This group includes cells of three different origins:
         (1) Cells formed in the tissue itself from undifferentiated *mesenchymal cells* of the embryonic mesenchyme or from undifferentiated mesenchymal cells that persist as reserve stem cells in the connective tissues of the adult. This group includes:
            (a) *Fibroblasts*: These are the basic cells of the connective tissues themselves. They synthesise most of the extracellular matrix.
            (b) *Reticular cells*: These are specialised forms of fibroblasts present in connective tissues of certain lymphatic organs.
            (c) *Myofibroblasts*: Fibroblasts that can contract because they possess a well-developed *actin microfilament* cytoskeleton. They resemble smooth muscle cells.
            (d) *Adipocytes*: These are cells in whose cytoplasm accumulate large lipid inclusions (fat droplets). They may be found singly or in small groups in connective tissue, or they may be the most abundant cell type, in which case they form *adipose tissue*.
            (e) *Mesenchymal cells*: These are the stem cells that remain in the connective tissue to produce the cells mentioned above.
            (f) *Pericytes*: These are cells that surround the smallest blood vessels (capillaries).
         (2) Cells produced from *embryonic haemopoietic* cells and installed in tissue during embryonic development.
            (a) This may include *resident macrophages*. They settle in connective tissue at the beginning of embryonic development and the population is renewed by cell divisions.

(3) Cells produced from *haemopoietic cells in the bone marrow* that reach the connective tissue by circulating in the blood.
   (a) Once they have reached the tissue, they act in it with little movement and do not leave it.[1]
   (b) Three cell types are included in this group:
      1. *Monocyte-derived macrophages*: These are cells derived from blood monocytes that penetrate and differentiate in connective tissue.
      2. *Dendritic cells (DC)*, also called *interdigitating dendritic cells (IDC)*: Some authors identify them with macrophages, but they have different functions. There are several types, some originating from monocytes and others not.
      3. *Mast cells*: They are derived from unidentified blood cells.

2. *Transient cells*: This group includes cells of the *immune system* that travel through the blood and infiltrate the connective tissue when there is an infection to carry out an immune response. These include:
   a. Monocytes
   b. Plasma cells
   c. Lymphocytes
   d. Neutrophils
   e. Eosinophils
   f. Basophils

C. Connective tissue cells can have different origins:
   1. *Cells of mesenchymal origin*, i.e. originating from the connective tissue itself. These have already been mentioned above.
   2. *Cells of haemopoietic origin*: As we have seen, these are two types:
      a. *Cells from embryonic haematopoiesis*, which settle in the tissue at early stages of development.
      b. *Cells originating from postnatal haematopoiesis*, i.e. produced in the *bone marrow* as blood cells.
         (1) They subsequently leave the blood and enter the connective tissue.
         (2) They include some resident cells (monocyte-derived macrophages, DCs and mast cells) and all transient cells.

D. There is some confusion about the classification of connective tissue cells:
   1. Some authors identify resident cells with cells of mesenchymal origin and transient cells with cells of haemopoietic origin.
   2. However, traditionally, some cells of blood origin (macrophages, DCs and mast cells) have been considered resident because of their low mobility once they reach the tissue.
   3. On the other hand, the plasma cell remains in the tissue once installed and yet is considered transient because of its relationship to the immune response.

---

[1] Dendritic cells can leave the connective tissue and move into the lymph, but because of their similarity to macrophages, they are usually included as resident cells.

4.4 Fibroblast

A. It is the characteristic cell of connective tissue. Its name means *fibre-forming* or *fibre-producing*.

B. Some histologists use the word 'fibrocyte' because they consider that the suffix -cyte should be used to denote the differentiated cell of a tissue, as already explained in Chap. 1. However, this term is incorrect, since the connective tissue cell was named fibroblast because of its fibre-producing activity.

C. Two types of fibroblasts can be observed depending on whether they have more or less synthetic activity (Fig. 4.3):

   1. In mature connective tissue, well-differentiated fibroblasts with little synthetic activity are observed.
      a. They have a variable shape.
         (1) They are usually oriented along the fibre bundles of connective tissue and are therefore often seen as spindle cells with long extensions.
         (2) Other times they are seen as flat stellate cells with thin extensions.
      b. The nucleus is elliptical, with one or two nucleoli and dispersed chromatin.
      c. Their cytoplasm is not very abundant.
         (1) It is usually eosinophilic, such as the collagen fibres of the extracellular matrix, making it difficult to distinguish in histological sections stained with haematoxylin and eosin.
         (2) A small Golgi apparatus, some cisternae of rough endoplasmic reticulum and some mitochondria are usually observed.
      d. This is the cell that is mistakenly referred to as a fibrocyte in some books.

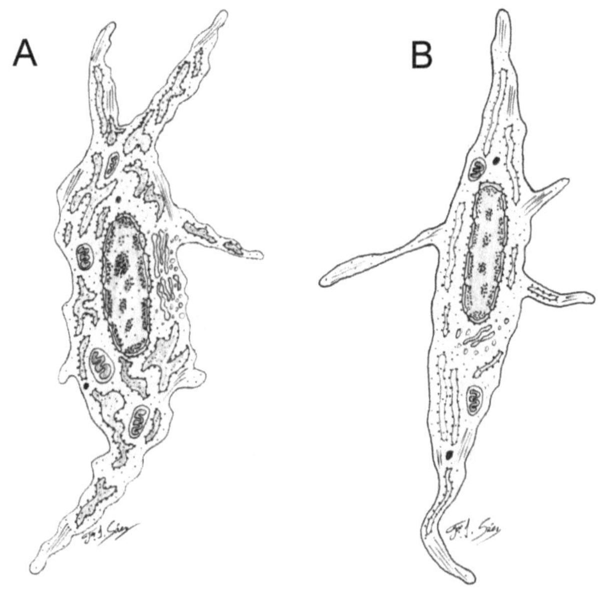

Fig. 4.3 Active fibroblast, with high synthetic activity (**A**), and inactive fibroblast, with low synthetic activity (**B**)

2. In developing connective tissues, or in tissues that are damaged and undergoing repair (e.g. wound healing), fibroblasts actively synthesise fibres and fundamental substance of the extracellular matrix.
   a. They are larger in size.
   b. Their cytoplasm is more voluminous and biosynthetic organelles are abundant:
      (1) It is somewhat basophilic due to the abundance of rough endoplasmic reticulum cisternae.
      (2) The Golgi apparatus is also more developed, as are the secretory vesicles that release the secreted products.

D. Each fibroblast is capable of producing all the components of the *extracellular matrix*, both the different types of fibres and the components of the ground substance, except for some molecules that leak from the blood.

4.5 Reticular Cell

A. Reticular cells are present in the variety of connective tissue called *reticular tissue*, which is present in some lymphatic organs and in bone marrow.
B. These are fibroblasts specialised in producing *reticular fibres* (to be described later).
C. They are stellate in shape, and their prolongations, which extend along the reticular fibres, join with those of other reticular cells and form a kind of network or lattice that allows other cells to be housed in their hollows.

4.6 Myofibroblast

A. It is a variant form of fibroblast, difficult to distinguish in routine haematoxylin and eosin stained preparations (Fig. 4.1), which has hybrid characteristics between *fibroblast* and *smooth muscle cell* that will be explained in the corresponding chapter.
B. Electron microscopy shows that these cells have bundles of *actin microfilaments* in their cytoplasm that, associated with *myosin* molecules, have the capacity to generate cell contraction.
C. Cell contraction can be transmitted to the extracellular matrix fibres due to the existence of specialised junctions in the plasma membrane.
D. The myofibroblast is clearly differentiated from the smooth muscle cell by two characteristics:
   1. It lacks a basal lamina (external lamina) surrounding it, whereas the smooth muscle cell has an external lamina.
   2. They are usually isolated, not associated with other similar cells. In contrast, smooth muscle cells are usually in clusters.

4.7 Adipocyte

A. *Adipocytes* are cells specialised in storing fat. They can be found scattered or in small groups in connective tissue, but in certain regions they can occupy the entire tissue, forming *adipose tissue*.
B. They are usually large, spherical or polyhedral in shape, and most of their cytoplasm is occupied by lipid inclusions (*fat droplets*).
C. During their formation from mesenchymal cells, lipid droplets accumulate in their cytoplasm and fuse together to form larger inclusions.
D. There are two types of adipocytes:
   1. The *unilocular adipocyte* has a large fat droplet that occupies almost the entire cell volume (Fig. 4.4).
      a. The rest of the cytoplasm and the nucleus are compressed into a thin peripheral layer attached to the plasma membrane.
      b. The adipose tissue formed by these cells is called *white* or *unilocular adipose tissue*.
   2. The *multilocular adipocyte* has multiple fat droplets.
      a. The nucleus is usually cell-centred and there are many mitochondria among the lipid droplets.
      b. The adipose tissue they form is called *brown* or *multilocular adipose tissue*.

4.8 Mesenchymal Cell

A. In adult connective tissues, embryonic *mesenchymal cells* persist and act as *reserve stem cells*.
B. They look like fibroblasts and are therefore difficult to identify. They are usually smaller and show fewer organelles in the cytoplasm.

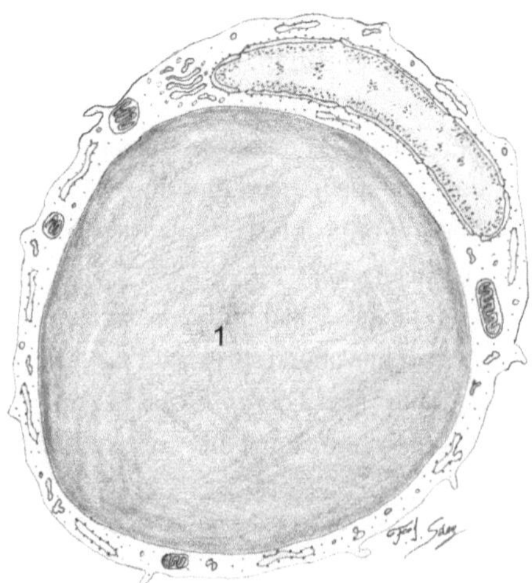

Fig. 4.4 Ultrastructure of the unilocular adipocyte. (1) Lipid droplet

C. Some are usually found close to blood vessels—especially small veins and venules—and are therefore referred to as *perivascular* or *adventitial cells*.
D. When stimulated, they can divide and give rise to fibroblasts and other connective tissue cells, such as adipocytes, and cells of other tissues, such as smooth muscle cells.

4.9 Pericyte

A. *Pericytes* (from the Greek περι- peri-, 'around') are cells that surround capillaries, the finest blood vessels that we will study in another chapter (Figs. 4.1 and and Fig. 12.5 in chapter 12).
B. They are derived from *mesenchymal cells* and can therefore be considered connective tissue cells, although they are separated from the connective tissue by a *basal lamina*.
C. They have *actin* and *myosin filaments* in their cytoplasm that allow them to contract.
D. In certain situations they can behave like *mesenchymal stem cells* and produce other cell types.

4.10 Melanocyte

A. *Melanocytes* (from the Greek μέλας, μέλανος mélas, mélanos 'black'), cells that synthesise *melanin*, are found in some connective tissues (iris and choroid of the eyeball).
B. Melanocytes are abundant in some epithelia, for example in the epithelium of the skin and the epithelium of the oral cavity, and will be discussed in the chapter on the oral mucosa.

4.11 Macrophage

A. *Macrophages* (from Greek μακρο- makro-, 'large' and Latin -phăgus, and this from Greek φαγο- phago- and -φάγος -phágos, 'eating') are, as their name suggests, the '*big eaters*' of connective tissue (Fig. 4.5).
B. They are specialised in *phagocytosis* of foreign agents and waste materials that need to be recycled, such as old or damaged erythrocytes.
C. In some tissues and organs, there are specialised macrophages with specific names (Kupffer cells in the liver, microglia cells in the central nervous system, alveolar macrophages in the lung, etc.).
D. Phagocytosis of pathogenic microorganisms is activated when they are coated with *antibodies*, as macrophages have *antibody receptors* on their surface.
E. An important function of macrophages is to present foreign molecules (*antigens*) to *T lymphocytes (T cells)*.

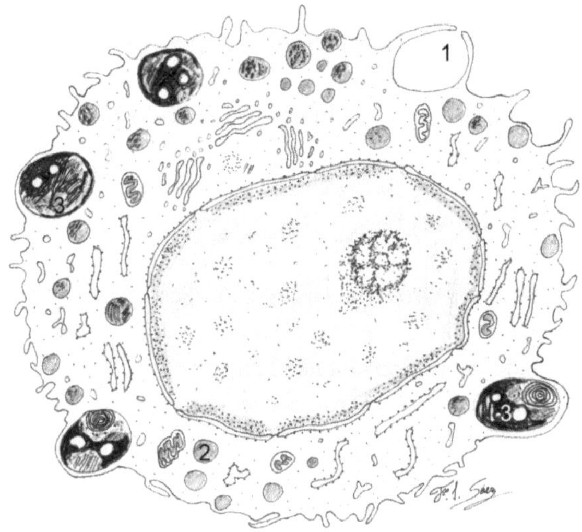

Fig. 4.5 Ultrastructure of the macrophage. Pseudopods involved in phagocytosis are seen on its surface (1) and many vesicles and lysosomes (2) in its cytoplasm. (3) Phagosomes with material to digest

1. T lymphocytes are cells that can recognise an antigen and react against it.
2. To recognise one of these antigens, other cells must present it to them bound to cell surface molecules called *major histocompatibility complexes (MHC)*.

F. There are two types of MHC molecules:
   1. *MHC I*, present in all cells of our body.
   2. *MHC II*, present only on *antigen-presenting cells* (*APCs*). The APCs in the human body are:
      a. Macrophages
      b. Dendritic cells
      c. B lymphocytes (B cells)
      d. Some epithelioreticular cells of the thymus

G. Since the 1960s, it has been accepted that all macrophages are formed from blood monocytes. However, new data allow us to distinguish between two types of macrophages:
   1. *Macrophages of embryonic origin* or *resident macrophages*:
      a. These macrophages settle in tissues during embryonic development.
         (1) In some tissues, they are established very early in embryonic development and originate from the first haemopoietic cells in the *yolk sac* wall.
         (2) In other tissues, they appear later in embryonic development from haemopoietic *liver* cells.
      b. They persist in adult tissues because they proliferate and the population is renewed.
   2. *Monocyte-derived macrophages*:
      a. They form in tissue when *inflammation* occurs.
      b. Monocytes circulating in the blood leave the blood vessels and turn into macrophages.

c. This population of monocyte-derived macrophages is maintained by the arrival of new monocytes that become macrophages, and because the macrophages themselves can proliferate.

H. The presence of phagocytic cells in connective tissues formed from blood precursors led to the proposal of a *mononuclear phagocytic system*, consisting of all cells related to macrophages and thought to be derived from monocytes.
   1. According to current knowledge, the cells of this system could be classified as follows:
      a. Originating during embryonic development:
         (1) Macrophages of many connective tissues (resident macrophages)
         (2) Liver Kupffer cells
         (3) Alveolar macrophages of the lung
         (4) Macrophages from lymphatic organs and bone marrow
         (5) Microglial cells of the central nervous system
         (6) Peritoneal macrophages
         (7) Perhaps some osteoclasts in the bone
      b. Originating in the adult from blood monocytes:
         (1) Monocytes (monocytes can persist in connective tissue)
         (2) Osteoclasts in bone, and probably chondroclasts in cartilage and odontoclasts in tooth
         (3) Breast macrophages
         (4) Macrophages of the intestinal wall
   2. Some authors include dendritic cells in this system.

4.12 Dendritic Cell (DC) or Interdigitating Dendritic Cell (IDC)

A. DCs have a clear, irregular nucleus and a cytoplasm with few differentiations and many branching prolongations, such as the branches of a tree, hence the name *dendritic cells* (from the Greek δενδρίτης dendrítēs, 'tree').

B. They are usually surrounded by lymphocytes.

C. They have long been related to macrophages. This is because these cells are also involved in the immune response, but have similarities and differences with them:
   1. *Differences*: They do not perform intense phagocytosis, and they do not usually have antibody receptors on their surface.
   2. *Similarities*: They are APCs, as they have MHC II on their surface, and their function is to capture antigens, process them and present them to T cells.

D. Antigen presentation can take place at two different sites:
   1. In situ, i.e. in the connective tissue itself.
   2. In the lymph nodes.
      a. DCs are highly mobile and can penetrate *lymphatic vessels* in connective tissues and travel through these vessels with the captured antigens.

b. When it reaches the *lymph nodes*, which are filled with lymphocytes, the antigen presentation takes place.

E. During their journey through the lymph, they are called *veiled cells.*

F. Several types of CD have now been distinguished:
   1. *Conventional* or *myeloid dendritic cells (cDC* or *mDC)*: Present in connective tissues as APCs.
   2. *Plasmacytoid dendritic cells (pDC)*: Present in lymphatic organs and until recently thought to be derived from the same lineage as lymphocytes.
   3. *Monocyte-derived dendritic cells (moDC)*: These are dendritic cells that are produced in tissues from monocytes.
   4. *Langerhans cells*: These are dendritic cells present in some epithelia, such as skin and mouth epithelia. The origin of these cells is highly debated, although it is currently believed that they are formed in the early stages of embryonic development.

G. The origin of these cells is controversial:
   1. From an unidentifiable bone marrow precursor that has been termed the *common dendritic cell precursor (CDP).* From this precursor, cells are formed and transported through the blood in immature form to the tissue where they would be formed:
      a. cDC
      b. pDC
   2. From *monocytes* that travel through the blood to the tissue where moDCs are formed.
   3. *From yolk sac* cells that migrate during embryonic development to the tissues of the embryo.
      a. This population is maintained and self-renewed in tissues.
      b. This is the origin of *Langerhans cells.*

4.13 Mast Cell

A. *Mast cells* (from the German mästen, 'to fatten') or *labrocytes* (from the Greek λαβρασ labras, 'voracious') are given these names because they were mistakenly thought to be cells with high phagocytic activity.

B. They can move through the tissue.

C. Cells are variable in size (10–30 μm), rounded or ovoid in shape (Fig. 4.6).
   1. Its surface has folds or small *microvilli*, the Golgi complex is well developed and there are few mitochondria and few endoplasmic reticulum cisternae.
   2. The small, round nucleus is difficult to see because the cytoplasm contains a large number of basophilic granules that are intensely coloured.
   3. The granules in the cytoplasm have a heterogeneous appearance:
      a. They have a rolled lamellar content.
      b. Metachromatically stained when using certain aniline dyes such as toluidine blue and thionine, two blue dyes that stain the granules purple (*metachromasia* is when a substance is coloured differently from the colour of the staining dye).

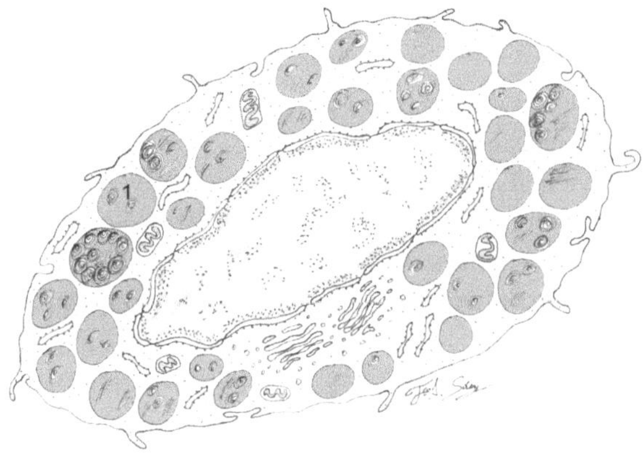

Fig. 4.6 Mast cell. In its cytoplasm there are many granules (1) with rolled lamellar material

   c. These granules accumulate inflammatory mediators such as *tryptase*, *heparin* and *histamine*, which are released into the extracellular matrix when the mast cell is stimulated.

D. When the mast cell is stimulated, other inflammatory mediators (*neosynthesised inflammatory mediators*) are synthesised and released as they are synthesised, without being stored in the granules.

E. Substances stored in the granules and neoformed substances cause tissue *inflammation*.

F. One factor that activates mast cells is *allergens*.
   1. Allergens are molecules that bind to *immunoglobulin type E* (*IgE*).
   2. As mast cells have IgE receptors on their surface, the presence of allergens stimulates mast cells and triggers the allergic reaction.

G. Mast cells are produced in the bone marrow from the haemopoietic stem cell.
   1. There, *mast cell precursors* (*MCp*) are formed, which are not morphologically differentiable, but can be identified with molecular markers.
   2. These precursors are incorporated into the bloodstream to the connective tissues.
   3. In tissue, the precursor can remain dormant or differentiate and develop its characteristic granules.

H. The entry of mast cells into tissues can be:
   1. *Constitutive*, which is produced in small quantities to maintain a stable population.
   2. *Induced* by an inflammatory process, which causes a massive influx.

I. Two subclasses of mast cells have been distinguished:
   1. *Connective tissue constitutive mast cells (CTMCs)*, which are usually located near venules and nerve endings.
   2. *Mucous mast cells (MMCs)*, present in the intestinal and pulmonary mucosa, normally infiltrating the lining epithelia. Their formation is immune-induced and they are T-cell-dependent.

J. As we shall see, they have similarities to basophils in the blood, but they are different cells.

4.14 Monocyte

A. Monocytes are blood cells that can leave vessels and settle in connective tissues.
B. They can have two functions:
   1. In *tissues in a steady state*, monocytes maintain an appearance similar to that in the blood. They can phagocytose *antigens* and pass into the lymph capillaries to carry the antigens to the lymph nodes.
   2. In *inflamed tissues*, monocytes turn into macrophages.

4.15 Plasma Cell

A. *Plasma cells* differentiate in tissue from activated B lymphocytes that travel through the blood.
B. They are *antibody-producing* cells. Antibodies are *immunoglobulins*, proteins that diffuse through tissues and blood and are responsible for *humoral immunity* to infection.
C. The size of these cells is variable, ranging from 10 to 20 μm in diameter (Fig. 4.7).
D. Its nucleus is rounded, with the chromatin arranged in spaced masses that give the nucleus the appearance of a cartwheel.
E. In the cytoplasm, close to the nucleus, there is usually a sparsely stained area occupied by the Golgi apparatus, while the rest of the cytoplasm is *basophilic* due to the abundance of rough endoplasmic reticulum cisternae. These cisternae may be:
   1. Flattened and arranged parallel to each other.
   2. Dilated and filled with a flocculent material (i.e. looking like tangled strands). In some cases, there are eosinophilic inclusions in the lumen called *Russell bodies*, formed by protein chains of immunoglobulins, the significance of which is unknown.

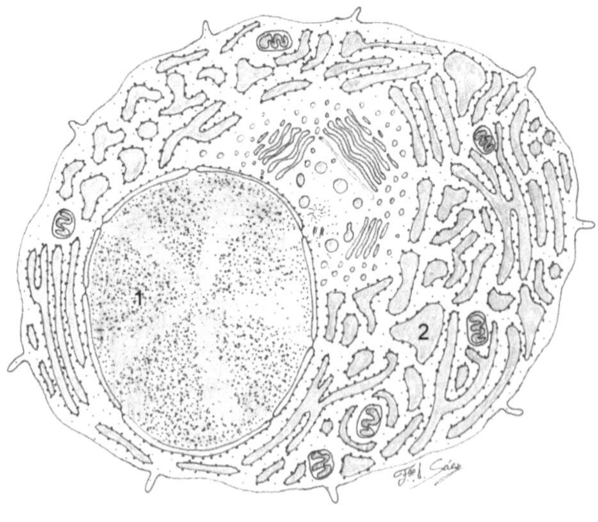

Fig. 4.7 Ultrastructure of the plasma cell. (1) Cartwheel-shaped heterochromatin; (2) cisternae of endoplasmic reticulum

4.16 Extracellular Matrix of Connective Tissue

A. The extracellular matrix is made up of three types of components (Fig. 4.1):
   1. *Protein fibres*
   2. An amorphous, viscous *ground substance* in which the protein fibres are embedded.
   3. *Tissue fluid*, which hydrates the ground substance.
B. Connective tissue fibres are usually classified into three different types:
   1. *Collagen fibres*
   2. *Reticular fibres*
   3. *Elastic fibres*

4.17 Collagen Fibres

A. They are formed by the polymerisation of a protein called *collagen*[2] (from the Greek κόλλα kólla 'glue' and -γενής -genḗs, 'that generates or produces something').
B. They got their name because when boiled they form a sticky substance, *gelatine*.
C. The collagen that forms these fibres is a protein of which there are many variants,[3] which are named with Roman numerals in the order in which they are discovered (type I collagen, type II collagen, type III collagen, etc.).
D. Some variants form fibres:
   1. *Fibre-forming collagens*. For example, type I and type II.
   2. *Collagens that do not form fibres*. For example, type IV collagen, which is one of the constituents of the *basal laminae* of epithelia.
E. The most abundant fibres in the connective tissue proper are formed by *type I collagen*. They are fibres of variable thickness and length, oriented in different directions and show a smooth undulation.
F. The thicker fibres (from 1 to 10 or 20 μm) are made up of thinner fibrils, but even the thinner fibres visible with the light microscope are made up of thinner *submicroscopic fibrils*, 20–100 nm thick, visible only with the electron microscope.
G. The submicroscopic fibrils show a *periodic transverse striation* every 67 nm, with alternating light and dark bands.
H. Type I fibres are flexible but very *resistant to traction*. They resist without stretching or breaking. The force required to break these fibres is several hundred kilograms per square centimetre, and in such situations few fibres stretch.

---

[2] It is also called tropocollagen.

[3] At least 29 forms of collagens have been described.

4.18 Reticular Fibres

A. They are so called because they are fine fibres that intertwine and form fine meshes (reticuli).
B. They are also characterised by the fact that they are stained by *silver impregnation* methods using *silver salts*, which allows them to be distinguished from collagen fibres.
C. In the past, these fibres were described as consisting of a protein called *reticulin*. Today, we know that this reticulin is a variety of fibre-forming collagen, *type III collagen*.
D. When viewed with the electron microscope, the submicroscopic fibrils are thinner than typical collagen fibrils (20–40 nm), but show the same transverse striation every 67 nm.
E. These are the first fibres to appear in the mesenchyme and are gradually replaced by type I collagen fibres.
F. They are relatively abundant in:
   1. The *reticular lamina* of the basal membrane.
   2. Around adipocytes and small blood vessels.
   3. In the nerve *endoneurium*.
   4. Around the muscle cells.
   5. In the stroma of the liver and other glands.
G. They are the predominant fibres in the *reticular connective tissue* variety.

4.19 Elastic Fibres

A. They are 0.2- to 1.0-µm-thick fibres that usually branch and anastomose.
B. In some tissues they form thick bundles of 4–5 µm.
C. They are difficult to distinguish in routine stains because they stain-like collagen fibres, but can be revealed with special stains (Verhoeff, Weigert resorcin-fuchsin or others).
D. If they are very abundant, they give a yellowish hue to the fresh tissue. In the walls of the arteries, the elastic fibres associate and form *elastic membranes*.
E. When viewed under the electron microscope, these fibres can be seen to have two zones:
   1. A homogeneous central structure. It is made up of *elastin*.
   2. A peripheral zone with 10 nm microfibrils. These microfibrils consist of *fibrillin*.
F. During embryonic development, fibrillin microfibrils appear before elastin.
G. In some tissues, other fibre varieties appear, which may be immature forms of the typical elastic fibres:
   1. Fibres consisting exclusively of fibrillin, which are called *oxytalan fibres*.
   2. Fibres with little elastin, called *elaunin fibres*.
H. Elastic fibres have a high *elasticity*, i.e. they are stretchable when pulled and spring back into shape when the stretching force ceases.

1. They can be extended by 150% of their initial length without breaking.
2. This elasticity seems to be mainly due to elastin molecules.

4.20 Ground Substance

A. The main component of the ground substance is polysaccharides called *glycosaminoglycans*. Glycosaminoglycans are long, unbranched, linear polymers formed by the polymerisation of certain special disaccharides (Fig. 4.8).
B. The disaccharides that make up glycosaminoglycans contain acidic groups (*carboxyl groups*), and may also be *sulfated*. The result is that glycosaminoglycans have a *net negative charge*.
C. The negative charge attracts *cations*, such as sodium, which is osmotically active.
D. The accumulation of electrically charged groups and the *osmotic pressure* create a hydrophilic environment that tends to concentrate the water in the tissue fluid.
E. As glycosaminoglycans are long linear molecules that fold little or not at all, they take up a lot of volume for their mass, and as this volume concentrates water, *turgor pressure* is generated, resulting in the formation of a viscous, gelatinous, highly hydrated substance (Fig. 4.9).
F. These characteristics of glycosaminoglycans give tissues two properties:
   1. It enables connective tissues to resist *compressive forces*.
   2. It provides an aqueous medium that allows the diffusion of water-soluble molecules dissolved in the matrix.
G. Two types of glycosaminoglycans can be distinguished:
   1. Glycosaminoglycans attached to some protein chain, forming *proteoglycans*.
      a. These glycosaminoglycans have tens or hundreds of disaccharides forming their molecule.
      b. This is the case of glycosaminoglycans such as *dermatan sulfate*, *chondroitin sulfate*, *heparan sulfate* and *keratan sulfate*.
   2. *Free glycosaminoglycans*, which do not form progeoglycans. There is one that is very abundant, *hyaluronic acid* or *hyaluronan*:
      a. Hyaluronic acid is a very long molecule of thousands of disaccharides, and it takes up a lot of space and, although it is not sulfated, it retains a lot of water.
      b. Although it does not covalently bind to proteins, it can indirectly bind to proteoglycans through special binding proteins forming proteoglycan aggregates, as we will see in cartilage tissue.
H. Proteoglycans are a variety of *glycoproteins*. They are distinguished from other glycoproteins by two characteristics (Fig. 4.8):
   1. Most glycoproteins have carbohydrate chains covalently linked to the peptide chain, which are characterised by the following:
      a. They are *short, branched oligosaccharide chains*.
      b. They consist of various types of *monosaccharides* (xylose, mannose, galactose, etc.).

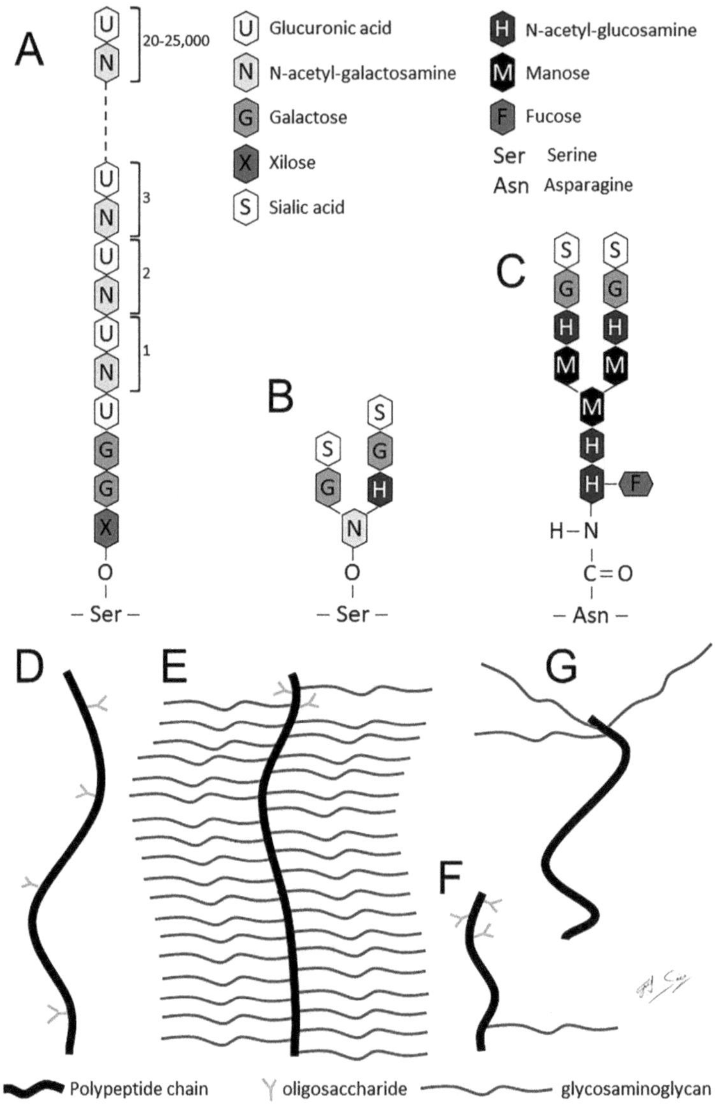

Fig. 4.8 Glycosaminoglycans (**A**) are unbranched linear polymers of 20,000–25,000 disaccharide molecules, depending on the case. In contrast, the usual oligosaccharides of glycoproteins (**B**, **C**) consist of short, branched chains made up of different types of monosaccharides that are attached to serine (**B**) or asparagine (**C**) residues of the peptide chain. Typical glycoproteins (**D**) consist of a polypeptide chain to which one or more oligosaccharides are covalently attached, whereas proteoglycans (**E**–**G**) consist of a polypeptide chain to which one or more glycosaminoglycans are attached. Proteoglycans, in addition to glycosaminoglycans, may have oligosaccharides (**E**, **F**) attached to them

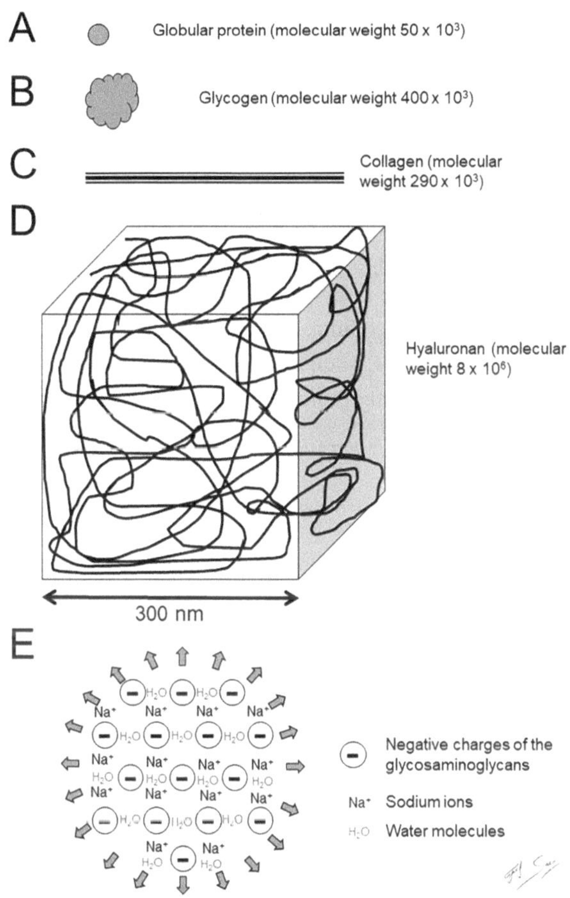

Fig. 4.9 Properties of the glycosaminoglycans of the ground substance. Glycosaminoglycans occupy a large volume in proportion to their molecular weight. Shown for comparison are a globular protein (**A**), a glycogen molecule (**B**), collagen (**C**) and a glycosaminoglycan (**D**). In addition to occupying a large volume (**E**), glycosaminoglycans have many negative charges that attract cations, mainly sodium. As sodium is osmotically active, large amounts of water are concentrated, which creates a turgor pressure (arrows). This turns the ground substance into a hydrated porous gel

2. A proteoglycan is a protein to which at least one glycosaminoglycan is covalently bound:
   a. Some proteoglycans have a single glycosaminoglycan chain.
   b. Others may own many chains, up to 200.

I. Some of the best known proteoglycans are:
   1. *Versican*, which binds to collagen fibrils and hyaluronan.
   2. *Aggrecan*, *decorin*, *biglycan* and *fibromodulin*, which are associated with collagen.
   3. *Syndecane*, which is on the surface of the plasma membranes of cells.
   4. *Perlecan*, which is present in epithelial basal laminae.

J. Another component of the ground substance are *glycoproteins* that perform very important functions, for example:
   1. *Fibronectin*, a protein that has regions on its molecule that can bind to other matrix molecules or to the surface of cells. This allows fibres to bind to cells.
   2. *Nidogen* or *entactin*, present in the basal lamina.
   3. *Laminin*, present in the basal laminae, which binds matrix molecules to cells.
   4. *Tenascin*, involved in cell attachment to the extracellular matrix in embryonic tissues and in healing processes.

K. Most of these glycoproteins are *adhesive proteins*, as they serve to bind cells to the fibrous components of the extracellular matrix.

L. In *specialised connective tissues*, such as bone tissue, *specific glycoproteins* such as osteonectin, osteocalcin and osteopontin may be involved in matrix mineralisation processes.

4.21 Tissue Fluid

A. Tissue fluid is the aqueous phase of the extracellular matrix in which fibres and ground substance are embedded.

B. Its main component is *water*.

C. It is formed by filtration of *blood plasma* (Fig. 4.10), so it may contain plasma components, although many plasma proteins do not usually pass into tissues.

D. It contains amino acids, sugars, fatty acids, coenzymes, hormones, neurotransmitters, mineral salts and cellular waste products.

E. In this aqueous medium, gases such as *nitric oxide* (NO, a messenger between cells), *oxygen* (O_2) and *carbon dioxide* (CO_2) diffuse.

F. Its composition is variable in each tissue, as it depends on the products secreted by the cells and the exchanges that take place between the blood and the tissues.

4.22 Varieties of Connective Tissue

A. Connective tissues can be classified into two groups: connective tissues proper and specialised connective tissues. The latter will be dealt with in other chapters. The characteristics of the connective tissues proper will be briefly summarised here.

B. *Embryonic connective tissue* or *mesenchyme*:
   1. Mesenchyme is the first connective tissue to form in the embryo.
      a. Most of it comes from the *mesoderm*.
      b. In cephalic regions, the mesenchyme (*ectomesenchyme*) originates from *neural crest* cells (*neuroectoderm*).
   2. It is a delicate tissue, with few thin collagen fibres, mainly *reticular fibres*, and a lot of *hyaluronan*.

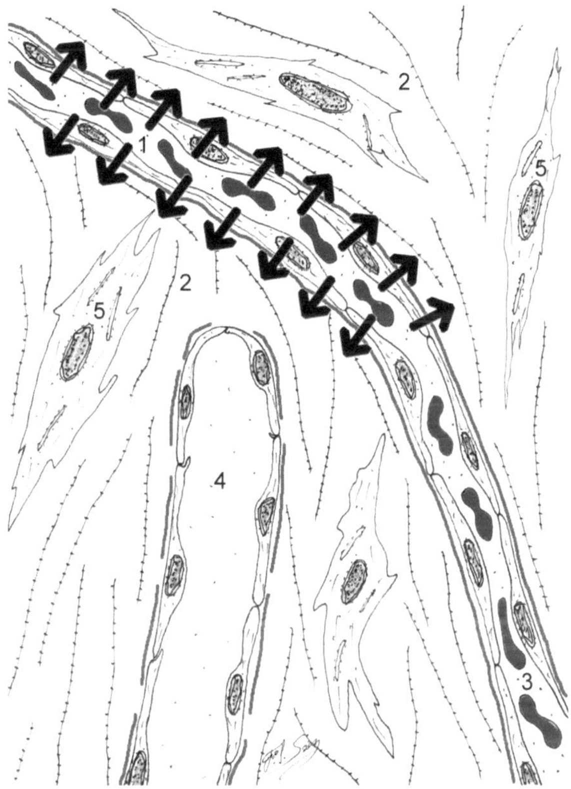

Fig. 4.10 Filtration of blood plasma to produce tissue fluid. At the initial part of the capillaries (1) the outflow of fluid (arrows) from the plasma into the extracellular matrix of the connective tissue (2) occurs. The outflow is due to the difference in hydrostatic pressure between the blood and the connective tissue. At the end of the capillary (3) where the hydrostatic pressure is lower, no filtration occurs. (4) Lymphatic capillary; (5) fibroblasts

3. The cells are poorly differentiated (*mesenchymal cells*), with many extensions that contact those of neighbouring cells by *gap junctions*. These cells are motile and able to divide. In some areas the cells are scattered, in others they are clustered.
4. In the umbilical cord, the embryonic connective tissue is very gelatinous, so it is called *mucous connective tissue* or *Wharton jelly*, and has few cells and few thin wavy fibres.

C. *Loose connective tissue*:
   1. Loose connective tissue is very common in certain locations close to *epithelia*:
      a. Beneath the epithelia that line the surface of the body and our body cavities.
      b. Surrounding the epithelia of the glands and the smallest blood vessels.
   2. It has few *fibroblasts* separated from each other with a lax matrix, consisting of *fine collagen fibres* that do not form large fibre bundles, *elastic fibres* and *reticular fibres*.
   3. The fibres leave large spaces occupied by ground substance and tissue fluid. As these materials are not stained in routine preparations, these spaces appear empty, and therefore this tissue was formerly called *areolar tissue* (from the Latin *areŏla* 'small area or space').
   4. All types of *resident* and *transient cells* can be found in this tissue variety, as this is where immune activity takes place.

D. *Dense connective tissue*:
   1. In dense connective tissue, fibres predominate over cells and ground substance.
   2. The cells, which are not abundant, are usually fibroblasts, macrophages and mesenchymal reserve cells.
   3. All other cell types are rare.
   4. There are two varieties of dense connective tissue:
      a. *Dense irregular connective tissue*:
         (1) It has thick, tightly woven bundles of collagen fibres.
         (2) It is widespread (dermis, organ capsules, tendon and nerve sheaths and other sites).
      b. *Dense regular connective tissue*:
         (1) The collagen fibre bundles are all oriented in the same direction, parallel, to meet special physical requirements.
         (2) Only fibroblasts are distinguishable and no other cell types are visible.
         (3) It is a highly specialised type of tissue that occurs, for example, in tendons.

E. *Reticular connective tissue*:
   1. In this tissue, the *reticular fibres* are the most abundant and are arranged in a mesh, with large spaces between the intertwined fibres.
   2. It has stellate-like *reticular cells*, with their elongated prolongations following and enveloping the reticular fibres.
   3. The spaces in the meshwork are occupied by cells of other types, so this tissue variety occurs in organs where cells need space to develop and a three-dimensional supporting meshwork.
   4. Reticular tissue is typical of *lymphatic organs* (except thymus) and *bone marrow*.

F. *Fusocellular connective tissue*:
   1. Fusocellular connective tissue is typical of the *ovary* and part of the uterine *endometrium*.
   2. It consists of many spindle-shaped fibroblasts, closely packed together, sometimes giving the appearance of a *shoal*.
   3. There is very little extracellular space with little matrix, where *type III collagen* predominates.

G. *Elastic connective tissue*:
   1. Elastic connective tissue is dominated by *elastic fibres*, which are thick and branched.
   2. Between them there are fine collagen fibres and few fibroblasts.
   3. It occurs in the yellow ligament of the vertebrae and in the vocal cords.
   4. In some hollow structures, such as arteries, there are areas of elastic connective tissue forming *fenestrated membranes*, which are perforated sheets of elastin of varying thickness arranged in concentric layers around the lumen of the blood vessel.

H. *Adipose tissue*:
   1. Adipose tissues are connective tissues in which *adipocytes* predominate. These cells leave little space between them, occupied by a sparse extracellular matrix that is usually traversed by *blood capillaries*.
   2. There are two types of adipose tissue:
      a. *White* or *unilocular adipose tissue*, consisting of *unilocular adipocytes*. It has several functions:
         (1) *Store fats* used as reserve substances that are released when there is a need for energy, e.g. during periods of fasting.
         (2) *Hormone production*, for example:
            (a) *Leptin*, which inhibits food intake and is involved in appetite regulation.
            (b) *Angiotensinogen*, which raises blood pressure.
            (c) *Adiponectin*.
            (d) *Resistin*.
            (e) *Steroid hormones* (testosterone, estrogens, glucocorticoids).
      b. *Brown* or *multilocular adipose tissue*, consisting of *multilocular adipocytes*:
         (1) There are many mitochondria in which oxidative phosphorylation does not occur (no ATP is produced):
            (a) The energy released by the reactions of the respiratory chain is dissipated as *heat*.
            (b) This heat production (*nonshivering thermogenesis*) is the main function of the tissue.
         (2) It is more abundant in *newborns* than in adults, and in winter than in summer.

I. *Cartilaginous tissue*: It forms the *cartilages* and will be studied in the next chapter.

J. *Osseous tissue*: It forms the *bones* and will be examined in a specific chapter.

K. Other varieties of connective tissue
   1. Other tissues that can be considered varieties of connective tissue are:
      a. *Blood* (which will be discussed in another chapter).
      b. The *vitreous body* (a connective tissue with very few fibres).
      c. The *nucleus pulposus* of the intervertebral discs (a connective tissue with few fibres).
      d. The *synovial fluid* of the joints (which we will study with the joints).

Take-Home Message

Connective tissues provide support for the rest of the body's tissues. They originate from embryonic mesenchyme (from the mesoderm), although in the cephalic region they may originate from ectomesenchyme (from the neural crests). They consist of many cell types, which are usually classified into resident cells and transient cells. The typical resident cell of connective tissue is the fibroblast. Other resident cells include adipocytes, reticular cells, mesenchymal cells, myofibroblasts, macrophages, mast cells and dendritic cells. Transient cells are cells of the immune system (neutrophils, plasma cells, eosinophils, lymphocytes, basophils and monocytes) that travel through the blood and infiltrate the connective tissue. In general, connective tissue cells can have three origins: mesenchyme (or ectomesenchyme in the cephalic region), embryonic haemopoiesis and postnatal haemopoiesis. In the extracellular matrix of connective tissue, there are collagen fibres that provide resistance to traction and elastic fibres that provide elasticity. The fibres are embedded in the ground substance, which contains glycosaminoglycans, hydrated by tissue fluid, providing resistance to compression. Connective tissues are classified into connective tissues proper (embryonic or mesenchymal, loose, dense, reticular, elastic, adipose) and specialised connective tissues (cartilaginous and osseous).

Further Reading

Dillac L, El Dika L, Ullah R, Kubiak J, Kloc M (2024) Macrophage cell cycle. Results Probl Cell Differ 74:119–134

Epelman S, Lavine KJ, Randolph GJ (2014) Origin and functions of tissue macrophages. Immunity 41:21–35

Frossi B, Mion F, Sibilano R, Danelli L (2018) Is it time for a new classification of mast cells? What do we know about mast cell heterogeneity. Immunol Rev 282:35–46

Gartner LP (2017a) Connective tissue. In: Textbook of histology, 4th edn. Elsevier, Philadelphia, pp 127–148

Gartner LP (2017b) Extracellular matrix. In: Textbook of histology, 4th edn. Elsevier, Philadelphia, pp 81–98

Goldberg BD, Rabinovitch M (1988) Connective tissue. In: Weiss L (ed) Cell and tissue biology – a textbook of histology, 6th edn. Urban & Schwarzenberg, Baltimore, pp 155–188

Gómez de Ferraris ME, Campos Muñoz A (2009) Histología general humana. In: Histología, Embriología e Ingeniería Tisular Bucodental, 3rd edn. Médica Panamericana, pp 57–78

Gurish MF, Austen F (2012) Developmental origin and functional specialization of mast cell subsets. Immunity 37:25–33

Kierszenbaum AL, Tres LL (2019) Connective tissue. In: Histology and cell biology – an introduction to pathology, 5th edn. Elsevier, Philadelphia, pp 135–176

Kloc M, Kubiak JZ, Zdanowski R, Ghobrial RM (2023) Memory macrophages. Int J Mol Sci 24:38

Komuro T (1990) Reevaluation of fibroblasts and fibroblast-like cells. Anat Embryol 182:103–112

Krstić RV (1985) Connective and supporting tissues. In: General histology of the mammal. An atlas for students of medicine and biology. Springer Verlag, Berlin, pp 99–235

Lendahl U, Muhl L, Betsholtz C (2022) Identification, discrimination and heterogeneity of fibroblasts. Nat Commun 13:3409

Li XD, Li N, Wang YJ, Han QX, Sun BS (2024) Research progress of fibroblasts in human diseases. Biomolecules 14:1478

Lowe JS, Anderson PG (2015) Support cells and the extracellular matrix. In: Stevens & Lowe's human histology, 4th edn. Elsevier, Philadelphia, pp 55–70

O'Dowd G, Bell S, Wright S (2023) Supporting/connective tissues. In: Wheater's functional histology. Elsevier, Amsterdam, pp 65–81

Pawlina W (2024) Connective tissue. In: Histology a text and atlas with correlated cell and molecular biology, 9th edn. Wolters Kluwer, Philadelphia, pp 176–215

Schuster R, Rockel JS, Kapoor M, Hinz B (2021) The inflammatory speech of fibroblasts. Immunol Rev 302:126–146

Varol C, Mildner A, Jung S (2015) Macrophages: development and tissue specialization. Annu Rev Immunol 33:643–675

Yu J, Qian BZ (2025) Cellular stress and macrophage activation. Semin Immunol 78:101953

Cartilaginous Tissue

5

What You Will Learn in This Chapter

Cartilage is made up of a specialised connective tissue called cartilaginous tissue. In this chapter you will learn that cartilage originates from mesenchyme and can grow by apposition or intussusception, and in many cases is surrounded by a connective tissue sheath called perichondrium. You will see that the cells of the cartilage are the chondrocytes, and that the extracellular matrix consists mainly of collagen fibres (type II collagen), and large aggregates of a specific proteoglycan (aggrecan). You will study that there are different varieties of cartilage (foetal, hyaline, elastic, fibrous and articular cartilage).

5.1 General Characteristics

A. The skeletal structures we call *cartilage* (from the Latin cartilāgo, 'cartilage, firm flesh, flesh of some fruits') are made up of *cartilaginous tissue* (from the Latin cartilaginōsus). The term cartilage is often used as a synonym for cartilaginous tissue.
B. Cartilaginous tissue is a variety of *specialised connective tissue*. It is characterised by its extracellular matrix consisting of fibres and a gel-like ground substance that produces a solid, firm but malleable tissue, giving it a certain elasticity.
C. Its rigidity allows it to form some skeletal structures in the human body. Its activity in the skeleton is possible because:
   1. *Collagen fibres* resist tension.
   2. Its abundant, highly hydrated *proteoglycans* allow it to withstand pressure.
D. Despite its relative stiffness, cartilaginous tissue has the capacity to grow, so the *embryonic skeleton* is largely made up of cartilage tissue.
E. Cartilage lacks blood vessels[1] and nerves.

[1] Exceptions exist, as there may be blood vessels in the laryngeal and foetal cartilages.

F. J. Sáez, I. Badiola, *Essential Oral Histology*, Learning Materials in Biosciences,
https://doi.org/10.1007/978-3-032-03165-5_5

F. Tissue nutrition occurs by *diffusion* of substances from the blood vessels of the tissues surrounding the cartilage.
G. In some cartilage, the surrounding tissue is a specialised connective tissue sheath called *perichondrium*.
H. Cartilaginous tissue can regenerate if tissue is lost or damaged:
   1. In some animals, cartilage tissue can regenerate itself.
   2. In mammals and humans, cartilage repair requires the involvement of perichondrium, in two phases:
      a. Pre-growth of connective tissue from the perichondrium.
      b. Transformation of connective tissue into cartilage in a process of *metaplasia* (transformation of one tissue into another).

5.2 Embryonic Origin and Growth of Cartilage

A. Cartilage develops from *embryonic mesenchyme* (Fig. 5.1).
   1. Cartilage originates in embryonic regions where *mesenchymal cells* accumulate and develop into *chondrification centres* (from the Greek χόνδρος, chondros, 'cartilage').
   2. Mesenchymal cells become *chondrogenic cells*.
   3. These cells synthesise a hyaline and metachromatic matrix including collagen fibres, which are barely distinguishable in the matrix.
   4. As matrix synthesis persists and the volume of matrix increases, the cells become separated. Each cell occupies a space devoid of matrix, which under the microscope looks like compartments resembling *lacunae*.
B. The cartilage thus formed can grow by two mechanisms (Fig. 5.2):
   1. *Appositional growth*:

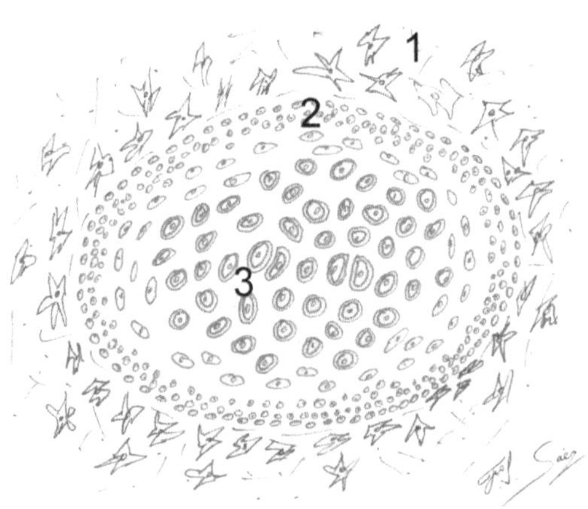

Fig. 5.1 Illustration representing embryonic cartilage (foetal cartilage). (1) Mesenchymal cells; (2) chondroblasts; (3) chondrocytes

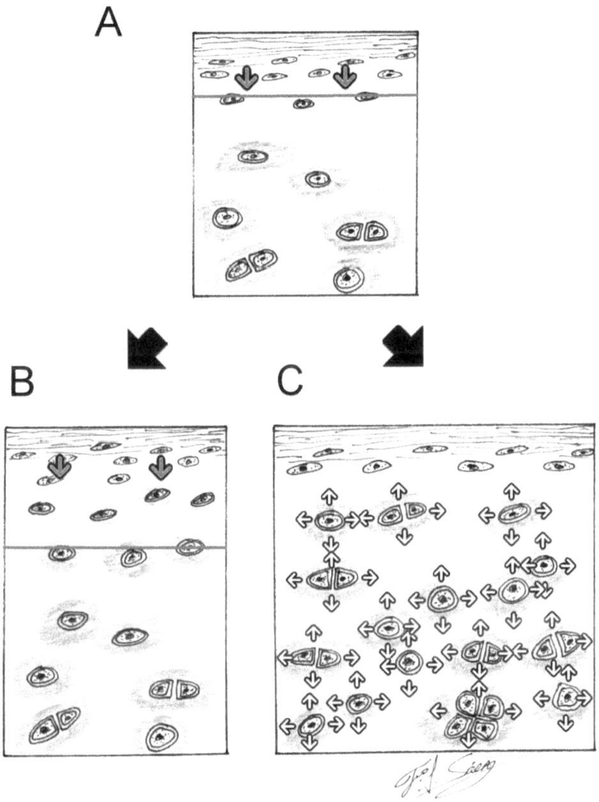

Fig. 5.2 A cartilage (**A**) can grow by apposition (**B**) or by intussusception (**C**). Although both types of growth can occur simultaneously, their effects have been shown separately in the illustration for clarity. When growth occurs by apposition, new tissue is produced from the surface perichondrium in the direction indicated by the grey arrows. The new tissue is deposited on top of the existing tissue, so the cartilage grows from the grey line indicated in **A** and **B**. If growth is by intussusception, the cartilage grows from the inside in all directions (white arrows) as the chondrocytes proliferate by mitosis and all synthesise new extracellular matrix between them

   a. The mesenchyme surrounding the cartilage condenses into a connective tissue called *perichondrium* (from the Greek περι- peri-, 'around').
   b. Cells on the inner side of the perichondrium synthesise cartilage matrix that accumulates on the surface of the previously formed cartilage (Fig. 5.2A, B).

2. *Growth by intussusception* (from Latin intus 'inwardly' and susceptio, -ōnis 'action of receiving') or *interstitial growth*:
   a. The inner cartilage cells continue to divide by mitosis and synthesise new matrix between them.
   b. The increase in cell number and extracellular matrix volume results in growth from within (Fig. 5.2A, C).

C. Cartilage growth is evident during embryonic and foetal development and is maintained in childhood and adolescence.
D. The ability of cartilage to grow persists in the adult individual, albeit in a latent form.

5.3 Perichondrium

A. Some varieties of cartilage are lined by a connective tissue sheath called perichondrium.
B. The perichondrium of adult cartilage is a connective tissue territory with blurred boundaries (Figs. 5.3 and 5.4). Two layers are usually distinguished:
   1. A more external, *fibrous layer*, consisting of dense connective tissue with few cells.
   2. Another more internal one, which has more cells and has the capacity to produce cartilage during development (*cellular* or *chondrogenic layer*).
C. Cartilage nutrition occurs by diffusion from the blood vessels of the perichondrium.

5.4 Cartilage Cells

A. The following cells can be distinguished in cartilage or related tissues such as perichondrium (Figs. 5.1, 5.3, 5.4 and 5.5):
   1. *Chondrogenic cells*: These are cells of the cellular layer of the perichondrium that can begin to synthesise cartilaginous matrix.
      a. During embryonic cartilage formation, they are derived from *embryonic mesenchymal cells*.

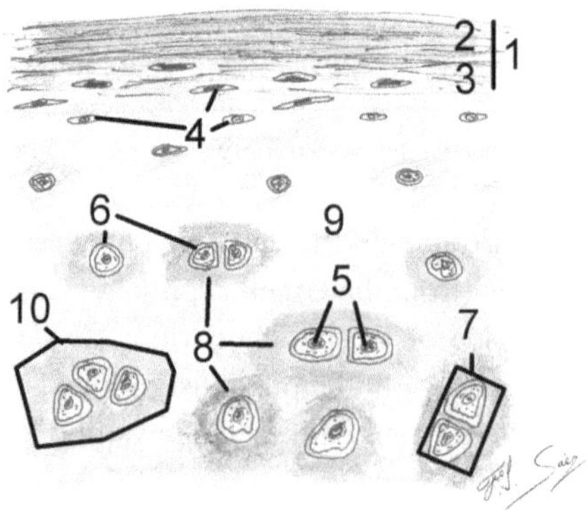

Fig. 5.3 Histological organisation of hyaline cartilage. (1) Perichondrium; (2) fibrous layer; (3) chondrogenic or cellular layer; (4) chondroblasts; (5) chondrocytes; (6) chondrocyte capsule (capsular matrix); (7) isogenous groups; (8) territorial matrix; (9) interterritorial matrix; (10) chondrone

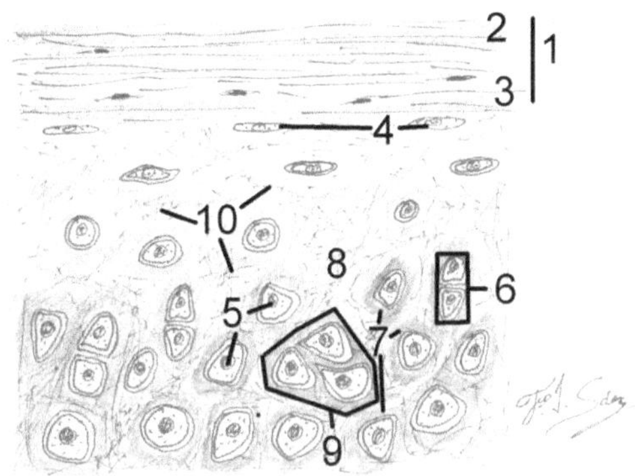

Fig. 5.4 Elastic cartilage is similar to hyaline cartilage, but its matrix has a fibrous appearance. (1) Perichondrium; (2) fibrous layer; (3) cellular or chondrogenic layer; (4) chondroblasts; (5) chondrocytes; (6) isogenous groups; (7) territorial matrix; (8) interterritorial matrix; (9) chondrone; (10) elastic fibres

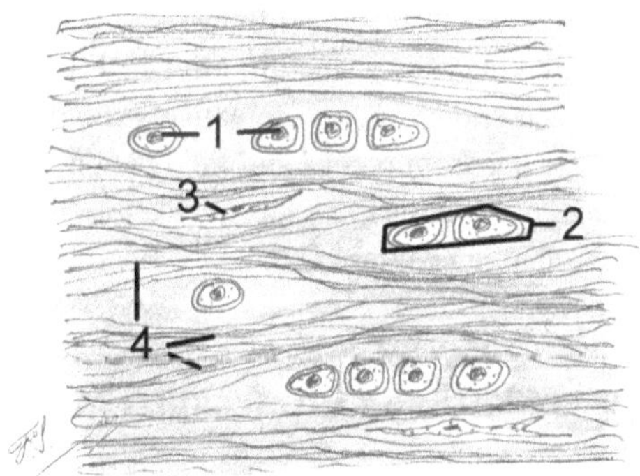

Fig. 5.5 Fibrous cartilage or fibrocartilage. (1) Chondrocytes; (2) isogenous groups; (3) fibroblasts; (4) type I collagen fibres

   b. After embryonic development, there are chondrogenic cells in the inner layers of the perichondrium that are derived from *reserve mesenchymal cells*.

2. *Chondroblasts*: These are chondrogenic cells that gradually become surrounded by cartilage matrix and increase in volume.
   a. Their shape varies from fusiform to oval, and as they develop they increase in size.
   b. They are basophilic cells that possess protein-synthesising organelles.
   c. During appositional growth, they are located between the innermost layer of the perichondrium and the cartilage surface.
   d. As they synthesise matrix, they move into more internal positions and become larger and develop into chondrocytes.
3. *Chondrocytes*: The cells of mature cartilage tissue.
   a. They are derived from the development of chondroblasts.

   b. The peripheral ones are ovoid, the deep ones are more rounded, 10–30 μm in diameter.
   c. They occupy a space from which they exclude the homogeneous extracellular matrix of the cartilage. On histological preparations, the cell retracts, so that it looks like a small cell inside a larger cavity called a *lacuna*, *chondroplast* or *chondrocele*.
   d. The chondrocyte shows a rounded *nucleus* with one or more *nucleoli*.
   e. When there is active matrix synthesis, the cytoplasm shows many cisternae of *rough endoplasmic reticulum* and a large *Golgi apparatus*. When there is no active matrix synthesis, the rough endoplasmic reticulum and Golgi apparatus are less developed.
   f. Elongated mitochondria, some lipid droplets and glycogen inclusions are observed.
4. *Fibroblasts*: Fibroblasts may be present in:
   a. The outer layers of the perichondrium.
   b. In one variety of cartilage, namely *fibrous cartilage*.

B. The concept of chondroblast and chondrocyte is not very clear, as each author uses different criteria, and different criteria are applied in different books.

C. When chondrocytes are actively dividing, clusters of two to four chondrocytes close together from recent mitotic divisions can be observed. These groups are called *isogenous groups* (from the Greek ἰσο- iso-, 'equal' and -γενής -genḗs, 'generating or producing something').

5.5 Cartilage Extracellular Matrix

A. The extracellular matrix of all varieties of cartilage has common characteristics.

B. The cartilage matrix may look very homogeneous in certain varieties, but not in all. When it appears homogeneous, it is because both the ground substance and the very small fibres have the same *refractive index*.

C. The matrix is basophilic:
   1. Stain with specific dyes for complex carbohydrates, such as periodic acid Schiff stain (*PAS*).
   2. This is because the matrix is dominated by *glycosaminoglycans*, especially various forms of *chondroitin sulfate*.

D. Sometimes a thin region of the matrix surrounding the chondrocele is often distinct from the rest of the matrix and is seen as if it were a *chondrocyte capsule* (*capsular matrix*).

E. After water, the most abundant components of the matrix in all varieties of cartilage are:
   1. Collagen fibres formed by *type II collagen*.
      a. These type II collagen fibres are characteristically very thin.

   b. Other types of collagen, such as type X and type XI, are involved in the formation of these fibres.
   c. The fibres are masked by the ground substance.
2. *Hyaluronan*, which as we know is a glycosaminoglycan that does not form proteoglycans.
3. *Aggrecan*, a proteoglycan:
   a. It contains 100–150 molecules of the glycosaminoglycans *chondroitin sulfate* and *keratan sulfate* bound to the core protein.
   b. It can bind to hyaluronan via *binding proteins* (Fig. 5.6).

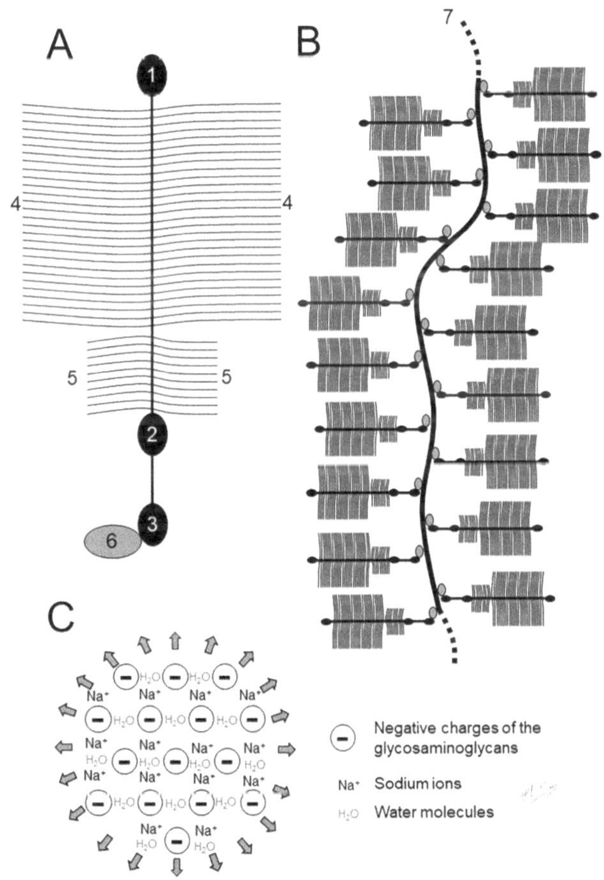

Fig. 5.6 Aggrecan is the most abundant proteoglycan in most cartilage tissues (**A**). It consists of a core fibrillar protein chain with some globular domains (1, 2 and 3), to which numerous chains of the glycosaminoglycans chondroitin sulfate (4) and keratan sulfate (5) are covalently attached. The peculiarity of this proteoglycan is that it can bind hyaluronic acid (7) by means of a binding protein (6). In cartilage, numerous aggrecan molecules bind to large molecules of hyaluronic acid (**B**). It is worth recalling the properties of glycosaminoglycans discussed in Chap. 4 (**C**), whose numerous negative charges attract sodium ions and retain water, creating a turgor pressure (grey arrows) that turns the ground substance into a hydrated porous gel. The fact that these large aggregates of aggrecan and hyaluronic acid are created in the matrix of cartilage tissue explains their consistency

4. *Chondronectin*, a fibronectin-like adhesion protein, which binds the surface of chondrocytes to collagen fibres and glycosaminoglycans.

F. In addition, in some varieties they can be found:
   1. *Elastic fibres*, present in *elastic cartilage*. They are visible under the microscope because they are not masked.
   2. *Type I collagen fibres*, found in *fibrous cartilage*.
   3. *Versican*, a proteoglycan:
      a. It contains several oligosaccharides and 12–15 *chondroitin sulfate* chains attached to the core protein sequence.
      b. It is present in *fibrous cartilage*.

G. The most characteristic feature of the cartilaginous matrix is that aggrecan forms *complexes* (Fig. 5.6):
   1. Numerous aggrecan molecules are non-covalently associated with large molecules of hyaluronan via binding proteins.
   2. Giant macromolecular aggregates, up to 3–4 mm in size, are formed.
   3. These macromolecular complexes contain a large amount of glycosaminoglycans (those of aggrecan and hyaluronic acid) and are responsible for the gel-like appearance of the tissue and its morphological stability.

5.6 Cartilage Varieties

A. In adult humans there are three varieties of cartilage: *hyaline*, *elastic* and *fibrous* (*fibrocartilage*).

B. Some authors consider a fourth variety, *articular cartilage*, which other authors consider a special form of hyaline cartilage. In any case, articular cartilage will be explained in the chapter dealing with the joints between bones.

C. There is another variety of cartilage, *foetal cartilage*, which is developing cartilage that is no longer found in the adult.

5.7 Foetal Cartilage

A. In *foetal cartilage*, cells are very abundant:
   1. They are fusiform, rounded or even stellate.
   2. They are evenly distributed throughout the matrix, singly or in pairs (Fig. 5.1).

B. There are no differences in the staining of the extracellular matrix, i.e. the chondrons typical of other strains, which are explained below, are not visible.

C. Occasionally *blood vessels* can be seen.

5.8 Hyaline Cartilage

A. *Hyaline cartilage* (from the Greek ὑάλινος hyalinos, 'translucent') is a semi-transparent, glassy, somewhat elastic, bluish-grey tissue with an opalescent appearance.[2]
B. It is found at the anterior ends of the ribs, and in the tracheal, bronchial and laryngeal rings.
C. The articular surface of many bones is often covered with hyaline cartilage, which is sometimes considered another variety (*articular cartilage*).
D. In histological preparations, the extracellular matrix of hyaline cartilage has a very homogeneous appearance. However, several zones can be distinguished (Fig. 5.3):
   1. *Capsular matrix*: The part of the matrix in intimate contact with the chondrocytes (or chondroplasts) is distinguishable as a thin, more stained sheet than the rest. It contains a fine network of collagen fibres embedded in a material similar to that of the basal lamina.
   2. *Territorial matrix*: Regions around chondrocytes and isogenous groups that stain more intensely because they have more sulfated glycosaminoglycans.
      a. Haematoxylin and eosin stains usually show intense violet blue.
      b. The set of chondrocytes and their territorial matrix is called a chondron.
   3. *Interterritorial matrix*: This is the rest of the matrix, located between the chondrons, which is stained in paler tones.
E. The hyaline cartilage is surrounded by *perichondrium*.
F. With ageing, hyaline cartilage can undergo degeneration due to several factors:
   1. Water loss.
   2. Alterations in proteoglycans.
   3. Formation of *asbestos fibres* (unmasked collagen fibres that are distinguishable under the microscope).
   4. Calcifications.
   5. Cell destruction.
G. *Calcification* of cartilage can also occur in physiological situations. Specifically, calcification is a necessary process as a step prior to the formation of many bones (ossification), as we will study in the corresponding chapter.

5.9 Elastic Cartilage

A. It is found in the ear, external auditory canal, Eustachian tube, epiglottis and some bronchial cartilages.
B. When fresh, its colour is yellowish and it is more opaque, more flexible and more elastic.
C. Histologically, it is very similar to hyaline cartilage (Fig. 5.4):

[2] Opalescent: appearing iridescent like opal.

1. It has rounded chondrocytes, with a capsule and isogenic groups of two to four cells.
2. Microscopically, the matrix differs from that of hyaline cartilage in that it is traversed by branched fibres. These fibres are *elastic fibres* that surround the chondrons, pass through the interterritorial matrix and contact the perichondrium.
3. Chondrocytes may be more abundant than in the hyaline variety.

D. The elastic cartilage is surrounded by perichondrium.

5.10 Fibrous Cartilage (Fibrocartilage)

A. *Fibrous cartilage* or *fibrocartilage* is found in regions of *dense connective tissue*, such as the intervertebral discs, in certain articular cartilages—for example, in the *temporomandibular joint*, which will be discussed in Chap. 16—in the pubic symphysis, in the round ligament of the femur and at the bony insertion points of some tendons.
B. Fibrous cartilage could be defined as a dense connective tissue with some regions of cartilaginous tissue between its collagenous fibre bundles.
C. It resembles dense connective tissue because there are large areas where bundles of *type I collagen fibres* are abundant (Fig. 5.5). These fibres are not masked by the extracellular matrix.
D. *Fibroblasts* are observed in the fibrous regions.
E. In the gaps between the bundles of type I collagen fibres, there are *chondrons* formed by isolated chondrocytes or isogenous groups in a row, with a cartilaginous matrix with *type II collagen*.
F. In fibrous cartilage, *versican*, secreted by fibroblasts, is more abundant than *aggrecan*, secreted by chondrocytes.
G. No perichondrium is visible around the fibrous cartilage:
   1. Sometimes, fibrocartilage may be surrounded by regions of dense connective tissue with diffuse border zones, without a defined boundary.
   2. Other times, fibrous cartilage continues into hyaline cartilage or bone.

Take-Home Message

Cartilaginous tissue is a specialised connective tissue that forms cartilage. It is formed in embryonic chondrification centres from mesenchymal cells that develop into cartilage matrix-producing chondroblasts. The cells become trapped in the matrix and develop into chondrocytes. Cartilage can grow by apposition of new matrix and cells on the surface or by intussusception (production of matrix and cells inside). Some varieties of cartilage are lined with a connective tissue sheath called perichondrium. In the extracellular matrix are type II collagen fibres and a proteoglycan called aggrecan because it forms large complexes. There are different varieties of cartilage. Foetal cartilage is found during prenatal development. Hyaline cartilage is a precursor of many bones during development and is also found in the adult. A variety of hyaline cartilage is articular cartilage, which is present in the joints between bones. Elastic cartilage is characterised by elastic fibres in its matrix. Fibrous cartilage has regions with type II collagen and chondrocytes and areas with type I collagen, a proteoglycan (versican) and fibroblasts.

Further Reading

Archer CW, Francis-West P (2003) The chondrocyte. Int J Biochem Cell Biol 35:401–404

Aubin JE, Liu F, Malaval L, Gupta AK (1995) Osteoblast and chondroblast differentiation. Bone 17:S77–S83

Cancedda R, Cancedda FD, Castagnola P (1995) Chondrocyte differentiation. Int Rev Cytol Surv Cell Biol 159:265–358

Gartner LP (2017) Cartilage and bone. In: Textbook of histology, 4th edn. Elsevier, Philadelphia, pp 149–178

Gómez de Ferraris ME, Campos Muñoz A (2009) Histología general humana. In: Histología, Embriología e Ingeniería Tisular Bucodental, 3rd edn. Médica Panamericana, pp 57–78

Goodwin DW (2001) Visualization of the macroscopic structure of hyaline cartilage with MR imaging. Semin Musculoskelet Radiol 5:305–312

Gvaramia D, Kern J, Jacob Y et al (2022) Regenerative potential of perichondrium: a tissue engineering perspective. Tissue Eng Part B Rev 28:531–541

Hickok NJ, Haas AR, Tuan RS (1998) Regulation of chondrocyte differentiation and maturation. Microsc Res Tech 43:174–190

Jee WSS (1988) The skeletal tissues. In: Weiss L (ed) Cell and tissue biology – a textbook of histology, 6th edn. Urban & Schwarzenberg, Baltimore, pp 211–254

Kierszenbaum AL, Tres LL (2019) Connective tissue. In: Histology and cell biology – an introduction to pathology, 5th edn. Elsevier, Philadelphia, pp 135–176

Kronenberg HM (2007) The role of the perichondrium in fetal bone development. Skelet Biol Med Part A 1116:59–64

Lin Z, Willers C, Xu JA, Zheng MH (2006) The chondrocyte: biology and clinical application. Tissue Eng 12:1971–1984

Malinin GI, Malinin TI (1999) Microscopic and histochemical manifestations of hyaline cartilage dynamics. Prog Histochem Cytochem 34:169

O'Dowd G, Bell S, Wright S (2023) Skeletal tissues. In: Wheater's functional histology. Elsevier, Amsterdam, pp 180–196

Pawlina W (2024) Cartilage. In: Histology a text and atlas with correlated cell and molecular biology, 9th edn. Wolters Kluwer, Philadelphia, pp 216–237

Osseous Tissue (I): Macroscopic and Microscopic Structure

6

What You Will Learn in This Chapter

In this chapter you will learn about the characteristics of bone tissue, a specialised connective tissue that forms bones. The tissue can give rise to different bone structures distinguishable with the naked eye, and different microscopic organisations of the tissue can also be distinguished. In addition, bones are often lined by a connective tissue called periosteum. You will study that the bone matrix is made up of organic components (type I collagen, glycosaminoglycans, proteins) and inorganic components (hydroxyapatite). Several cell types can be found, some of mesenchymal origin and others of blood origin.

6.1 General Characteristics

A. *Osseous tissue* (from Latin osseus, 'bone') is the main and characteristic component of the *bones* of the skeleton. The term bone is often used as a synonym for osseous tissue.
B. Osseous tissue is a variety of *specialised connective tissue*. As such connective tissue, it consists of *cells* and an abundant *extracellular matrix* composed of *fibres* and *ground substance* with *tissue fluid*.
C. The peculiarity of bone tissue is that the extracellular matrix is *mineralised*, *calcified*, which makes the bone a hard and very resistant substance in order to perform its function in the *skeleton*.
D. When we talk about *hardness*, we must understand the term from a geological point of view.
   1. In geology, hardness is the degree of resistance of one mineral to being scratched by another.
   2. The softest mineral is *talc* and the hardest is *diamond*.

F. J. Sáez, I. Badiola, *Essential Oral Histology*, Learning Materials in Biosciences,
https://doi.org/10.1007/978-3-032-03165-5_6

E. Human tissues are usually very soft materials, but bone and some tooth tissues are mineralised, which makes them harder.
F. The hardness of mineralised human tissues is difficult to measure and contradictory data have been published.
   1. The following table shows the hardness value of various minerals and mineralised tissues measured on two of the scales used in geology.
   2. The value shown for tissues is approximate and may vary.

	Mohs scale	Knoop scale
Talc	1	1
Gypsum	2	32
Gold	2.5	69
Dental cementum	*2.5–3*	*40*
Bone	*2.5–3*	*40*
Dentin	*3–4*	*64–68*
Fluorite	4	163
Enamel	*5*	*343*
Apatite	5	430
Quartz	7	800–900
Diamond	10	8000–8500

G. Among the properties of bone tissue are the following:
   1. *Tensile* strength.
   2. *Compressive* strength.
   3. Relative *elasticity.*
   4. Relatively *light* weight. It can be stated that the organisation of the bone allows for the greatest strength with the least weight.
H. Despite being an apparently very resistant material, bone is a very *dynamic* tissue:
   1. It is renewed and reformed throughout an individual's life and is under metabolic, nutritional and endocrine influence.
   2. These properties are of utmost importance in order to be able to act on it by means of surgical or orthodontic procedures.
I. The functions of bones include:
   1. Internal body *support.*
   2. Place of insertion of muscles and tendons, which makes *locomotion possible.*
   3. *Protection* of vital organs (brain, heart, etc.).
   4. Housing of the *bone marrow*, which is the tissue that produces blood cells.
   5. Calcium deposition and *homeostatic regulation* of blood and tissue calcium levels.

6.2 Macroscopic Structure of Bone Tissue

A. With the naked eye or a low magnification magnifying glass, two different organisations of bone tissue can be distinguished in a bone:
   1. *Compact* or *dense bone*: a solid mass with no gaps or voids visible to the naked eye. Microscopic cavities are present.
   2. *Spongy* or *cancellous bone*: it is made up of:
      a. *Spicules* (from Latin spicŭla, diminutive of spica 'spike') or *trabeculae* (from Latin trabecŭla 'small beam') of bone tissue that interlock with each other and form a mesh or lattice of bony pieces.
      b. Between the trabeculae there are interconnected spaces that form a labyrinthine system.
      c. The spaces are occupied by soft tissue called *bone marrow*.

B. In a bone we can find regions of compact bone and regions of spongy bone in a variable distribution according to each bone (Fig. 6.1):
   1. In the *long bones* of our body, we can distinguish several zones:
      a. The *diaphysis* (from Ancient Greek διά dia 'separation, through' and φύσισ phýsis 'area of growth'), which is the central cylindrical region.
         (1) The diaphysis is a hollow cylinder formed by a wall of compact bone tissue surrounding a central *medullary cavity*, which houses bone marrow.
         (2) Spongy bone is usually absent or very sparse.

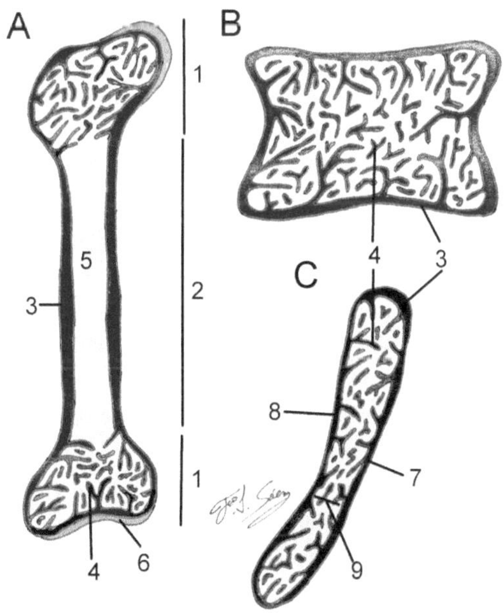

Fig. 6.1 Distribution of compact bone and spongy bone in a long bone (**A**), a short bone (**B**) and a flat bone (**C**). (1) Epiphyses; (2) diaphysis; (3) compact bone; (4) spongy bone; (5) medullary cavity; (6) articular cartilage. (7) external table; (8) internal table; (9) diploe

b. The *epiphyses* (singular *epiphysis*, from Renaissance Latin *epiphysis*, and this from Greek ἐπί epí 'on' and φύσισ phýsis 'zone of growth'), which are the dilated regions found at both ends of the diaphysis. The epiphyses possess:
   (1) A thin covering of compact bone on the outside.
   (2) A central region of spongy bone.

c. Between the epiphysis and diaphysis of growing bones is an internal region with columns of cancellous bone called *metaphysis* (from the Greek μετά meta 'in the middle of').

2. *Short bones*, such as vertebrae, usually have a thin outer layer of compact bone surrounding inner spongy bone.
3. In the *flat bones* of the skull, scapula and the sternum, we observe peripheral compact bone forming the so-called external and internal tables, between which is situated spongy bone forming the *diploe* (from the Greek διπλός diplós 'double').

6.3 Bone Surfaces

A. The outer surface of the bones is covered by a tissue called the *periosteum*:
   1. It is a connective tissue with *osteogenic capacity* (from the ancient Greek ὀστεο osteo, 'bone').
   2. There is no periosteum at the insertion points of tendons and ligaments and in regions covered with articular cartilage.
   3. In some cases, the periosteum is given special names, although it is still the same tissue. For example, in the inner tables of the skull bones, the periosteum is called *dura mater*.
   4. The structure of the periosteum is variable.
      a. The outermost layer of the periosteum is a *dense connective tissue* with few cells and blood vessels.
      b. The innermost layer of the periosteum contains cells in contact with bone. These cells are osteogenic in growing bones, but remain inactive in the rest.
   5. Some collagen fibres of the periosteum can penetrate the superficial bone matrix forming the so-called *perforating* or *Sharpey's fibres*.

B. The inner surfaces of bones are lined by the *endosteum*, a very thin tissue with *osteogenic capacity*.
   1. It lines the surface of the bone tissue in the medullary cavity of the diaphyses of long bones and the trabeculae of spongy bone.
   2. The endosteum also lines some microscopic cavities that are organised as ducts or canals in the compact bone.
   3. It appears as a thin pavement layer consisting of cells and few fibres.

6.4 Bone Extracellular Matrix

A. The *bone matrix* can be considered as consisting of two closely related components: the organic component or *organic matrix* and the inorganic component or *mineral matrix*.
   1. If the bone is treated with a *weak acid*, the mineral salts are removed and the bone loses its hardness and becomes a tough, flexible material that retains its macroscopic and microscopic organisation.
   2. If the organic components are removed and the inorganic components are left in place, the macroscopic shape of the bone and partly the microscopic organisation is maintained, but the bone loses its tensile and compressive strength and is very brittle.

B. In short, the organic matrix gives the bone its strength and elasticity, while the mineral matrix is responsible for its hardness.

C. The *organic matrix* of bone tissue does not differ much from that of other connective tissues.
   1. *Collagen* makes up 90% of the organic matrix. It is *type I collagen*.
   2. There are other proteins, including:
      a. *Osteopontin*: It is involved in cell adhesion, similar to fibronectin.
      b. *Osteonectin*: It is involved in providing the growth nucleus for hydroxyapatite crystals during mineralisation.
      c. *Bone sialoprotein*: It appears to be involved in mineralisation.
      d. *Bone morphogenetic protein* (*BMP*):
         (1) It is a local chemical mediator involved in cell-to-cell communication and promotes cell proliferation.
         (2) It plays an important role in bone formation.
      e. *Osteocalcin* or *Bone Gla protein*[1] (*BGP*):
         (1) It binds calcium to collagen.
         (2) It is produced by bone cells and can travel in the blood, so osteocalcin may be involved in the mineralisation of bone elsewhere in the body.
      f. *Matrix Gla protein (MGP)*: It is involved in the regulation of calcium levels in the body.
      g. *Alkaline phosphatase*: It is involved in mineralisation, releasing phosphate groups from organic molecules so that they can form phosphates from the mineral matrix.
   3. In addition, there are *proteoglycans* with their characteristic *glycosaminoglycans*.
      a. Among the proteoglycans, decorin and biglycan.
      b. Glycosaminoglycans include hyaluronic acid, chondroitin sulfate, dermatan sulfate and heparan sulfate.

---

[1] Gla proteins contain a sequence of amino acids that bind Ca^{2+}. Among these amino acids is γ-carboxyglutamate, which is glutamate carboxylated by the action of vitamin K.

D. The *mineral matrix* of bone tissue can account for about 75% of the dry weight of a bone.
   1. It consists mainly of *calcium phosphate* crystals in the form of *hydroxyapatite*, $Ca_{10}(PO_4)_6(OH)_2$.
   2. These crystals are acicular and measure 40–60 nm long by 1.5–3 nm wide.
   3. The crystals are arranged in the 40 nm gaps between the ends of the collagen molecules of the fibrils and in the spaces between the fibrils (Fig. 6.2).
   4. The newly formed bone matrix is not mineralised, is called *osteoid* (from the Greek ὀστεο osteo, 'bone' and ειδ-ής/ές -eid(és) 'having the appearance of') and consists exclusively of organic matrix.
   5. Bone matrix synthesis always occurs on a surface of bone tissue or mineralised cartilage tissue, so the osteoid is always coating mineralised matrix.
   6. The mineralisation process starts 10–20 days after the synthesis of the organic matrix. It is explained in detail in Chap. 7.

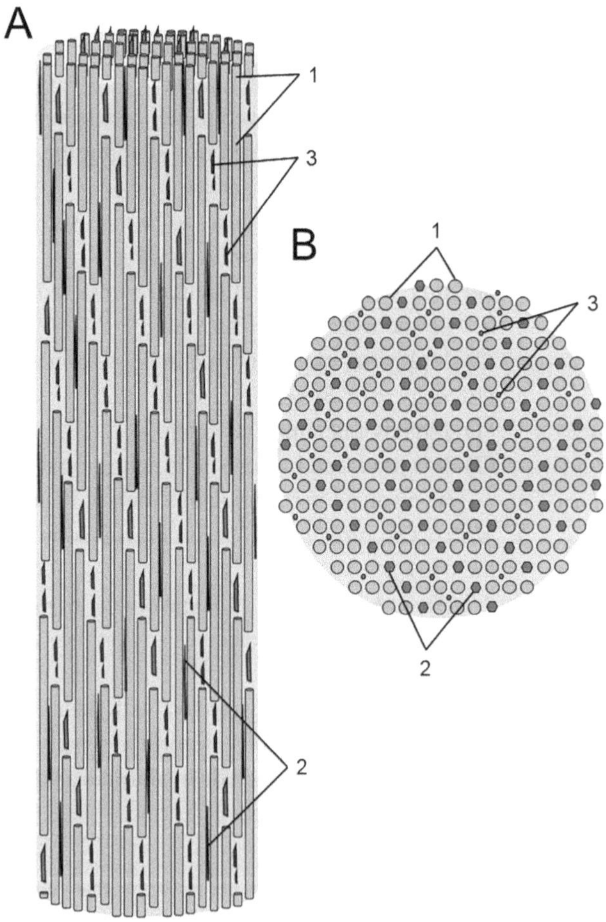

Fig. 6.2 Longitudinal view (**A**) and cross-section (**B**) of a collagen fibril. Hydroxyapatite crystals are deposited on the surface and within the fibrils formed by collagen or tropocollagen molecules (1), either next to these molecules (2) or in the gaps between them (3)

7. *Citrate* and *carbonate* ions can be found in the mineral matrix.
8. The hydroxyl groups of apatite can be replaced by *fluoride*, forming *fluorapatite*.
9. Other minerals stored in bone include *magnesium*, *potassium* and *sodium*.

6.5 Bone Survey Method

A. There are two different techniques for bone studies with the brightfield microscope.
   1. *Histological sections of decalcified tissue*:
      a. When preparing the tissue, *decalcification* is carried out, which allows the bone to be cut into very thin sections. Without this decalcification, it would not be possible to cut the bone.
      b. After histological sections have been made, they can be stained with any of the usual stains.
      c. It allows the study of cellular components and organic components of the extracellular matrix.
   2. *Ground sections*:
      a. They are made from clean, macerated bone, which is cut with a saw and then ground to a thin film that can be observed under the microscope, with or without the use of dyes.
      b. This technique allows the microscopic architecture of the extracellular matrix to be appreciated and other morphological information to be obtained.

6.6 Microscopic Structure of Bone Tissue

A. Microscopically, there are two types of bone tissue. The main difference between the two is the way the extracellular matrix is organised (Fig. 6.3):
   1. *Woven*, *non-lamellar*, *bundle* or *immature bone* is the first bone tissue to form when a bone is created or when a fractured bone is repaired.
      a. Its extracellular matrix is homogeneous and no special organisation is discernible.
      b. Its collagen fibres are arranged in a randomly woven pattern.
      c. The tissue cells are more abundant than in lamellar bone and are arranged without apparent organisation by the matrix.
   2. *Lamellar* or *mature bone* is characterised by the following features:
      a. Its bone matrix is organised in the form of *bony lamellae* or *lamellae*, 3–7 μm thick.
      b. The cells, which are less abundant than in woven bone, are regularly arranged at different levels, as most of them are between the lamellae.

B. The first bone tissue to form is always *non-lamellar woven bone*, and when it is remodelled it is replaced by *lamellar bone*, so the bone we usually see is lamellar bone.

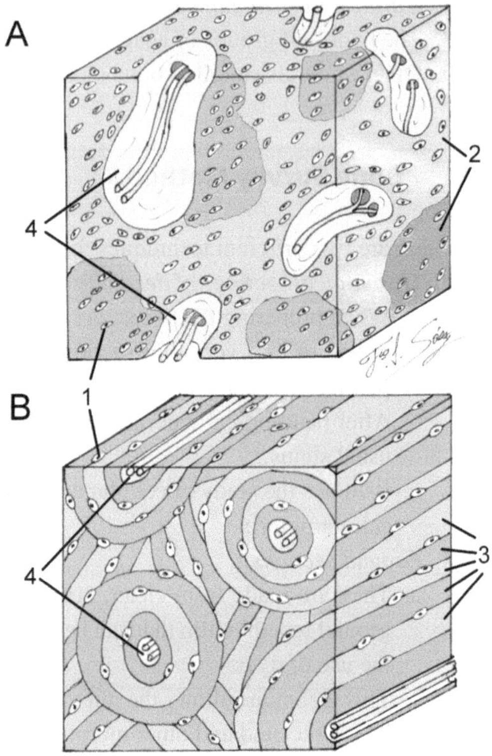

Fig. 6.3 Microscopic structure of woven or immature bone (**A**) and lamellar bone (**B**). In woven or immature bone the matrix is homogeneous and the cells are abundant, whereas in lamellar bone the matrix is arranged in lamellae and the cells are more sparse. (1) Bone cells (osteocytes) in lacunae; (2) homogeneous bone matrix of woven bone; (3) lamellae of bone matrix of lamellar bone; (4), vascular channels with blood vessels

C. Immature bone may persist in some places in the adult, such as the bones of the *dental sockets*.

D. As almost all bone in adults consists of lamellar bone, its microscopic structure will be explained as follows:

1. In each of the lamellae of lamellar bone, all the collagen fibres are arranged in the same orientation and are therefore parallel to each other. In contrast, the orientation changes in each lamella.
2. Between the lamellae, and sometimes within them, there are small cavities, called *osteoplasts*, *osteoceles* or *lacunae*, in which bone cells, *osteocytes*, are found.
3. From the lacunae, small canals run in all directions, connecting osteoceles to osteoceles (*calcophore ducts*).
4. *Microscopic organisation of compact bone*: In compact bone, the organisation of the lamellae is complex and several structures can be distinguished (Fig. 6.4):
   a. *Circumferential lamellae*: These are of two types:
      (1) *External circumferential lamellae*: These are lamellae that are arranged in the peripheral region of the compact bone, parallel to the surface, below the periosteum and surrounding the entire bone.

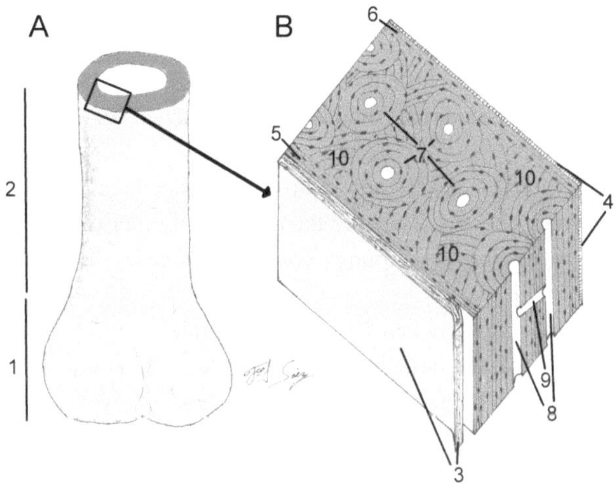

Fig. 6.4 Representation of a long bone (**A**), cut transversely in the diaphysis, and of the microscopic structure of the compact bone (**B**) of the area squared in **A**. (1) Epiphysis; (2) diaphysis; (3) periosteum (partially separated from the bone surface to better distinguish it from the external circumferential lamellae); (4) endosteum; (5) external circumferential lamellae; (6) internal circumferential lamellae; (7) osteons or Haversian systems; (8) Haversian canal; (9) Volkmann canal; (10) interstitial lamellae

(2) *Internal circumferential lamellae*: Lamellae parallel to the inner surface of the compact bone covered with endosteum. Sometimes these inner circumferential lamellae are not very visible or not complete.

b. *Osteons* (from the German osteon, from the Ancient Greek ὀστεο osteo, 'bone') or *Haversian systems*: They are cylindrical structures.
   (1) They are made up of:
      (a) Several lamellae (4–20) arranged concentrically around a small central canal.
      (b) A central channel of 22–110 µm surrounded by lamellae. This is called *Haversian canal.*
      (c) Occasionally, Haversian canals branch laterally, pass through the wall of the osteon and contact other Haversian canals in neighbouring osteons. These lateral branches are called *Volkmann canals*.
   (2) In the osteonal canals it can be observed:
      (a) *Endosteum*, which lines the bony surfaces of the canals.
      (b) A small amount of *loose connective tissue.*
      (c) Blood vessels circulate between the connective tissue of the canals, usually a *capillary* and a *postcapillary venule*, although occasionally *arterioles* may be found.

(3) Blood vessels enter this system of channels from the periosteum, pass through the compact cortical bone, reach the bone marrow and re-form branches that penetrate the cortical bone.

(4) It should be noted that osteons can be branched, so they are not as simple as a cylinder.

c. *Interstitial lamellae*. These are groups of lamellae located in the spaces or interstices between osteons. They are the remains of older osteons.

5. *Microscopic organisation of spongy bone*: In this case, the microscopic organisation is simpler:
   a. The *trabeculae* are between 50 and 400 μm thick.
   b. In each trabecula, the lamellae are distributed as flat discs approximately 60 μm thick and 600 μm long.
      (1) Each disc consists of about 20 lamellae parallel to the disc surface.
      (2) These discs have been called *trabecular osteons*, although this term is not often used in order not to confuse them with the true Haversian systems of compact bone.
   c. When viewing trabeculae under the microscope, the lamellae are seen parallel to the surface of the trabeculae.

E. Nutrition of bone tissue:
   1. *Compact bone nutrition*:
      a. Compact bone can be very thick, so that the inner regions are distant from the nearest vascularised tissues (periosteum on the outside and bone marrow on the inside).
      b. Osteons serve to bring blood vessels into the inner regions, allowing nutrients to be brought closer to the tissue.
   2. *Nutrition of spongy bone*:
      a. Trabeculae are relatively thin structures, so nutrients can enter the tissue by diffusion from the blood vessels in the bone marrow.
      b. Cancellous bone trabeculae usually lack true osteons and blood vessels do not penetrate into them.
      c. Exceptionally, Haversian systems can be observed in very large trabeculae.

6.7 Bone Tissue Cells

A. Several cell types can be found in osseous tissue or on its surface (Fig. 6.5). Most of them are related and originate from mesenchymal cells:
   1. *Osteoprogenitor cells*: Derived from mesenchymal cells and give rise to other bone cells.
   2. *Osteoblasts*: They are derived from osteoprogenitor cells. They are coating bone matrix and synthesise new bone matrix on it.
   3. *Osteocytes*: They are derived from osteoblasts that are enveloped by the matrix they are synthesising, so they are found within the bone tissue.

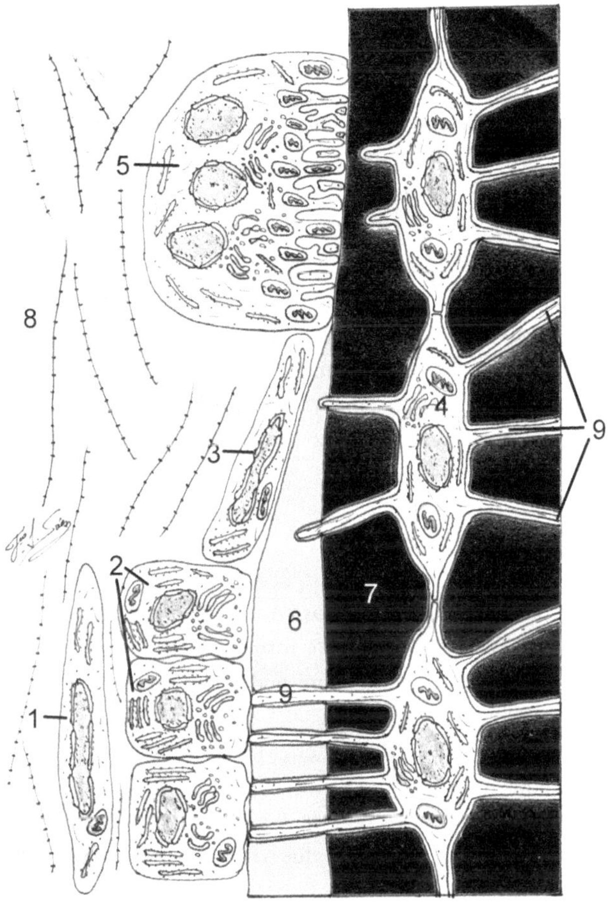

Fig. 6.5 Cells of bone tissue. In the connective tissues located on the surface of the bone tissue (periosteum and endosteum), there are osteoprogenitor cells (1) which are activated and converted into osteoblasts (2) which synthesise bone matrix. When osteoblasts are inactivated they become bone-lining cells (3). Some osteoblasts are trapped in the bone matrix that they themselves synthesise and become osteocytes (4). Multinucleated osteoclasts can be found on the bone surface (5), which degrade the bone matrix. (6) Non-calcified bone matrix or osteoid; (7) calcified bone matrix; (8) connective tissue matrix (periosteum or endosteum); (9) calcophore ducts with osteocyte extensions

4. *Bone-lining cells*: Derived from osteoblasts that become inactive. They are found on bone surfaces.

B. However, there are other cells that originate from blood precursors:
   1. *Osteoclasts*: They are derived from blood monocytes, although it is also thought that some osteoclasts (or their precursors) may be tissue residents installed there from embryonic development.
   2. *Bone macrophages*: These are macrophages derived from monocytes. They are located next to osteoblasts on the surfaces of bone tissue and cause their death.

6.8 Osteoprogenitor Cells

A. *Osteoprogenitor cells* are produced from *mesenchymal cells*.
B. They are found on the surfaces of bone tissue.
C. They are elongated, spindle-shaped, with an elongated oval nucleus and an eosinophilic or weakly basophilic cytoplasm, as they have little rough endoplasmic reticulum and some free ribosomes.
D. They can be found in two forms:
   1. *Periosteal cells*, present in the inner layer of the *periosteum*.
   2. *Endosteal cells*, which line the bone surfaces in the medullary cavities and vascular ducts (Haversian and Volkmann canals) forming the *endosteum*.
E. When bone is formed, i.e. during embryonic development or in other bone remodelling or repair processes, these cells divide to produce osteoblasts that synthesise bone.

6.9 Osteoblasts

A. *Osteoblasts* are the cells produced by *osteoprogenitor cells* when bone matrix synthesis is activated. They are responsible for forming bone matrix.
B. They are found in the inner periosteum and endosteum, always in contact with the surface of the bone matrix.
C. When active, osteoblasts form an epithelial layer of cuboidal cells:
   1. They have a nucleus at the pole furthest from the bone and a large Golgi apparatus oriented towards the opposite pole.
   2. The cytoplasm is highly basophilic due to its high content of cisternae of the *rough endoplasmic reticulum*.
D. As the osteoblasts are always on the surface of the bone tissue, the bone tissue always grows by *apposition*.
E. The matrix they synthesise at the beginning is not mineralised and forms the *osteoid*. After some time, the innermost areas of the osteoid become mineralised.
F. During its lifetime, an osteoblast can have three fates:
   1. Approximately 15% of the osteoblasts are enveloped by the bone matrix they have synthesised and become *osteocytes*.
   2. About 30% become inactive and form *bone-lining cells*.
   3. The rest die by *apoptosis* and are eliminated by bone macrophages. In this process, the macrophage produces substances that stimulate the formation of new osteoblasts.

6.10 Osteocytes

A. *Osteocytes* are the characteristic cells of bone tissue, which are located in *osteoceles*.
   1. In *mature bone*, osteoceles are between lamellae and sometimes within a lamella.
   2. In *immature bone*, osteoceles are scattered throughout the bone matrix.

B. The shape of the cell is adapted to the shape of the lacuna in which it is found, which is usually *lenticular*.
C. The osteocytes emit numerous fine extensions that travel inside the *calcophore ducts* until they contact the extensions of neighbouring osteocytes by means of *gap junctions*. In this way, the bone cells communicate with each other.
D. The cytoplasm is somewhat *basophilic*. Electron microscopy shows variable development of the Golgi apparatus and rough endoplasmic reticulum:
   1. The cells closest to the surface (the youngest) have these organelles more developed.
   2. Cells in deeper positions (older) show less development of biosynthetic organelles, indicating that older cells have less synthetic activity.
E. The osteocyte is involved in the maturation of the bone matrix, but is also capable of resorbing the surrounding mineral matrix (*osteocytic osteolysis*), which allows calcium to be released into the blood.
F. Osteocytes can live for a long time, between 10 and 20 years.
   1. In the newborn, 1% of osteocytes may be dead. In the elderly, 75% of osteocytes may be dead.
   2. It appears that the lacunae and canaliculi of dead osteocytes are filled with mineralised matrix.

6.11 Bone-Lining Cells

A. *Bone-lining cells* are produced from osteoblasts that stop synthesising matrix and become apparently inactive.
B. These cells become flattened, spindle-shaped cells, with the organelles less evident, resting on a thin layer of osteoid.
C. *Gap junctions* are observed at the points of contact between them.
D. They may also emit extensions that are inserted into the superficial *calcophore ducts*.
   1. There they contact the cytoplasmic extensions of nearby osteocytes, where they form *gap junctions*.
   2. These contacts serve to regulate the nutrition of the osteocytes and to regulate the movement of calcium and phosphate between the blood and the bone tissue.
E. The bone-lining cells of the periosteum are called *periosteal cells* and those of the endosteum are called *endosteal cells*.
F. In order for the bone to be resorbed, it is essential that the mineral matrix is not covered by non-mineralised matrix:
   1. Prior to bone resorption, bone-lining cells produce *collagenase*, an enzyme that degrades collagen, which destroys osteoid.
   2. The bone-lining cells then retract so that the mineralised matrix is exposed to the surface and can be degraded by the cells in charge of the process (*osteoclasts*).

6.12 Osteoclasts

A. *Osteoclasts* (from the Greek -κλάστης -klástēs 'breaking') are very large *multinucleated cells*, up to 150 μm in diameter, found, like osteoblasts, on the surface of bone tissue.
B. Osteoclasts often create cavities in which they are located. These spaces are called *Howship lacunae*.
C. They have a very specialised structure:
   1. The *nuclei* are reddish violet when stained with haematoxylin and eosin and are located at the pole opposite the bone matrix.
   2. The *cytoplasm* is *basophilic* in young cells and *eosinophilic* in mature cells. Numerous Golgi apparatus, many mitochondria, rough endoplasmic reticulum cisternae and lysosomes are observed in the cytoplasm.
   3. The *plasma membrane in* contact with the bone matrix is highly folded:
      a. They form a *ruffled border* that changes shape continuously.
      b. In the cytoplasm of this region there are many *mitochondria* and *lysosomes*.
D. The periphery of the osteoclast surface that contacts the bone matrix adheres strongly to the matrix *osteopontin*:
   1. This seals the space between the cell and the bone matrix.
   2. A microenvironment is created between the osteoclast and the bone matrix in which substances produced by the osteoclast are released.
E. For osteoclasts to attach to bone tissue, it is essential that the bone tissue is not coated with osteoid, which must be degraded by *bone-lining cells*.
F. The osteoclasts then take over the resorption of the bone matrix, both the mineral matrix and the organic matrix. This is done with *lysosomal and non-lysosomal enzymes*.
G. The osteoclast acidifies the medium by releasing *protons* (H^+), which, in addition to creating the optimal environment for *acid hydrolases*, dissolves *calcium salts*.
H. Origin of osteoclasts:
   1. Osteoclasts are derived from blood *monocytes*, which leave the blood circulation and fuse on the bone surface and form part of the *mononuclear phagocytic system*.
   2. It is thought that there may also be a population of *resident osteoclasts* that settle into the tissue during embryonic development.

6.13 Bone Macrophages

A. *Bone macrophages* are produced from blood *monocytes*, which are activated by substances produced by osteoblasts.
B. They are located on *osteoblasts* and cause cell death by *apoptosis* of many of them.
C. These macrophages eliminate dead osteoblasts.
D. During the process of removing dead osteoblasts, they produce substances that stimulate the production of new osteoblasts.

Take-Home Message

Bones are made up of osseous tissue. Macroscopically, compact bone and cancellous bone can be distinguished. The outer surfaces of the bone are covered by periosteum. The internal spaces are lined by endosteum. The extracellular matrix of osseous tissue contains type I collagen fibres, proteins (osteopontin, osteonectin, bone sialoprotein, osteocalcin, etc.) and glycosaminoglycans, as well as high levels of minerals, mainly calcium phosphate crystals (hydroxyapatite). The newly formed bone matrix lacks minerals and is called osteoid. Under the microscope, immature bone tissue is seen to be non-lamellar, but this tissue is replaced by lamellar bone tissue. In compact lamellar bone, lamellae can form concentric structures called osteons, inside which there is a vascular channel. Several cell types can be found in bone: osteoprogenitor cells, derived from mesenchymal cells, are found on the surfaces of bone (periosteum and endosteum) and develop into osteoblasts, which produce bone matrix. When osteoblasts are trapped by the matrix they produce, they become osteocytes. If they stop their osteogenic activity, they become bone-lining cells. There are also cells of blood origin, such as osteoclasts, multinucleated cells formed by the fusion of monocytes, which destroy the bone matrix, and bone macrophages, also produced by monocytes on the bone surface, which are responsible for controlling the osteoblasts.

Further Reading

Arnett TR, Dempster DW (1990) Protons and osteoclasts. J Bone Miner Res 5:1099–1103

Ascenzi MG, Roe AK (2012) The osteon: the micromechanical unit of compact bone. Front Biosci-Landmark 17:1551–1581

Aubin JE, Liu F, Malaval L, Gupta AK (1995) Osteoblast and chondroblast differentiation. Bone 17:S77–S83

Chang B, Liu XH (2022) Osteon: structure, turnover, and regeneration. Tissue Eng Part B Rev 28:261–278

Chen X, Wang ZQ, Duan N et al (2018) Osteoblast-osteoclast interactions. Connect Tissue Res 59:99–107

Gartner LP (2017) Cartilage and bone. In: Textbook of histology, 4th edn. Elsevier, Philadelphia, pp 149–178

Gómez de Ferraris ME, Campos Muñoz A (2009) Histología general humana. In: Histología, Embriología e Ingeniería Tisular Bucodental, 3rd edn. Médica Panamericana, pp 57–78

Jee WSS (1988) The skeletal tissues. In: Weiss L (ed) Cell and tissue biology – a textbook of histology, 6th edn. Urban & Schwarzenberg, Baltimore, pp 211–254

Kierszenbaum AL, Tres LL (2019) Connective tissue. In: Histology and cell biology – an introduction to pathology, 5th edn. Elsevier, Philadelphia, pp 135–176

Kitase Y, Prideaux M (2023) Targeting osteocytes vs osteoblasts. Bone 170:116724

Lian JB, Stein GS (1995) Development of the osteoblast phenotype: molecular mechanisms mediating osteoblast growth and differentiation. Iowa Orthop J 15:118–140

Matsuo K, Irie N (2008) Osteoclast-osteoblast communication. Arch Biochem Biophys 473:201–209

McDonald MM, Kim AS, Mulholland BS, Rauner M (2021) New insights into osteoclast biology. JBMR PLUS 5:e10539

O'Dowd G, Bell S, Wright S (2023) Skeletal tissues. In: Wheater's functional histology. Elsevier, Amsterdam, pp 180–196

Park-Min KH, Lorenzo J (2022) Osteoclasts: other functions. Bone 165:116576

Pawlina W (2024) Bone. In: Histology a text and atlas with correlated cell and molecular biology, 9th edn. Wolters Kluwer, Philadelphia, pp 238–279

Peck WA, Rifas L (1982) Regulation of osteoblast activity and the osteoblast-osteocyte transformation. Adv Exp Med Biol 151:393–400

Pierce AM, Hammarstrom L (1991) Osteoclasts – structure and function. Electron Microsc Rev 4:1–45

Sanchez C, Gabay O, Henrotin YE, Berenbaum F (2008) Osteoblast: a cell under compression. Biomed Mater Eng 18:221–224

Sun Y, Li JB, Xie XP et al (2021) Recent advances in osteoclast biological behavior. Front Cell Dev Biol 9:788680

Thiel A, Reumann MK, Boskey A et al (2018) Osteoblast migration in vertebrate bone. Biol Rev 93:350–363

Osseous Tissue (II): Ossification, Bone Remodelling and Joints

7

What You Will Learn in This Chapter
The two mechanisms of bone formation are explained here: intramembranous ossification and chondral ossification. In addition, osseous tissue has a high mineral content, so the organic matrix undergoes a mineralisation process, the mechanism of which you will learn about. The mechanisms of bone growth, the mechanisms of bone modelling and remodelling and fracture repair are also explained. Finally, you will learn about the different types of joints between bones, and you will see the complex organisation of synovial joints or diarthrosis.

7.1 Bone Histogenesis (Ossification)

A. Bone formation always takes place by replacement or modification of a previous connective tissue, i.e. by *ossification* of an existing tissue.
B. Two types of ossification are usually distinguished:
   1. Intramembranous or desmic ossification, which takes place in an embryonic connective tissue 'membrane' (mesenchyme). It is a direct ossification because:
      a. Mesenchymal cells convert directly into osteoblasts.
      b. The mesenchymal matrix is converted directly into bone matrix.
   2. *Chondral ossification*, which takes place on previously formed cartilaginous *tissue*. *It is* therefore an *indirect ossification*:
      a. Mesenchymal cells first develop into chondroblasts that produce cartilage.
      b. Subsequently, the cartilage degrades and new mesenchymal cells turn into osteoblasts and deposit bone tissue on its remains.
C. In both cases, the first bone formed is spongy bone (*primary* or *primitive spongy bone*).
D. In some areas, the growth of trabeculae occupies the space between them, which fuse, and *compact bone* is formed.
E. In other regions of the future bone, the mesenchyme between the trabeculae develops into haemopoietic tissue (*bone marrow*).

F. J. Sáez, I. Badiola, *Essential Oral Histology*, Learning Materials in Biosciences,
https://doi.org/10.1007/978-3-032-03165-5_7

7.2 Intramembranous Ossification

A. *Intramembranous ossification* occurs in connective tissues (mesenchyme).

B. Through this mechanism, they are formed:

   1. Some bones called *intramembranous bones* or *membrane bones*: frontal, parietal and part of the occipital bones, the maxilla and the mandible.
   2. The bone tissue necessary for the growth of short bones and the increase in thickness of long bones.

C. In the case of intramembranous bones, intramembranous ossification follows these steps (Fig. 7.1):

   1. It begins with a condensation of mesenchyme in the embryo, i.e. a *blastema* (from the ancient Greek βλάστημα blástēma 'germ, sprout') is formed because cells accumulate in the future bone region.
   2. The cells acquire *osteogenic capacity* (*osteoprogenitor cells*) and begin to synthesise *bone matrix*. The first bone matrix produced is not mineralised and is called *osteoid*.
   3. Cells settle on this bone matrix, become cubic in shape, their cytoplasm becomes more basophilic and they become *osteoblasts* that synthesise more matrix on top of the already formed matrix.

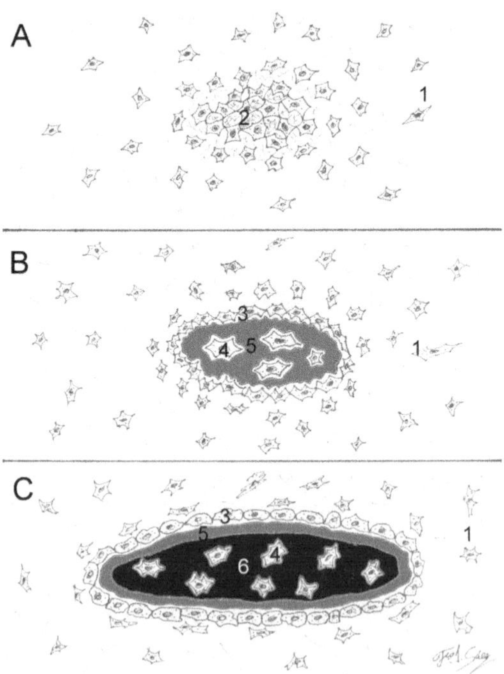

Fig. 7.1 Representation of the intramembranous ossification process in three consecutive phases, **A**, **B**, and **C**. (1) Mesenchyme with mesenchymal cells; (2) condensation of mesenchymal cells forming a blastema with osteogenic capacity; (3) osteoblasts; (4) osteocytes; (5) osteoid; (6) bone matrix

4. The bone matrix formed occurs in small areas producing *trabeculae* (*primary* or *primitive spongiosa*), separated from each other by regions of unossified mesenchyme with blood vessels.
5. As the bone matrix calcifies, osteoblasts are observed on a thin layer of *osteoid* on top of already mineralised bone matrix.
6. When any osteoblast is surrounded by bone matrix, it becomes an *osteocyte*.
7. The surrounding mesenchymal cells divide by mitosis and provide new osteoprogenitor cells and new osteoblasts to replace those that have become osteocytes.
8. The trabeculae grow and converge. In some areas they fuse and thus produce the first areas of compact bone (*primary* or *primitive compact*).

D. A bone can form from several areas of mesenchyme that ossify:
   1. Each area is a *centre of ossification*.
   2. Gradually, the bone tissue formed in the different ossification centres comes together to form the complete bone.
   3. The fontanelles of the neonatal skull are soft regions where connective tissue has not yet been replaced by bone.

7.3 Chondral Ossification

A. In this case, the outline of the bone is formed as a structure of hyaline cartilage. This is why they are called *cartilaginous bones*.

B. This is the process by which most of the bones in our bodies are formed, especially the short and long bones.

C. Once a cartilaginous outline of the bone is formed, *chondral ossification occurs*, in a mechanism that has been well studied in long bones, where it takes place in two successive phases:
   1. *Perichondral ossification*:
      a. A mesenchymal envelope forms around the central region of the cartilage outline, the future diaphysis (Fig. 7.2).
         (1) This envelope, shaped like a *periosteal cuff*, forms on the outside of the skeletal piece and becomes bone (*bony collar*).
         (2) Strictly speaking, it is an *intramembranous ossification* that occurs on the surface of the cartilage.
      b. The cartilage surrounded by the bony cuff undergoes several changes.
         (1) Their cells become *hypertrophic*, with many mitochondria.
         (2) They cause *calcification* of the cartilage matrix.
         (3) The process leads to *cell death*.
      c. Blood vessels and mesenchyme penetrate from the periphery into the spaces produced by cell death. This penetration occurs at a single site of the cuff, the *periosteal bud*, by osteoclast activity.
      d. The following process of chondral ossification, described below, can then begin.

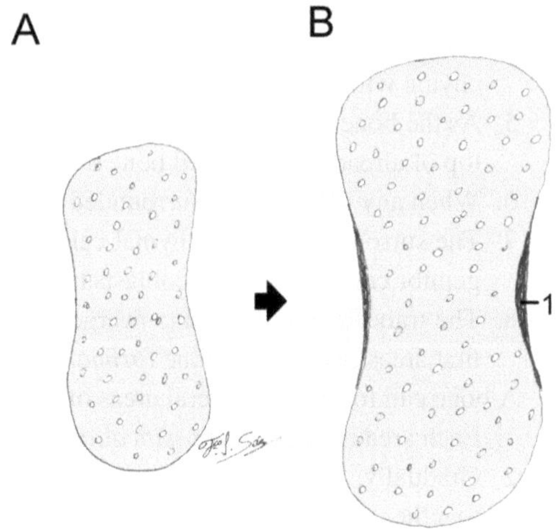

Fig. 7.2 Early stages of the chondral ossification process. A sketch of the future bone is formed by hyaline cartilage (**A**), around which a bony collar is formed by perichondral ossification (**B**). (1) Bony collar

2. *Endochondral ossification*: This process is caused by the appearance of ossification centres within the cartilage outline. Two centres of endochondral ossification are usually distinguished:
   a. *Primary ossification centre*: It occurs in the centre of the *diaphysis*.
      (1) In this region the chondrocytes proliferate and organise themselves in columns, following the longitudinal axis of the future bone.
      (2) The cartilage cells in this area *hypertrophy*.
         (a) This process causes the lacunae occupied by cells to become very large and to merge with each other.
         (b) This forms large spaces separated by thin regions of cartilaginous matrix (Fig. 7.3).
      (3) The matrix calcifies and cell death by *apoptosis* occurs:
         (a) This results in a *calcified cartilaginous matrix* with numerous empty spaces.
         (b) The spaces are occupied by mesenchyme and blood vessels penetrating through the periosteal bud (Fig. 7.4).
      (4) Some mesenchymal cells develop into osteoprogenitor cells that give rise to *osteoblasts*.
      (5) A layer of bone matrix is deposited on the acellular, calcified cartilage matrix trabeculae, first without calcification (*osteoid*), which then gradually calcifies.
      (6) As it grows in thickness, some osteoblasts become trapped inside the bone matrix and turn into *osteocytes*.
      (7) This forms trabeculae of bone tissue with calcified cartilage matrix within them, which are arranged longitudinally along the major axis of the bone (Fig. 7.5).

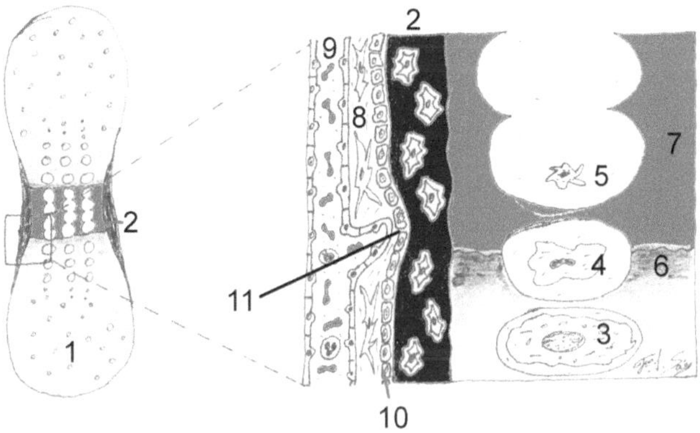

Fig. 7.3 Representation of the process of chondral ossification (continuation of Fig. 7.2). (1) Hyaline cartilage mould; (2) bony collar; (3) hypertrophic chondrocyte; (4) regressing chondrocyte; (5) dead chondrocyte; (6) calcifying cartilage matrix; (7) calcified cartilage matrix; (8) mesenchyme; (9) blood vessel; (10) osteoblasts; (11) periosteal bud formation

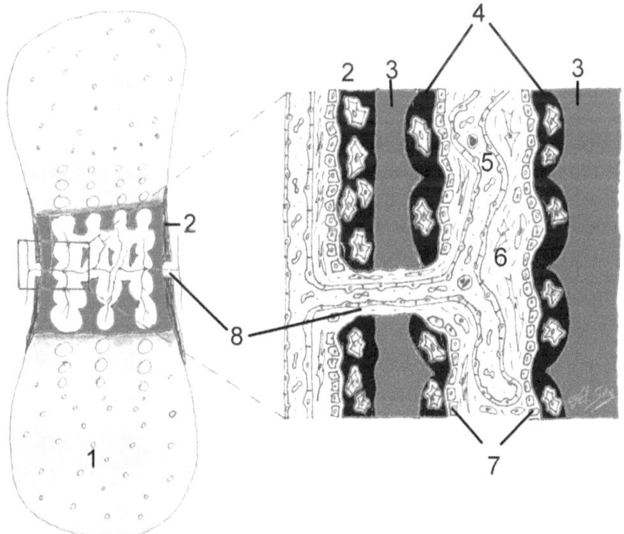

Fig. 7.4 Representation of the process of chondral ossification (continuation of Fig. 7.3). (1) Hyaline cartilage mould; (2) bony collar; (3) calcified cartilage trabecula; (4) bone matrix with osteocytes; (5) blood vessels; (6) mesenchyme; (7) osteoblasts; (8) periosteal bud penetrating the diaphysis

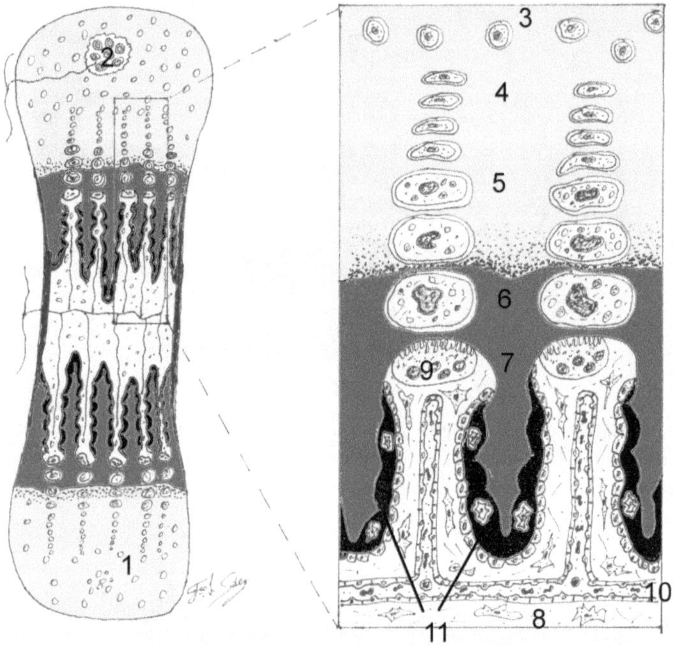

Fig. 7.5 Representation of the process of chondral ossification (continuation of Fig. 7.4) and the growth of a long bone in the epiphyseal plate. (1) Hyaline cartilage mould; (2) secondary ossification centre; (3) zone of reserve cartilage; (4) zone of proliferation; (5) zone of hypertrophy; (6) zone of calcification; (7) zone of resorption; (8) mesenchyme; (9) osteoclast; (10) blood vessel; (11) bone matrix with osteocytes

(8) *Osteoclasts* then appear, degrading some of the newly formed bone trabeculae, so that the bone tissue is remodelled as it is synthesised.

b. *Secondary ossification centre*: They are produced in the *epiphyses* by mechanisms similar to those of the primary ossification centre (Fig. 7.5).
   (1) They are formed in the central zone of the epiphyseal cartilage, which means that ossification begins in the centre while the peripheral zones are still formed by hyaline cartilage.
   (2) They usually appear later than the primary ossification centres, so that much of the epiphysis of embryonic bone is cartilaginous, while the diaphysis is already bone tissue.
   (3) The trabeculae that form are arranged radially, as the chondrocytes grow in all directions.
   (4) When the entire cartilage mould is replaced by bone, two regions of cartilage remain:
      (a) *Articular cartilage*: A layer of cartilage that remains in the articular regions of the epiphysis.
      (b) *Epiphyseal plate* or *epiphyseal disc*: This is a cartilaginous region between the epiphysis and the diaphysis (*metaphysis*), which is responsible for the growth in length of the bone in infancy.

7.4 Bone Matrix Mineralisation

A. Matrix *mineralisation* is important not only because it occurs in bone tissue but also because it occurs in other tissues, such as cartilage and various dental tissues (*dentine*, *enamel* and *cementum*).

B. In all mineralised tissues, except enamel, the matrix has collagen fibres and ground substance.

   1. The mineral substances are located on the collagen fibrils and within them, between the collagen molecules (Fig. 6.2).
   2. The whole process is regulated by the cells of each tissue.

C. The mineralisation process in bone, cartilage, cementum and dentine is very similar and requires a certain concentration of *calcium* (Ca^{2+}) and *phosphate* (PO_4^{3-}) to be exceeded.

D. In bone, the process occurs on the first, non-mineralised bone matrix, the *osteoid*.

E. Mineralisation starts 10–20 days after the synthesis of the organic matrix.

F. Mineralisation occurs according to this scheme:

   1. Increased Ca^{2+} concentration in the extracellular matrix due to ion binding by *sialoproteins* and *osteocalcin*.
   2. Action of *alkaline phosphatase*, an enzyme present in osteoblasts, which releases phosphate groups from various organic molecules and increases the concentration of PO_4^{3-} in the extracellular matrix.
   3. It has been described that mineralisation begins in the osteoid matrix by two mechanisms:
      a. *Mineralisation points*: These are places in the matrix where crystallisation of hydroxyapatite occurs. It is not clear whether these crystallisation points actually occur in the matrix.
      b. Within the *matrix vesicles* or *mineralisation vesicles*, small vesicles of 50–200 nm in diameter are detached by budding from the surface of the osteoblast. In other words, they are *ectosomes*.
   4. The matrix vesicle-mediated mineralisation process occurs in two phases:
      a. *Matrix vesicle-mediated mineralisation phase*: It occurs within the *matrix vesicles*.
         (1) Matrix vesicles take up Ca^{2+} and PO_4^{3-} from the extracellular matrix and deliver them into the matrix.
         (2) In the lumen of these vesicles, *nanospheres* of 10 nm in diameter are formed by *amorphous calcium phosphate* [$Ca_9(HPO_4)(PO_4)_5(OH)$].
         (3) Amorphous calcium phosphate crystallises to form *octocalcium phosphate* [$Ca_8H_2(PO_4)_6 \cdot 5H_2O$].
         (4) Due to the high concentration of calcium and phosphate in the matrix vesicles, octocalcium phosphate is transformed into acicular *hydroxyapatite* crystals [$Ca_{10}(PO_4)_6(OH)_2$], which accumulate in the vesicles.

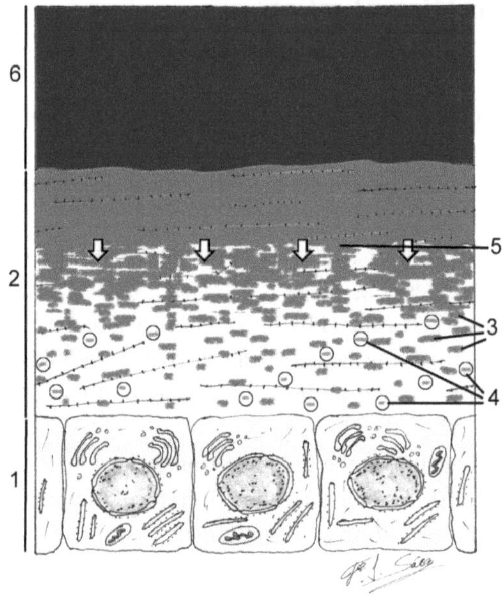

Fig. 7.6 Progress of bone matrix mineralisation. The first matrix synthesised by osteoblasts (1) is osteoid (2), which is not mineralised. Osteoblasts increase the concentration of calcium and phosphate in the osteoid, and hydroxyapatite crystals begin to form, either free (3) or within matrix vesicles (4), which eventually release the crystals. As more matrix is synthesised, the matrix further away from the osteoblasts is more mineralised than the matrix closer to the osteoblasts. Mineralisation occurs as a mineralising front (5) advancing from the matrix furthest from the osteoblast to the one closest to the osteoblast (white arrows). (6) Calcified bone matrix

(5) *Phospholipases* open pores in the matrix vesicle membrane through which crystals pass into the extracellular matrix.

b. *Collagen mineralisation phase*: It occurs outside the matrix vesicles.
   (1) The hydroxyapatite crystals form *mineralised nodules*, where the crystals continue to grow between the collagen fibres due to the high concentration of calcium and phosphate and the action of matrix proteins.
   (2) The nodules grow and contact each other, until the entire matrix is mineralised. At this point, the matrix vesicles disintegrate.
   (3) The fusion of the mineralised nodules forms a *mineralising front* that propagates through the osteoid (Fig. 7.6).

7.5 Bone Growth

A. The bone can grow in thickness and length.
B. In short, flat bones, growth occurs by apposition from the periosteum.
C. In the long bones:

1. Thickening occurs by apposition of new bone from the periosteum.
2. The growth in length is produced by the activity of the *epiphyseal plate* or *epiphyseal disc*, which as mentioned above is located in the *metaphysis*.
3. In this epiphyseal plate, we can observe, from the epiphysis towards the diaphysis, the following regions (Fig. 7.5):
    a. *Zone of reserve cartilage (resting zone)*: epiphyseal cartilage, not yet ossified, with little cellular proliferation.
    b. *Zone of proliferation*: an area of cartilage in which chondrocytes are actively proliferating and are organised in well-defined columns.
    c. *Zone of hypertrophy*: chondrocytes increase in size, still arranged in columns, between which remain thin cartilage matrix septa.
    d. *Zone of calcification*: as chondrocyte hypertrophy continues, calcification of the cartilage matrix occurs.
        (1) The growth of the chondrocytes leads to the growth of the lacunae they occupy, causing lacunae in the same column to come into contact with each other, forming a single elongated cavity.
        (2) The boundaries between this region of calcification and the previous zone of hypertrophy are blurred, as both processes (hypertrophy and calcification) are interrelated and some authors consider them to be a single zone.
    e. *Zone of resorption*: chondrocytes die by *apoptosis*.
        (1) The elongated cavities occupied by the chondrocytes are invaded by mesenchymal tissue (in the embryo) or connective tissue (in the child), actually *haemopoietic tissue* (in post-embryonic periods) with *blood vessels*.
        (2) This zone consists of vertical *trabeculae* of *calcified hyaline cartilage* matrix surrounding elongated cavities (parallel to the trabeculae) filled with *bone marrow* (haemopoietic tissue with blood vessels).
    f. *Zone of ossification*: this zone is highly vascularised and is formed by trabeculae with an inner region of calcified cartilage matrix on which bone matrix is being deposited, which progressively calcifies. Between the trabeculae is the bone marrow.

D. When the epiphyseal plate is fully ossified at the end of puberty, neither cartilage nor the areas described above are visible, and the bone stops growing in length.

7.6 Bone Modelling

A. During growth, bones maintain a similar shape, but may undergo small changes in shape to adapt to the developing body.
B. At the same time, the shape of the bone retains certain proportions and modifies others.
C. To achieve this, there is synthesis of bone matrix in some areas and resorption in others. In this case, the processes of synthesis by the osteoblasts and degradation by the osteoclasts are *independent* and not coupled.
D. This process is called *bone modelling*.

7.7 Bone Remodelling

A. Throughout the life of a bone, during growth and in adult individuals, the bone tissue of a bone is constantly being replaced and reorganised.
   1. In this process, bone degradation is *coupled* to the synthesis of new bone tissue to replace it.
   2. This process is called *bone remodelling*.
B. From the age of 30–40 years, less bone tissue is produced than is resorbed, which can lead to loss of bone tissue.
C. The first bone tissue to form in both intramembranous and endochondral processes is always *immature bone*, without lamellae.
D. This bone is later replaced by *lamellar bone*. The first generation of lamellar bone is subsequently replaced by new generations of lamellar bone, and so on throughout the life of the individual.
E. The reorganisation of compact bone differs from that of spongy bone.
   1. *Remodelling of the compact bone*:
      a. Cavities are produced by the action of osteoclasts, which degrade bone and move in one direction.
         (1) These resorption cavities become enlarged and cylindrical in shape, such as a tunnel, and are occupied by blood vessels and connective tissue.
         (2) These cavities are referred to as *bone-remodelling units* (Fig. 7.7).
         (3) Two zones can be distinguished:
            (a) *Cutting cone* or *perforating cone*, also called a *resorption canal*.

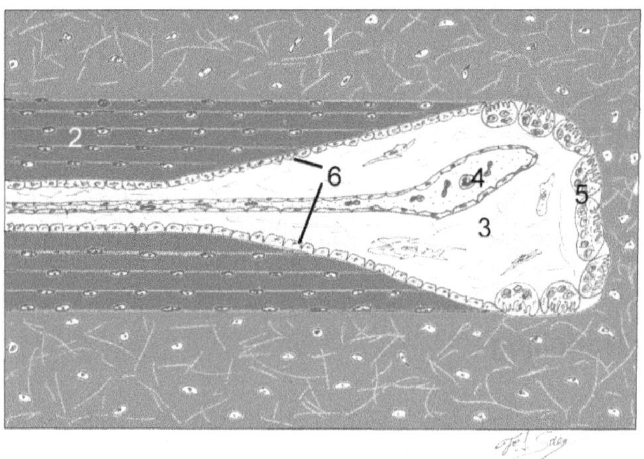

Fig. 7.7 Compact bone remodelling. The illustration shows a bone-remodelling unit. (1) Reticular or immature bone; (2) newly formed lamellar bone; (3) connective tissue; (4) blood vessel; (5) cutting cone or perforating cone with osteoclasts; (6) closing cone with osteoblasts

        1. It is the advancing front of the cavity where *osteoclasts* are found covering the bone, which they degrade, so the cavity grows in that direction.
        2. The diameter of the cutting cone, about 200 μm, marks the diameter of the future osteon.
    - (b) *Closing cone*, which follows the cutting cone and forms new bone. The closing cone is longer than the cutting cone.
        1. Here the cavities have an *endosteal* lining in contact with the bone that borders them. This endosteum contains *osteoblasts*, which form new bone.
        2. The synthesis of new bone tissue occurs by depositing new lamellae centripetally, i.e. towards the interior of the cavity.
        3. Finally, the resorption cavity is filled with concentric bone lamellae and a small central vascular channel remains.
        4. The result is the formation of a new *osteon*.
  - b. The formation of a new osteon can take several months.
  - c. The bone remodelling unit does not form coincidentally with existing osteons, but occurs in any region of the compact bone. Therefore, new osteons are produced by 'passing through' existing osteons.
  - d. The remnants of the partially degraded osteons go on to form the *interstitial lamellae* (Fig. 7.8).
  - e. Between the newly formed osteon and the surrounding bone matrix, a *cementation line* can be observed, which forms the edge of the new osteon.
2. *Spongy bone remodelling*:
  - a. In spongy bone, the bone tissue of the trabeculae is also constantly renewed.
  - b. The procedure is as follows:
    - (1) Degradation of the trabecula from its surface by the action of osteoclasts, which create a *pit* or *resorption groove*.
    - (2) When the action of the osteoclasts ceases, the bone surface is covered with *endosteum*, which at some point may re-form bone on the surface of the old bone.

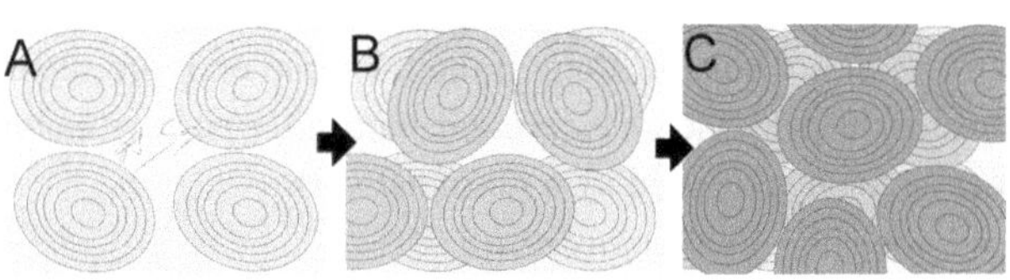

Fig. 7.8 Compact bone remodelling. The illustration shows three consecutive phases (**A–C**) of formation of new osteons by bone-remodelling units. The new osteons formed do not coincide with the existing ones

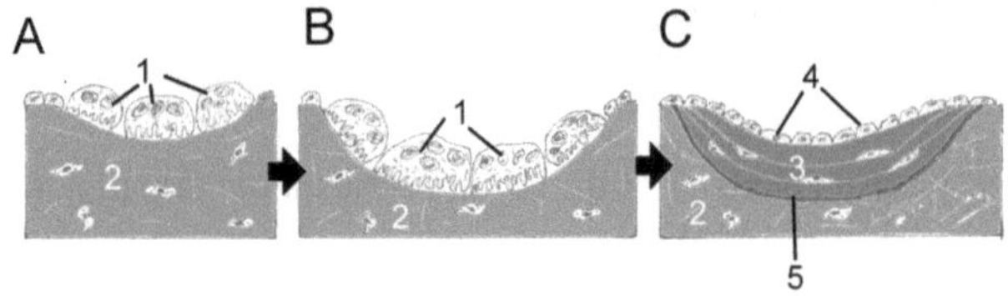

Fig. 7.9 Remodelling of spongy bone. The illustration shows three consecutive phases of resorption (**A**, **B**) and formation of new bone tissue (**C**). (1) Osteoclasts; (2) spongy bone; (3) newly formed bone; (4) osteoblasts; (5) cementation line

(3) The newly formed tissue constitutes what has been termed *trabecular osteona.*[1]

c. A *cementation line* can be observed at the boundary between old and newly formed bone (Fig. 7.9).

d. In older people, where the destruction of bone tissue is greater than bone formation.

(1) This can lead to progressive thinning of the trabeculae and increases the risk of excessive resorption destroying a trabecula.

(2) This can be serious, as bone matrix can only be synthesised on top of an existing one, and if a trabecula is lost, a new one cannot be formed.

7.8 Bone Repair

A. When a fracture occurs, it can be repaired due to the dynamism of the bone tissue.

B. This repair depends on the position of the fractured ends and their mechanical stability, which makes it possible to distinguish between two types of healing:

1. *Primary* or *direct bone healing*: if the fracture ends are in direct contact by compression.

   a. This occurs when the broken bone is surgically stabilised with plates that prevent movement between the bone fragments.

   b. In ruptured Haversian ducts, osteoclasts are produced and open resorption cavities at the neighbouring ends that cross the fracture line.

   c. The cavities are filled with new bone, thus establishing a firm union between the fracture ends.

2. *Secondary* or *indirect bone healing*: if the fractured ends are separated by a gap.

   a. This space is filled with *haemorrhagic material (fracture haematoma)*, consisting of erythrocytes trapped by a fibrin network.

   b. Broken bone ends undergo *necrosis*.

[1] We use the term here as a reminder, but it should be stressed that we do not recommend its use in order to avoid confusion with the true osteon.

c. An *acute inflammatory* process occurs, with invasion of *neutrophils* and *macrophages*.
d. The area is invaded by blood vessels and *fibroblasts*, and the haemorrhagic material is gradually replaced by highly cellular and vascularised tissue (*granulation tissue*) with type I and type III collagen fibres.
e. If the fractured ends of the bone are well stabilised:
   (1) The granulation tissue is covered with *fibrocartilage* that progresses inwards and replaces it. This produces a *cartilaginous callus* or *soft callus*.
   (2) Finally, the cartilaginous callus ossifies from the periphery inwards and a *hard callus* is formed.
f. If there is some instability, the cartilaginous callus may take time to ossify, although it may be a mechanically weak region for a long time.
g. If there is too much instability, bone does not form to repair the fracture, so the bone ends are held together by dense connective tissue that forms a *pseudarthrosis* (something like a false joint).
h. Over time, the bone formed in the fracture remodels and the bone regains its original appearance, without showing a scar.

7.9 Joints

A. *Classification of joints*:
  1. Joints are points of contact between two bones in which there are specialisations of some connective tissues to hold them together.
  2. The classification of joints varies according to the authors. We can consider two types of joints:
     a. *Synarthrosis* (from Ancient Greek συν- syn-, 'link' and ἄρθρωσις árthrōsis, 'articulation') or false joints (plural synarthroses).
     b. *Diarthrosis* (from Ancient Greek δια- dia-, 'separation' or 'through' and ἄρθρωσις árthrōsis, 'joint'), or true joints, which allow movement between bones (plural diartrhoses).

B. *Synarthrosis*:
  1. In synarthroses, the two bones are joined by *connective tissue* (fibrous or cartilaginous). There is little or no movement of one bone in relation to the other.
  2. When there is some capacity for movement in a synarthrosis, some authors call it an *amphiarthrosis*.
  3. Synarthroses can be:
     a. *Syndesmosis*: bones are joined together by a *ligament* of dense connective tissue. In the flat bones of the skull, these joints are called *sutures*.
     b. *Gomphosis*: this is a variety of the previous one that is found between the roots of the teeth and the alveolar bone. It will be studied in the chapter dedicated to the insertion periodontium.

c. *Synchondrosis*: the bones are held together by *cartilage*. There are two types of synchondroses:
   (1) *Primary synchondrosis*: there is only cartilage between the bones. This is what occurs at the junction between ribs and sternum.
   (2) *Secondary synchondrosis* or *symphysis*: each bone is lined with *hyaline cartilage*, and there is *fibrous connective tissue* between the two. There is usually a transition zone between the hyaline cartilage and the fibrous connective tissue formed by *fibrocartilage*. Sometimes the entire central fibrous tissue is fibrocartilage.

d. *Synostosis*: when bone growth ceases, most syndesmoses and synchondroses ossify and become synostoses, as in the bones of the skull.

C. *Diarthrosis*:
1. The *true joint* is very complex with several peculiar structures (Fig. 7.10).
   a. The articular surface of the bone is covered by *articular cartilage*.
   b. The entire region of contact between the bones is lined by a *joint capsule*, which leaves an inner cavity between the articular surfaces of the bones, called the *articular cavity* or *synovial cavity*, filled with a fluid, called *synovial fluid*.
2. *Articular cartilage* is a variety of hyaline cartilage, although in certain joints it is fibrous cartilage (Fig. 7.11).
   a. It borders the bone on one side and the articular cavity on the other.
      (1) Both surfaces lack perichondrium.
      (2) The surface of the articular side, which borders the articular cavity, is smooth.

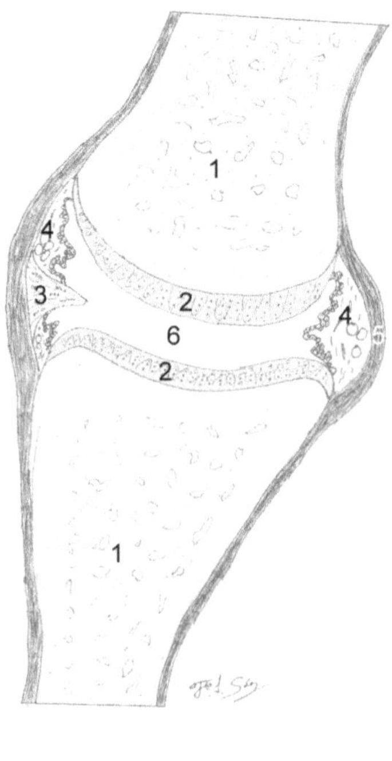

Fig. 7.10 Medial section of a true joint (diarthrosis). (1) Bone; (2) articular cartilage; (3) meniscus; (4) synovial membrane; (5) joint capsule; (6) articular cavity

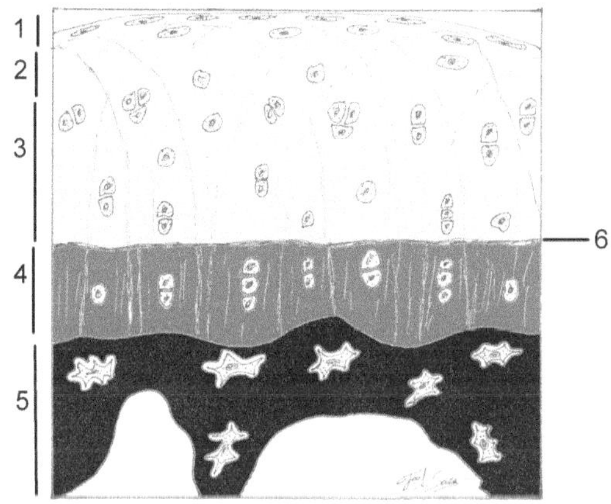

Fig. 7.11 Section of the articular surface of a diarthrosis. (1) Superficial zone or stratum I; (2) transitional zone or stratum II; (3) radial zone or stratum III; (4) zone of mineralised cartilage or stratum IV; (5) bone; (6) boundary line or tidemark

b. It is between 1 and 7 mm thick and is organised in a peculiar way, in zones or strata located between the articular cavity and the bone.

c. Its peculiar organisation is mainly due to the *type II collagen fibres*, which have a round trip:
   (1) They protrude perpendicularly from the surface of the bone.
   (2) They form an arch near the articular surface.
   (3) They are again directed perpendicular to the surface of the bone.

d. The areas to be observed are from the articular surface to the bone surface:
   (1) *Superficial zone* or *stratum I*. Type II collagen fibres are almost parallel to the surface or are seen in arcs. The chondrocytes are spindle-shaped and their major axis is parallel to the surface.
   (2) *Transitional zone* or *stratum II*. The fibres are cross-linked and the chondrocytes are isolated or in pairs.
   (3) *Radial zone* or *stratum III*. The fibres are arranged perpendicular to the surface (*radial fibres*) and the chondrocytes form elongated isogenic groups, which are arranged parallel to each other and to the fibres.
   (4) *Zone of mineralised cartilage* or *stratum IV*. This is the zone of contact with the bone. The collagen fibres are radial and do not penetrate the bone. There are few chondrocytes and many are regressing and dying.

e. Between strata III and IV there is a *boundary line* or *tidemark*, 2–5 μm thick, in which calcium is abundant.

3. The *joint capsule* is made up of connective tissue that forms a sheath surrounding the contact area of the bones. It consists of two layers (Fig. 7.12):
   a. *Fibrous membrane* of dense connective tissue that is a continuation of the periosteum of the two bones.
   b. *Synovial membrane*, which is a tissue that is attached internally to the fibrous membrane.

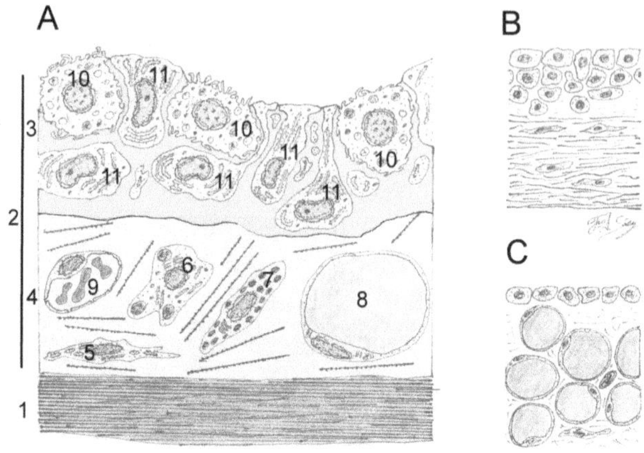

Fig. 7.12 Illustrations depicting the structure of the joint capsule. A joint capsule is depicted with a synovial membrane that has an areolar subintima (**A**), a fibrous subintima (**B**) and an adipose subintima (**C**). (1) Fibrous membrane; (2) synovial membrane; (3) intima or synovial layer; (4) subintima or subsynovial layer; (5) fibroblast; (6) macrophage; (7) mast cell; (8) adipocyte; (9) fenestrated blood capillary; (10) Synoviocyte A; (11) Synoviocyte B

4. The *synovial membrane* has an irregular surface and forms folds and villi (*synovial villi*) that protrude into the articular cavity.
   a. It consists of two layers:
      (1) *Subintima (subsynovial layer)*, the outermost layer.
         (a) It is a vascularised connective tissue, with adipocytes in variable quantity.
         (b) The capillaries are usually fenestrated, which facilitates the diffusion of substances into the tissues.
         (c) Elastic and collagen fibres are abundant.
      (2) *Intima (synovial layer)*, a region that looks epithelial in appearance but is not a true epithelium.
         (a) It consists of one to four layers of cells called *synoviocytes*, which are tightly bound together, but between which there is an amorphous extracellular matrix (*afibrillar*).
         (b) These cells lack a basal lamina.
         (c) There are two cell types:
            1. *Synoviocytes A*, similar to macrophages.[2]
            2. Fibroblast-like *synoviocytes B* with secretory activity that produce synovial fluid materials that fill the articular cavity.
   b. Three types of synovial membranes can be distinguished, which differ in the organisation of their subintima:
      (1) *Areolar*: It is formed by a subsynovial layer of loose connective tissue and a thick synovial layer.

[2] Type A synoviocytes are derived from monocytes and belong to the mononuclear phagocytic system.

(2) *Fibrous*: Its subintima is of dense connective tissue and its intima is thin.
(3) *Adipose*: It has a subsynovial layer with abundant adipocytes, which increase with age and certain pathological situations. The synovial layer is flat and simple.

c. There may be intermediate situations between these three types of synovial membranes.

5. The *articular cavity* contains the *synovial fluid* or *synovia*.
   a. It is made up of:
      (1) Substances filtered from blood plasma (water, proteins, glucose).
      (2) Molecules produced by B-type synoviocytes (*hyaluronan* and glycoproteins such as *lubricin*), many of which have lubricating capacity.
      (3) Free cells, mainly lymphocytes, but also monocytes, macrophages, free synoviocytes and granulocytes.
   b. This fluid has a *lubricating* (lubricin, hyaluronan) and *nourishing* (water, glucose) function.
   c. Projections of the joint capsule may be present in the articular cavity.
      (1) These projections may be incomplete, with their free edge, in which case they are called *menisci* (sing. meniscus).
      (2) In some cases, as in the *temporomandibular joint*, this projection forms a continuous disc that divides the articular cavity into two cavities. This disc is called the *articular disc*.

Take-Home Message

Bone can be formed directly from mesenchyme (intramembranous ossification) or by replacement of a cartilage outline (chondral ossification). Chondral ossification requires calcification of cartilage tissue, on which bone matrix is produced. The formed organic matrix (osteoid) is mineralised either from points of mineralisation in the matrix or by the action of matrix vesicles produced by cells, in which hydroxyapatite crystals are produced. Long bones grow in length by endochondral ossification occurring at the epiphyseal plates. Bones are modelled during development to adapt their shape to the growing body. Bones are also constantly remodelled, so that bone tissue is destroyed and new tissue is generated to replace it. This bone remodelling capacity allows bone fractures to be repaired. The bones of our skeleton contact each other at the joints. In some cases there is no possibility of movement between the bones (synarthrosis), but in others there is (diarthrosis). One type of synarthrosis is the gonphosis (joint between the root of the tooth and the alveolar bone). Diarthroses have a complex structure. The articular surfaces of both bones are lined with articular cartilage. In addition, these joints are protected by a joint capsule, the inner part of which is the synovial membrane, a tissue that produces the synovial fluid that fills and lubricates the articular cavity between the two bones.

Further Reading

Cohen MM (2000) Merging the old skeletal biology with the new. I. Intramembranous ossification, endochondral ossification, ectopic bone, secondary cartilage, and pathological considerations. J Craniofac Genet Dev Biol 20:84–93

Fitzgerald O, Bresnihan B (1995) Synovial-membrane cellularity and vascularity. Ann Rheum Dis 54:511–515

García-Aznar JM, Nasello G, Hervas-Raluy S et al (2021) Multiscale modeling of bone tissue mechanobiology. Bone 151:116032

Gómez de Ferraris ME, Campos Muñoz A (2009) Histología general humana. In: Histología, embriología e ingeniería tisular bucodental, 3rd edn. Médica Panamericana, pp 57–78

Kierszenbaum AL, Tres LL (2019) Osteogenesis. In: Histology and cell biology – an introduction to pathology, 5th edn. Elsevier, pp 177–198

Krstić RV (1991) Locomotor apparatus. In: Human microscopic anatomy. An atlas for students of medicine and biology. Springer Verlag, pp 439–445

Moskalewski S, Osiecka-Iwan A, Jankowska-Steifer E, Hyc A (2014) Synovial membrane asks for independence. Folia Morphol (Warsz) 73:395–398

Muehleman C, Arsenis CH (1995) Articular-cartilage.1. The normal joint. J Am Podiatr Med Assoc 85:277–281

Nanci A (2013) Bone. In: Ten Cate's oral histology, 8th edn. Elsevier, pp 95–121

O'Dowd G, Bell S, Wright S (2023) Skeletal tissues. In: Wheater's functional histology. Elsevier, pp 180–196

Robling AG, Turner CH (2009) Mechanical signaling for bone modeling and remodeling. Crit Rev. Eukaryot Gene Expr 19:319–338

Seeman E (2009) Bone modeling and remodeling. Crit Rev. Eukaryot Gene Expr 19:219–233

Simmons DJ (1974) Chronobiology of endochondral ossification. Chronobiologia 1:97–109

Sinder BP, Pettit AR, McCauley LK (2015) Macrophages: their emerging roles in bone. J Bone Miner Res 30:2140–2149

Souteyra JD, Amouroux J (1972) Anatomy, histology and pathology of synovial membrane and fluid. 2. Functions of synovial membrane. Nouv Presse Med 1:401

Wallis GA (1996) Bone growth: coordinating chondrocyte differentiation. Curr Biol 6:1577–1580

Blood

8

What You Will Learn in This Chapter
This chapter shows the components of blood, which consists of a liquid matrix, blood plasma, and various cell types. The functions of blood and methods of studying its organisation are introduced. The composition of blood plasma, the characteristics of erythrocytes, the various types of leukocytes (neutrophils, eosinophils, basophils, monocytes and lymphocytes) and the structure of platelets are explained.

8.1 General Characteristics

A. Blood is a red, viscous and slightly alkaline (pH 7.4) fluid that flows through the *circulatory system*.
B. In a human being, the blood volume is about 5–6 L, representing 7–8% of body weight.
C. It is considered a *variety of connective tissue* consisting of cells and a liquid extracellular matrix. Strictly speaking, blood is a fluid with suspended cells that circulates inside blood vessels and is closely related to connective tissues for several reasons:
   1. Blood cells are produced in a special connective tissue that we call *haemopoietic* or *haematopoietic tissue* (from the Ancient Greek words αἱμο- haimo- and αἱματο- haimato-, 'blood' and ποίησις poíēsis, 'action, creation') and which is found in the *bone marrow*.
   2. With the exception of erythrocytes, all other blood cells use blood as a means of transport, but carry out their activities in the connective tissues they colonise.
   3. The *liquid extracellular matrix* of the blood is produced by secretion (mainly from the liver) and filtration of fluid from other tissues, and in turn contributes by filtrate to the *tissue fluid* of other tissues.
   4. Although blood is not strictly a structure that allows the mechanical support of the body, it can be considered a form of connective tissue because it acts as a connector and communicator of distant tissues.

F. J. Sáez, I. Badiola, *Essential Oral Histology*, Learning Materials in Biosciences,
https://doi.org/10.1007/978-3-032-03165-5_8

D. Cells and cell derivatives are found in the blood:
   1. True *cells* (leukocytes).
   2. *Anucleated cells* without cytoplasmic organelles, which have a relatively short life span in the blood (erythrocytes).
   3. *Cell fragments* surrounded by their plasma membrane, anucleate but with cytoplasmic organelles (platelets).
E. Blood begins to form very early in embryonic development, in several stages:
   1. The first blood originates from the *extraembryonic mesenchyme* shortly after gastrulation:
      a. Cells of the yolk sac wall become primitive erythroid progenitors that begin to produce primitive erythrocytes, exclusively.
      b. Other cells produce the lining of the first blood vessels.
   2. A few days later, a second wave of erythrocyte formation occurs from new yolk sac cells that circulate throughout the embryo and populate the liver.
   3. Subsequently, 5 weeks after fertilisation, other *stem cells* are produced in intra-embryonic positions. These cells are produced from *aortic endothelial* cells and multiply in the liver.

8.2 Blood Functions

A. We can attribute to blood the following roles in our organism:
   1. Transport of necessary substances to the tissues, such as *oxygen* (O_2) and *nutrients.*
   2. Transport of *waste* or unnecessary *substances*, such as *carbon dioxide* (CO_2) and other molecules, from tissues to systems that remove them from the body.
   3. Transport and distribution of *hormones* from the endocrine glands that produce them to all tissues of the body.
   4. Intervention in the processes of *homeostasis* (from the Greek ὁμοιο- homoio-, 'like' and στάσις stásis, 'position, stability'), i.e. in the conservation of conditions suitable for cellular activity and the maintenance of life. They include the following actions:
      a. Buffering, e.g. as a pH *buffer* medium.
      b. *Coagulation* of blood. When blood leaves a blood vessel, it turns into a jelly-like substance and forms a *blood clot* (from the Old English clott, and this from Proto-Germanic klutto- 'a round mass, lump'). When blood vessels break, the clotting process seals the breaks and prevents bleeding.
      c. *Thermoregulation*, to maintain body temperature close to 37 °C.
   5. Transport of the cells and humoral substances of the *immune system* responsible for the body's defence against external pathogens.

8.3 Extracellular Matrix

A. The liquid extracellular matrix of blood is called *blood plasma*.
B. If blood is withdrawn, a *blood clot* is produced because all the cells or cell derivatives of the blood together with some components of the plasma form a gelatinous mass.
   1. This clot is separated from a liquid phase that constitutes the *serum* (perhaps from the pre-Roman word sŏrus or the Latin serum, 'milk and blood serum').
   2. The composition of serum is not the same as that of plasma, as serum lacks plasma components that are included in the clot.
C. Therefore, if we want to study plasma, we have to collect the blood in tubes coated with an *anticoagulant*, such as *heparin*. In this situation, the blood does not clot and can be centrifuged. This causes the blood to separate into three phases (Fig. 8.1):
   1. The lower sediment is a thick, reddish-coloured layer, and is made up of blood cells called *erythrocytes*.
      a. Measuring the percentage of the tube occupied by this sediment gives the percentage of red blood cells per volume of blood, which is called the *haematocrit*.
      b. In healthy males, the haematocrit remains between 39% and 50%, while in healthy females it may be between 35% and 45%.
   2. The intermediate layer is much thinner. It is a sediment that is called the *buffy coat* because it is made up of the rest of the blood's cell derivatives (*platelets and leukocytes*). It is whitish in colour.
   3. The top layer is a yellowish, semi-transparent fluid, which is the blood *plasma*.
D. Plasma is 90–92% *water*.
E. Approximately 7–9% are proteins of various kinds.
   1. When histological sections are performed, plasma proteins from blood vessels are fixed and stained with eosin.

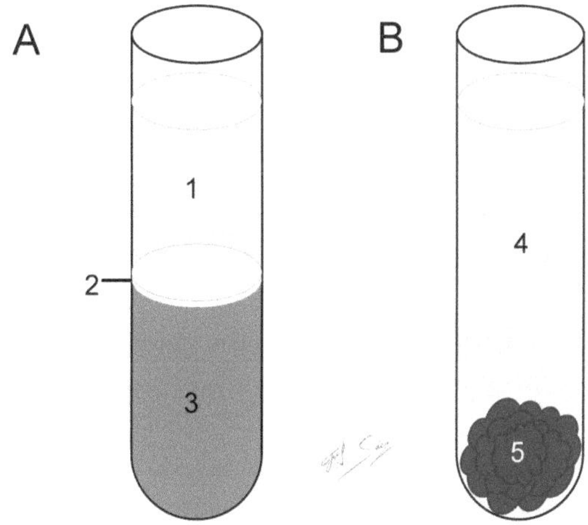

Fig. 8.1 Separation of blood components. In (**A**), blood is collected with anticoagulants and its components are separated by centrifugation. In (**B**), blood is collected without anticoagulants and a blood clot is produced. (1) Blood plasma; (2) buffy coat; (3) erythrocytes; (4) blood serum; (5) blood clot

2. There are proteins with different functions in the blood, the most abundant are:
   a. *Albumin*, a small protein (70 kDa) synthesised in the liver.
      (1) Maintains the colloidosmotic pressure between blood and tissues.
      (2) Transports:
         (a) Hormones, such as thyroxine.
         (b) Metabolites, such as bilirubin.
         (c) Drugs, such as barbiturates.
   b. *Immunoglobulins* or *γ-globulins*, produced by plasma cells and B-lymphocytes. They are the *antibodies* of the *humoral immunity (humoral response).*
   c. *Ceruloplasmin*, which transports copper.
   d. *Transferrin*, which transports iron.
   e. *Haptoglobin*, which transports haemoglobin to be destroyed in the liver.
   f. *Complement proteins*, produced in the liver, which are involved in defence against microorganisms and inflammation.
   g. Plasma *lipoproteins*, whose mission is to transport triglycerides and cholesterol between tissues. The most important are:
      (1) Chylomicrons
      (2) Very low density lipoproteins (VLDL)
      (3) Intermediate density lipoproteins (IDL)
      (4) Low-density lipoproteins (LDL)
      (5) High-density lipoproteins (HDL)
   h. *Fibrinogen*, a large protein (340 kDa), involved in clotting when converted to *fibrin* (323 kDa). It is produced in the liver.

F. The remaining components of plasma, up to a maximum of 2%, are various substances including:
   1. *Electrolytes*, such as Na^+, K^+, Ca^{2+}, Mg^{2+}, Cl^-, HCO^{3-}, PO_4^{3-} and SO_4^{2-}.
   2. *Nutrients*, such as glucose, lipids and amino acids.
   3. *Organic nitrogenous substances*, such as urea, uric acid, creatine[1], creatinine[2] and ammonium salts.
   4. *Gases*, such as nitrogen (N_2), O_2 and CO_2.
   5. Regulatory substances, such as certain *hormones*, some of which are proteins.

G. For some clinical test, plasma is required, for others serum is needed and in some cases serum or plasma can be used interchangeably.

8.4 Cells and Cell Derivatives of Blood

A. The cells and cell derivatives of blood are true cells, anucleated cells without organelles and cell fragments.

[1] Creatine is very important in muscle cells and neurons, where it is used to regenerate ATP.

[2] It is a breakdown product of creatine.

B. It is not easy to observe blood in the usual histological sections. In these, we can see the inside of the blood vessels with their contents, but the cellular structure and organisation is difficult to distinguish.
C. A special preparation technique is used to observe the blood cells by making *blood smears* (from Proto-Germanic smerwjan 'to spread grease on').
   1. A small drop of blood is deposited on a slide.
   2. With the help of another slide, the blood from the drop is spread on the surface of the glass slide.
   3. The blood is fixed by air-drying or flame-drying.
   4. Staining is carried out using special dye mixtures.
D. The dye mixtures used to dye blood extensions often contain:
   1. *Eosin*: It is an acidic dye.
   2. *Methylene blue*: It is a basic or alkaline dye.
   3. Blue dyes known as *azures* (azur A, azur B, methyl violet, methyl blue):
      a. Many of them are oxidation products of methylene blue.
      b. They can be *metachromatic*, i.e. they can dye blue or other colours, usually purple.
      c. They are alkaline dyes.
E. With these mixtures, *acidophilic* cell structures stain pink, *basophilic* cell structures blue and azure-fixing cell structures, called *azurophilic*, stain purple. The different staining characteristics of each cell allow different cell types to be distinguished.
F. The most commonly used blood stains are mixtures derived from *Romanowsky* stain, such as Giemsa, May-Grünwald, Pappenheim and Wright stains.

8.5 Erythrocytes

A. *Erythrocytes* (from Greek ἐρυθρο- erythro-, 'red') are also called *red blood cells*.
B. These cells lack a nucleus and cytoplasmic organelles and are therefore not true cells but cell debris (Fig. 8.2).
C. With the usual stains, they stain salmon pink.
D. Young erythrocytes may retain some ribosomes and are called *reticulocytes* or *polychromatophilic erythrocytes* (from the Greek πολυ- poly- 'much', χρῶμα, -ατος chrôma, -atos 'colour' and φιλο- philo-, -φιλος -philos, 'friend').
E. The number of red blood cells is $5.4 \times 10^6/\text{mm}^3$ in males and $4.8 \times 10^6/\text{mm}^3$ in females.
F. Erythrocytes, like all other blood cells, are produced in the bone marrow and enter the blood once formed but, unlike other blood cells, they do not leave the blood unless a blood vessel ruptures.[3]

[3] There are exceptions to this rule, as in some organs there may be areas where erythrocytes leave the blood vessels and pass into the tissues.

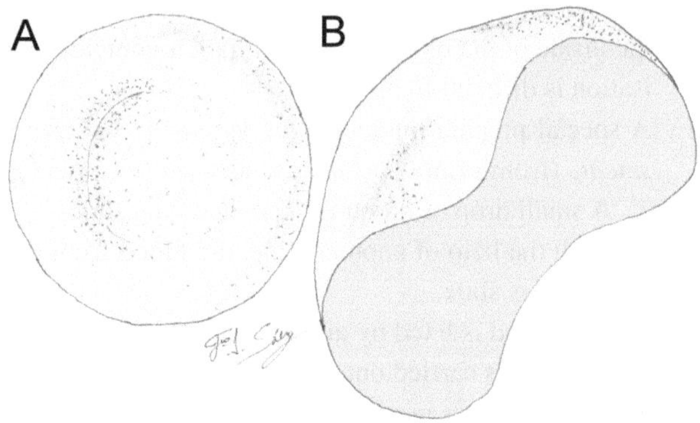

Fig. 8.2 Erythrocyte, general view (**A**) and cross-section (**B**)

G. Lacking organelles, they cannot replace their membrane molecules as they degrade, so once in the blood, their half-life is about 120 days.
H. They obtain energy from *glycolysis* and *lactic fermentation* occurring in their cytosol.
I. At the end of their life, their membrane is unstable and fragile. As a result, some erythrocytes disintegrate in the blood, but the vast majority are phagocytosed by *macrophages* in the spleen, liver and bone marrow.
J. They are biconcave disc-shaped, 7–8 μm in diameter, 2.6 μm thick at the periphery and 0.8 μm at the centre.
K. The biconcave shape gives them a large surface/volume ratio, which facilitates the exchange of gases (O_2 and CO_2) between the cytosol and the outside.
L. The biconcave shape is obtained by the organisation of the cytoskeleton, consisting of:
    1. Membrane-attached protein meshwork. Its main components are:
        a. *Spectrin*, tetrameric protein
        b. *Actin*
    2. Integral membrane proteins, such as *glycophorin* and *band 3 protein.*
    3. Proteins that link the protein lattice to integral membrane proteins, such as *ankyrin.*
M. Although they maintain this shape, they are not very rigid and can be deformed, which is useful when they pass through the smallest blood vessels (*capillaries*).
N. Much of the erythrocyte cytosol is occupied by the protein *haemoglobin.*
    1. Haemoglobin is a *tetramer*, i.e. it consists of four polypeptides. Various forms of haemoglobin can be found:
        a. *HbA* (96%). Made up of two α-chains and two β-chains.
        b. *HbA*$_2$ (1.5–3%). Two α-chains and two δ-chains .
        c. *HbF* (less than 1%). Two α-chains and two γ-chains. It is the most abundant haemoglobin in the foetus.
    2. Each polypeptide is attached to an *iron-containing heme group.*
    3. Haemoglobin is responsible for gas transport.

   a. The O_2-bound haemoglobin, *oxyhaemoglobin*, transports this iron-bound gas from the lungs to the tissues.
   b. Haemoglobin bound to CO_2, *carbaminohaemoglobin*, transports it from the tissues to the lungs.
   c. Haemoglobin can also transport *nitric oxide (NO)*, a gas that acts as a local mediator and causes blood vessels to dilate.
   d. If *carbon monoxide (CO)*, which binds to haemoglobin with greater affinity than O_2, is inhaled, *carboxyhaemoglobin* is produced, which cannot deliver oxygen to the tissues, leading to death.

O. The presence of iron makes the cytosol appear dense (dark) and finely granular under the electron microscope.

P. Erythrocytes have the enzyme *carbonic anhydrase*, which produces *carbonic acid* from carbon dioxide and water. Carbonic acid dissociates into *bicarbonate* and *hydrogen*. Bicarbonate is the main form in which carbon dioxide is transported to be expelled in the lungs.

8.6 Leukocytes

A. *Leukocytes* (from the Greek λευκός leukós, 'white') or *white blood cells* are a group of several cell types that when blood is centrifuged form part of the whitish *buffy coat*.

B. What they all have in common is that when isolated in large numbers, they form a white mass.

C. They have fine granules in their cytoplasm that stain with azure dyes. We call these granules *azurophilic* or *primary granules*, and they are vesicles with lysosome-like characteristics.

D. Some types of leukocytes have other granules in their cytoplasm that are capable of staining with blood staining dyes. We call these granules *specific* or *secondary granules*, which are characteristic of each type of leukocyte.

E. Several types of leukocytes have been differentiated and are classified into two groups:
   1. *Granular leukocytes* or *granulocytes*. These are leukocytes that have azurophilic granules and secondary granules. These are of three types, which are distinguished microscopically by the colour affinity of their specific granules:
      a. Neutrophils.
      b. Eosinophils.
      c. Basophils.
   2. *Agranular leukocytes* or *agranulocytes*. These are leukocytes that lack secondary granules. They only have azurophilic granules. These are of two types:
      a. Monocytes.
      b. Lymphocytes.

F. Granulocytes often have a lobed nucleus, which may give the impression that they are polynuclear cells. They have been called *polymorphonuclear* or *polynuclear cells*.

G. Lymphocytes and monocytes are also called *mononuclear cells*, as their nuclei are not lobed.[4]

H. All white blood cells leave the bloodstream by passing through the vessel wall (usually the *postcapillary venules*) in a process called *diapedesis* (from the Greek διαπήδεσις diapḗdesis 'filtration') (Fig. 8.3).
   1. The cells adhere weakly to the endothelium of the blood vessel.
   2. They begin to roll and move along the surface of the endothelium.

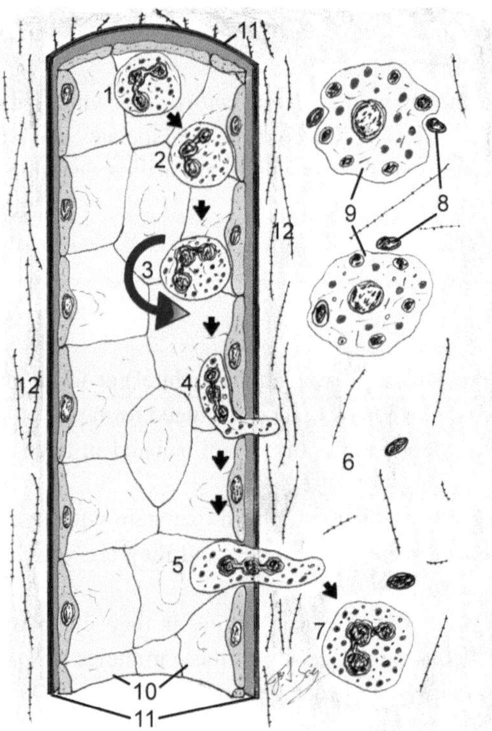

Fig. 8.3 Process of diapedesis of a leukocyte, in this case a neutrophil, through the endothelium of a postcapillary venule. Several phases are depicted and the short arrows indicate the evolution of the process. Neutrophils travel through the blood (1). Some (2) bind weakly to the surface of the endothelial cells (10) and begin to roll (curved arrow) and move (3), always attached to the endothelium. Finally, the neutrophil binds more tightly and emits pseudopods (4), which begin to penetrate the endothelial wall. The neutrophil moves through the endothelial perforation (5) and migrates (7) into the connective tissue (6), where it begins to phagocytose pathogenic microorganisms (8). Neutrophil adhesion and migration is triggered by substances produced by macrophages (9) that are phagocytosing infectious agents. (11) Basal lamina; (12) lamina reticularis

[4]The terms polymorphonuclear (polynuclear) and mononuclear are no longer in use, as all leukocytes have a single nucleus, although in some cases there are lobules.

3. They attach more strongly and emit *pseudopods* that create an opening in the endothelium (through a cell or between two cells), allowing the pseudopod and then the entire leukocyte to pass through.
4. The leukocyte migrates like an amoeba due to the action of its cytoplasmic cytoskeleton, namely *actin microfilaments* associated with *myosin*.

8.7 Neutrophils

A. *Neutrophils* are the most numerous leukocytes, accounting for 48–66% of all leukocytes.
B. Neutrophils are 12–15 μm in diameter in smears and 7 μm in suspension.
C. Its nucleus, with highly condensed chromatin, is multilobed (Fig. 8.4):
   1. It has three to four lobes (from the Greek λοβός lobós 'sack') joined by narrow nuclear regions.
   2. These segmented nuclei make the cell look multinucleated and are called *polymorphonuclear leukocytes*.
   3. The number of nuclear lobes increases with the age of the neutrophil and in some diseases.
   4. In the neutrophils of women, but not in the neutrophils of men, a drumstick-shaped nuclear appendage called *Barr body* is observed.
D. It has few synthetic organelles (endoplasmic reticulum, Golgi apparatus), as the mature cell is short-lived and has limited synthetic activity.
E. In the cytoplasm they have primary, secondary and tertiary granules. In humans, it is sometimes difficult to distinguish between the different types of granules.
   1. *Primary (azurophilic) granules* are 0.5 μm in diameter.
      a. They are azurophilic, staining red to purple.

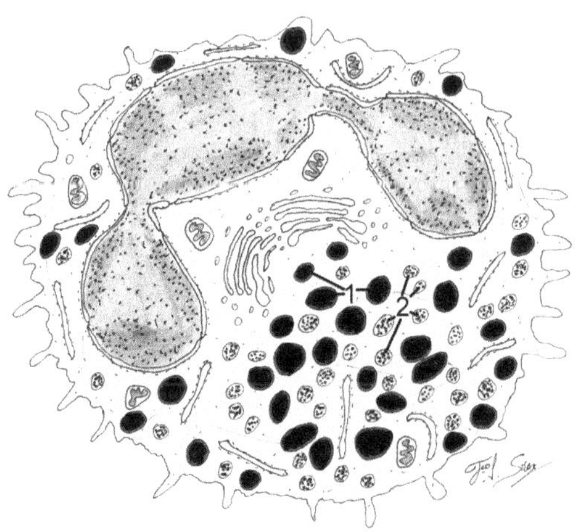

Fig. 8.4 Neutrophil granulocyte. (1) Azurophilic granules; (2) specific granules

b. They are lysosomes that possess:
   (1) Acid hydrolases
   (2) *Myeloperoxidase*, which produces *hydrogen peroxide* (H_2O_2)
   (3) Bactericidal agents:
      (a) *Defensins.*
      (b) *Lysozymes.*

2. *Secondary (specific) granules* are smaller but more abundant than azurophilic granules.
   a. They stain very little with mixed dyes, so they acquire a faint pink or violet colour, as they do not have a high affinity for a specific dye (they are neutral).
   b. Under the electron microscope, they are oblong in shape.
   c. They contain:
      (1) Antibacterial substances:
         (a) *Lactoferrin*, which sequesters iron.
         (b) *Lysozyme*, which breaks down the bacterial wall.
      (2) *Collagenase*, which degrades collagen fibres to facilitate tissue migration.
3. *Tertiary granules* have only recently been discovered.
   a. They are secretory. They release their contents into the extracellular matrix by *exocytosis*.
   b. They contain enzymes that assist the neutrophil in its migration through the tissues:
      (1) *Phosphatases*
      (2) *Metalloproteinases* (gelatinases, collagenases)

F. When the neutrophil phagocytes a bacterium, it inserts it into an intracellular vesicle called a *phagosome*. Primary and secondary granules attach to the phagosome to destroy the bacteria.

G. Neutrophils circulate for about 10 h in the blood, but some attach themselves to the walls of blood vessels.

H. When an infection occurs, they move into the tissues. In *acute inflammation* (the first phase of the inflammatory process), the number of neutrophils in a tissue increases.

I. These cells have several functions:
   1. They phagocytose infectious microorganisms, which is why they have also been called *microphages*. They have receptors for antibodies and complement proteins in their membrane, so their phagocytic activity is enhanced by *opsonisation*[5] of the bacteria.
      a. Phagocytosed microorganisms are destroyed or inactivated by various mechanisms:
         (1) Enzymatic degradation, due to its lysosomal enzymes.
         (2) Production of reagents toxic to bacteria, among others:

[5] Opsonisation is the binding of antibodies and complement components to the bacterial surface to facilitate phagocytosis by defence cells (macrophages and neutrophils).

(a) Toxic oxygen compounds, such as *hydrogen peroxide* (H_2O_2) and *superoxide* (O_2^-).
(b) Chlorinated compounds, such as *hypochlorous acid* (HClO).

b. When they finish their function of destroying microorganisms, they die and accumulate with other debris to form *pus*.

2. Destroy microorganisms by *NETosis* (neutrophil extracellular trap-associated cell death).
   a. When a neutrophil phagocytes a pathogen, a process of neutrophil cell death (*suicidal NETosis*) may be triggered:
      (1) The nuclear envelope is degraded and chromatin invades the cytoplasm.
      (2) The plasma membrane is ruptured, and the chromatin and the contents of the cytoplasm are spread across the extracellular matrix to form a lattice.
   b. The *extracellular chromatin lattice* becomes a trap for pathogenic microorganisms, which has two effects:
      (1) Pathogens are immobilised and prevented from spreading through the tissues.
      (2) Antimicrobial substances that have left the neutrophil cytoplasm along with the chromatin attack the microorganisms.
   c. NETosis can destroy *bacteria*, *viruses*, *fungi* and *parasites*.
3. They synthesise *leukotrienes* that help to trigger the inflammatory process.

J. As they do not have the capacity to regenerate their components, if they do not act against microorganisms they die after 1–4 days in a tissue.

K. Dead neutrophils can be degraded by macrophages or, in the gastrointestinal tract, they are released into the intestinal lumen and digested there.

L. They do not have the capacity to recirculate between tissues.

8.8 Eosinophils

A. *Eosinophils* (Fig. 8.5) constitute 1.4–4.8% of leukocytes.

B. They are similar in size to neutrophils or slightly larger and have a bilobed nucleus.

C. They also possess *azurophilic granules*, which are lysosomes. They contain *hydrolases* (acidic and other types) that degrade parasites and antigen–antibody complexes.

D. They have large *secondary eosinophilic granules* (1.0 μm or more):
1. Under the light microscope, they appear bright pink and are *birefringent*.[6]
2. Electron microscopy usually shows equatorial *crystalloid material* surrounded by a less dense matrix.
3. It has a varied content:
   a. Antiparasitic substances:

[6] In other words, it produces a double refraction of a light beam. The fact that a biological material is birefringent indicates an arrangement of molecules.

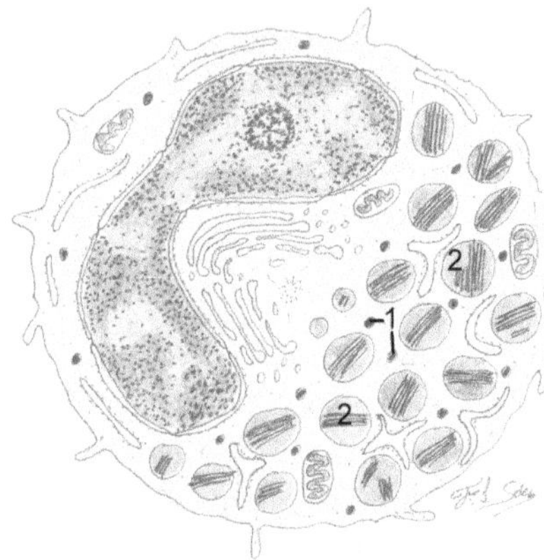

Fig. 8.5 Eosinophilic granulocyte. (1) Azurophilic granules; (2) specific granules

(1) *Major basic protein* (*MBP*): Rich in arginine. Found in the crystalline inclusion. In addition to being toxic to parasites, it causes degranulation of basophils and mast cells.
(2) *Eosinophil cationic protein* (*ECP*): It is in the amorphous matrix. Produces pores on the surface of the parasite.
(3) *Eosinophil-derived neurotoxin* (*EDN*): Attacks the nervous system of parasites. Also has antiviral function. It has been localised in the amorphous matrix.

b. Anti-inflammatory substances:
   (1) *Histaminase*, which degrades histamine.
   (2) *Arylsulfatase*, which degrades leukotrienes secreted by basophils and mast cells.
c. Antimicrobial substances: For example, *eosinophil peroxidase* (*EPO*) located in the matrix of the granules.

E. It has antibody receptors (IgG and IgE) and complement protein receptors on its surface.
F. The eosinophil is produced in the bone marrow, transported in the blood and carries out its functions in the tissues.
G. The eosinophil is involved in the following processes:
   1. *Phagocytosis*:
      a. Of *antigen–antibody complexes*, which are degraded within the cell.
      b. *Bacteria*, in small quantities.
   2. Defence against *parasites*, primarily helminth worms. When a parasite is coated with antibodies (IgG and IgE) and complement proteins, eosinophils attach to the parasite and release the toxic contents of their granules.
   3. They are involved in the induction and maintenance of *allergic reactions*.
   4. Regulation of the *inflammatory process*:
      a. They are attracted by mast cells and basophils to inflamed tissues.

b. They produce substances that deactivate inflammatory molecules and leukotrienes that increase inflammation.
5. Regulation of mast cells and basophils.

H. The number of eosinophils increases in the blood in the presence of parasitic infections or allergic processes.
I. Eosinophils are abundant in the intestinal mucosa and in sites of *chronic inflammation*, such as asthma.
J. It seems odd that eosinophils can recirculate in the blood.

8.9 Basophils

A. *Basophils* (Fig. 8.6) are the rarest of all leukocytes, accounting for less than 0.5% of the total.
B. They are similar in size to neutrophils.
C. The nucleus is irregular, twisted, S-shaped or bilobed, but is usually hidden because the secondary granules are stained with the same alkaline dyes, making it difficult to distinguish one from the other.
D. They possess azurophilic granules that are scarce. They are lysosomes that produce *acid hydrolases*.
E. They possess specific granules:
   1. They are stained with alkaline dyes. The staining is *metachromatic* and they stain from blue to purple.
   2. They are numerous.
   3. They are large (0.5 μm).

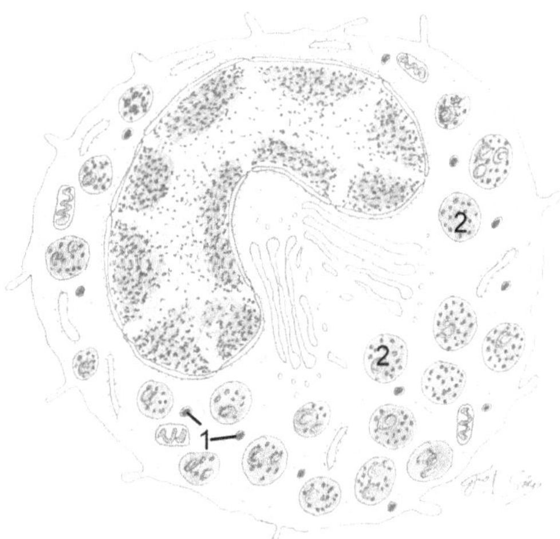

Fig. 8.6 Basophilic granulocyte. (1) Azurophilic granules; (2) specific granules

4. Under the electron microscope, they show a granular texture with *myelin figures*.[7]
5. They produce substances similar to those produced by mast cells in connective tissue:
    a. *Histamine*: Histamine is a vasoactive substance that produces:
        (1) Contraction of bronchial smooth muscle.
        (2) Smooth muscle relaxation of arterioles.
        (3) Loss of fluid in capillaries and venules. This increases the amount of fluid in the tissues, resulting in swelling or *oedema*.
    b. *Heparin*: A sulfated glycosaminoglycan that acts as an anticoagulant.
    c. *Heparan sulfate*: A sulfated glycosaminoglycan similar to heparin. It is a vasodilator, i.e. it causes dilation of the smallest blood vessels.
    d. *Leukotrienes*: These are lipids that cause:
        (1) Contraction of airway smooth muscle.
        (2) Increased vascular permeability and consequently oedema.
    e. *Interleukins*[8] that promote IgE synthesis.
    f. *Major basic protein (MBP)*, the same as that produced in eosinophils.

F. They secrete some *cytokines*[9] that regulate T-lymphocyte activity.
G. It has antibody receptors (IgE receptors) on its surface.
H. Like all other blood cells, basophils are produced in the bone marrow and pass into the bloodstream. The half-life of basophils in the blood is 1–2 days.
I. From the blood they pass into the tissues where they can be activated to perform various functions:
    1. *Intensify inflammation*: Basophils respond like mast cells. When an antigen binds to IgE molecules, basophil IgE receptors bind to this complex.
        a. This results in the release of the contents of the basophil granules into the extracellular matrix.
        b. The molecules released (mainly histamine and leukotrienes) cause:
            (1) Contraction of the bronchi
            (2) Dilation of blood vessels
            (3) Increased permeability of blood vessels
        c. This process occurs with *allergens*, the substances that cause allergic reactions.
    2. *Action against parasites*: For example, with the release of *MBP*.
    3. *Antibacterial action*, with low intensity.
J. For a long time it was thought that basophils and mast cells were two life forms of the same cell type.

---

[7] A myelin figure is a structure that shows the appearance of a spirally coiled sheet, as in the myelin sheath of a nerve fibre (see Chap. 11).

[8] Interleukins are a type of cytokine.

[9] Cytokines are small proteins involved in cellular communication processes, especially between immune cells.

1. According to this belief, basophils are blood-borne forms which, upon entering the connective tissues, are transformed into mast cells.
2. In reality, mast cells and basophils are related in origin, but different in the final differentiation process:
   a. *Mast cells*:
      (1) These are cells whose progenitors (*MCp*) travel through the blood as indistinguishable forms, probably similar to lymphocytes or monocytes.
      (2) They are differentiated once they have settled in the tissues.
      (3) In tissues they have a long life.
   b. *Basophils*:
      (1) These are cells that fully differentiate in the bone marrow before moving into the blood through which they travel.
      (2) They are attracted to the tissues where they are needed to support mast cells.
      (3) In tissues, the basophil is short-lived.
3. An important difference between the two cells is that the mast cell produces *tryptase*, but the basophil does not. The presence of tryptase is used as a marker of mast cell activity.
4. The characteristics of mast cells and basophils are given in the following table:

Feature	Mast cell	Basophil
Origin	Haemopoietic stem cell	Haemopoietic stem cell
Place of differentiation	Connective tissue	Bone marrow
Mitosis	Sometimes	No
Longevity	Weeks or months	Days
Size	20–30 μm	8–10 μm
Nuclear form	Rounded	Irregular, twisted, bi-lobed
Granules	Many, great	Few, somewhat smaller
Tryptase production	Yes	No

8.10 Lymphocytes

A. *Lymphocytes* (from the Latin lympha 'water', referring to a body fluid called *lymph*) are quite numerous, accounting for 25–27% of leukocytes.
B. They are the main cells of the *lymphatic system.*
C. Blood lymphocytes (Fig. 8.7) are circulating cells that have formed in the bone marrow and travel to other organs where they differentiate. They may subsequently recirculate to other tissues.
D. *Small* (6–10 μm), the most abundant, *medium* (12–15 μm) and *large* (15–18 μm and even larger) *lymphocytes* can be distinguished, the latter being rare in blood and more frequent in other tissues.

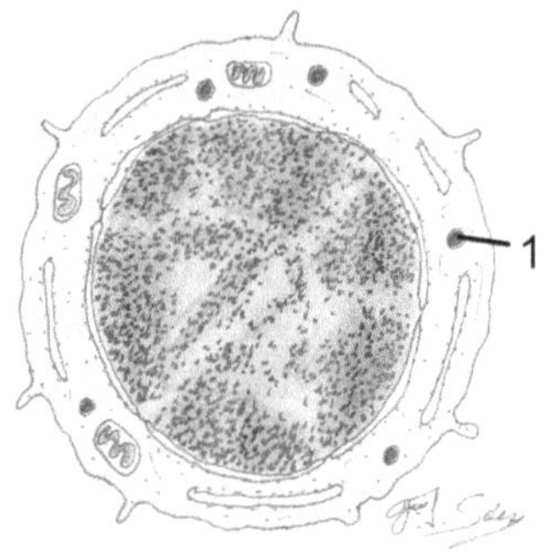

Fig. 8.7 Lymphocyte.
(1) Azurophilic granules

E. Its nucleus is hyperchromatic, spherical, slightly indented.
F. There is a very thin cytoplasm and very few organelles (mitochondria, ribosomes, little rough endoplasmic reticulum and azurophilic granules).
G. Each lymphocyte specifically recognises a foreign substance (*antigen*).
   1. When it divides, it forms a *clone* of cells capable of recognising that antigen.
   2. We have millions of different but indistinguishable lymphocyte clones in our bodies.
H. There are several types of lymphocytes, each specialising in a different function, but it is impossible to distinguish between them under the microscope, unless they are identified using immunohistochemical techniques to detect proteins specific to each cell type:
   1. *B lymphocytes* or *B cells* are produced in the bone marrow and mature there or in the connective tissues of the digestive tract.
      a. They are called B because in birds they undergo differentiation in an organ called the *bursa of Fabricius*.
      b. In the blood they make up 10–15% of circulating lymphocytes.
      c. Each B lymphocyte clone can make proteins of the globulin families (*immunoglobulins* or *antibodies*) that are able to specifically recognise its antigen (by recognising an antigen we mean that it binds to it).
      d. The antigen-recognising molecules on its surface are transmembrane immunoglobulins of the IgM and IgD types.
      e. When a B lymphocyte encounters an antigen that it recognises, it is activated and transformed into an antibody-producing *plasma cell.*
         (1) These cells colonise infected tissues or settle in the bone marrow.
         (2) Antibodies produced by plasma cells are soluble proteins that can travel through the blood to the site of infection, where they bind to the antigen and block it or cause its degradation by *macrophages*.
         (3) This soluble factor-mediated activity of the immune system is called *humoral immunity*.
   2. *T lymphocytes* or *T cells* make up 70–80% of circulating lymphocytes. There are two types of T lymphocytes:
      a. *Helper T lymphocytes* (*Th*), whose function is to help activate Tc lymphocytes.

(1) They are activated by *major histocompatibility complex type II (MHC II)* proteins, which are used by *antigen-presenting cells* (*APCs*).
(2) There are two varieties:
(a) *Th1*: Activate Tc cells, NK cells and macrophages.
(b) *Th2*: Cooperate with B lymphocytes.

b. *Cytotoxic T lymphocytes (Tc)*:
(1) They are activated when antigens are presented to them by an infected or tumour cell (i.e. they react to antigens from external agents or abnormal own cells).
(2) Antigens are presented on *major histocompatibility complex type I* (*MHC I*) molecules, which are present on the membranes of all cell types in the body.
(3) Tc lymphocytes produce substances that stimulate the death of cells that present antigens to them (*infected cells* and *tumour cells*).
(4) This Tc-mediated immunity is called *cellular immunity* or *cell-mediated response*.

c. *Regulatory (suppressor) T lymphocytes (Tr)*, which produce substances (*lymphokines*) that regulate the entire humoral and cellular response.

d. Other T lymphocyte subtypes have been described, but will not be described here.

3. *Natural killer lymphocytes* (*NK*) are 5–10% of the lymphocytes in peripheral blood.
   a. They are large, around 15 μm in diameter and have large granules, which is why they are called *large granular lymphocytes* (*LGL*).
   b. Their azurophilic granules are cytotoxic, but these cells cannot phagocytose.
   c. They destroy *virus-infected cells* and *tumour cells*.

I. All lymphocytes are produced in the *bone marrow*.
   1. B lymphocytes mature and acquire their immunological capacity in the bone marrow.
   2. Immature T lymphocytes migrate to the *thymus*, where they undergo differentiation.

J. When the cells have matured (*naïve mature lymphocytes*), they move to the organs of the *lymphatic system* and multiply slowly and form clones.

K. If the lymphocyte encounters its antigen, it is activated and proliferates rapidly. This proliferation produces two populations of lymphocytes from the clone:
   1. *Effector cells*: cells that act against the antigen.
      a. In the case of B lymphocytes, they produce antibodies and become *plasma cells*.
      b. In the case of T lymphocytes, they destroy infected or altered cells (Tc lymphocytes), or assist in the process (Th).
   2. *Memory cells*: cells that remain inactive, dormant, but if they encounter the antigen again, they react more quickly against it.

L. Lymphocytes can constantly recirculate in the blood, lymph and tissues.

8.11 Monocytes

A. The number of *monocytes* is between 2% and 8% of blood leukocytes.
B. Monocytes are the largest leukocytes, between 15 and 22 μm (Fig. 8.8).
C. Their nuclei may be spherical, oval, kidney-shaped or horseshoe-shaped and are usually lighter in colour than those of lymphocytes.
D. Next to the nuclear cleavage, the Golgi apparatus is visible in the cytoplasm. There are also smooth and rough endoplasmic reticulum, small mitochondria and few azurophilic granules.
E. Like the others cells, they are produced in the *bone marrow* and travel through the blood for 3 days until they colonise the connective tissues, where they form some of the cells of the *mononuclear phagocytic system* mentioned in Chap. 4.[10]
F. In particular, monocytes can be used to produce:
   1. *Osteoclasts* in the bone, and probably *chondroclasts* in the cartilage and *odontoclasts* in the tooth.
   2. *Macrophages* in the breast.
   3. *Macrophages* of the intestinal wall.
   4. *Monocytic dendritic cells* (moDC).
G. In addition to the cells mentioned above, monocytes can leave the vessels and settle in the connective tissues, where they maintain an appearance similar to that which they show in the blood. In this case, they can play two roles:
   1. In *steady-state tissues*, they perform a surveillance function. They can phagocytose antigens and enter lymphatic capillaries to carry antigens to lymph nodes.
   2. In *inflamed tissues* they can become *macrophages*.

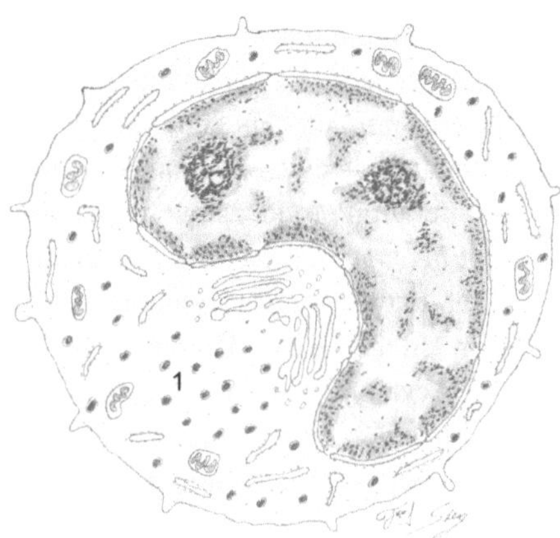

Fig. 8.8 Monocyte. (1) Azurophilic granules

[10] Other cells of this system settle in tissues during embryonic development and perpetuate themselves there.

8.12 Platelets

A. *Platelets* or *thrombocytes* (from Greek θρόμβος thrómbos 'clump, clot') are not true cells, but cytoplasmic fragments of large polyploid bone marrow cells called *megakaryocytes* (from Ancient Greek μεγα- mega-, 'large', κάρυον káryon 'nut', 'nucleus').
B. They are very numerous, approximately 300,000 platelets per mm^3.
C. They are biconvex discs 2–3 μm thick and have no nucleus (Fig. 8.9). They live in blood for about 20 days. In blood smears they are often seen in clusters (*agglutination*).
D. Under the light microscope, two areas are observed.
   1. *Chromomer* or *granulomer*, central area that stains very intensely.
   2. *Hyalomer*, very pale peripheral area.
E. Under the electron microscope, several zones can be distinguished according to the distribution of organelles. Some of these zones do not have well-defined boundaries and may overlap:
   1. Peripheral zone: It consists of the plasma membrane and its thick *glycocalyx*.
   2. Structural zone: A region beneath the plasma membrane, consisting of:
      a. A network of *actin microfilaments* associated with *myosin*.
      b. Beneath this network, at the margin of the platelet disc, is a ring of *microtubules* whose function is to maintain the shape of the platelet.
   3. Organelle zone: This is the centre of the platelet. It contain mitochondria, peroxisomes, glycogen and three types of granules:
      a. *Alpha (α) granules*, 300–500 nm.
         (1) They are azurophilic.
         (2) They contain factors involved in the initial phase of coagulation:
            (a) *Fibrinogen*.
            (b) *Von Willebrand factor*.

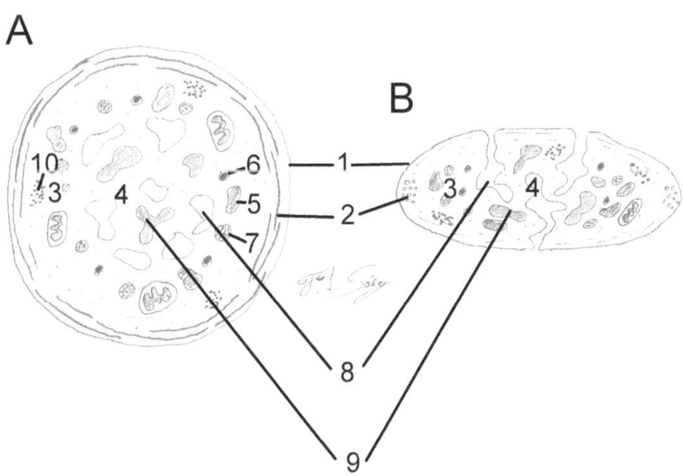

Fig. 8.9 Platelets in equatorial (**A**) and transverse (**B**) sections. (1) Peripheral zone; (2) structural zone; (3) organelle zone; (4) membranous zone; (5) α granules; (6) δ granules; (7) λ granules; (8) open canalicular system; (9) dense tubular system; (10) glycogen

b. *Dense granules* or *delta (δ) granules*, 250–300 nm. They are more abundant. They are involved in platelet adhesion and vasoconstriction. They contain:
   (1) *Histamine*, which causes vasodilation and increased vessel permeability.
   (2) *Serotonin*, which produces vasoconstriction.
c. *Lambda (λ) granules*, 200–250 nm. Similar to lysosomes, they contain enzymes that reabsorb the clot in the final stages of vascular repair.

4. Membranous zone: This is a region in which there are cavities surrounded by membrane. There are two types of cavities:
   a. *Open canalicular system (OCS)*: a remnant of the *demarcation channels* that occurs in *megakaryocytes* when platelets are formed.
      (1) They are invaginations of the plasma membrane.
      (2) Their cavities are channels that communicate with the cell exterior.
   b. *Dense tubular system*: cisternae of the *smooth endoplasmic reticulum*.
      (1) They serve as a calcium store and contain dense material.
      (2) These cavities do not communicate with the outside.

F. Platelets are involved in several functions:
   1. *Haemostasis*, by sealing vascular lesions, which prevents *bleeding*.
      a. When the blood vessel wall is injured, the collagen in the surrounding connective tissue is exposed to the blood.
      b. Platelets are attached to collagen by *von Willebrand factor*, which is produced by the injured endothelial cells and the platelets themselves.
      c. The wound triggers a blood *clotting* reaction that produces fibrous proteins, such as *fibrin*, formed from fibrinogen.
      d. The binding of platelets together with fibrous proteins produces a lattice that traps more platelets and erythrocytes at the site of injury.
         (1) This forms a *thrombus* or *clot* attached to the vessel wall that plugs the ruptured vessel.
         (2) The clot prevents bleeding.
      e. Finally, the platelets contract by the action of actin and myosin filaments, causing the clot to contract and facilitating closure of the blood vessel wound.
   2. *Tissue repair*, as platelets release *growth factors*.

Take-Home Message

Blood is a liquid connective tissue that performs many functions: transport of oxygen, nutrients, wastes and hormones; maintenance of homeostasis (pH buffer, coagulation, thermoregulation) and transport of immune system cells. Its liquid extracellular matrix is blood plasma, which can be obtained by centrifuging blood obtained with anticoagulants. If blood is collected without anticoagulants, a clot is produced and separated from the blood serum, whose composition is similar to that of plasma. Blood is usually studied microscopically in smears stained with special stains. Erythrocytes are anucleated, organelle-free, biconcave cells filled with haemoglobin that carry oxygen and carbon dioxide. There are several types of leukocytes: granulocytes (neutrophils, eosinophils and basophils) and agranulocytes (monocytes and lymphocytes), which travel through the blood and pass into the connective tissues where they perform their functions. Neutrophils phagocytose microorganisms and destroy or inactivate them by various mechanisms. They also inactivate microorganisms by NETosis, a cell suicide mechanism that uses cellular debris to trap them. Eosinophils possess substances that attack parasites, induce allergic reactions and regulate inflammation. Basophils make pro-inflammatory substances and also attack microorganisms and parasites. They are similar to mast cells. There are several types of lymphocytes, whose function is to identify and destroy pathogenic microorganisms and malignant cells. T lymphocytes (T cells) are cytotoxic (attack infected or transformed cells), while B lymphocytes (B cells) produce antibodies against foreign agents. Monocytes penetrate tissues and form different cell types of the mononuclear phagocytic system. Platelets are biconvex, anucleated cell fragments whose function is to plug breaks in blood vessels to prevent bleeding.

Further Reading

Burn GL, Foti A, Marsman G et al (2021) The neutrophil. Immunity 54:1377–1391

Fettrelet T, Gigon L, Karaulov A et al (2021) The enigma of eosinophil degranulation. Int J Mol Sci 22:7091

Gartner LP (2017) Blood and hemopoiesis. In: Textbook of histology, 4th edn. Elsevier, pp 251–286

Germic N, Hosseini A, Yousefi S et al (2021) Regulation of eosinophil functions by autophagy. Semin Immunopathol 43:347–362

Gigon L, Yousefi S, Karaulov A et al (2021) Mechanisms of toxicity mediated by neutrophil and eosinophil granule proteins. Allergol Int 70:30–38

Gómez de Ferraris ME, Campos Muñoz A (2009) Histología general humana. In: Histología, Embriología e Ingeniería Tisular Bucodental, 3rd edn. Médica Panamericana, pp 57–78

Kierszenbaum AL, Tres LL (2019) Blood and hematopoiesis. In: Histology and cell biology – an introduction to pathology, 5th edn. Elsevier, pp 199–236

Kim TS, Moutsopoulos NM (2024) Neutrophils and neutrophil extracellular traps in oral health and disease. Exp Mol Med 56:1055–1065

Krstić RV (1991) Blood. In: Human microscopic anatomy. An atlas for students of medicine and biology. Springer Verlag, pp 31–37

Lowe JS, Anderson PG (2015) Blood cells. In: Stevens & Lowe's human histology, 4th edn. Elsevier Mosby, pp 105–122

Miyake K, Ito J, Karasuyama H (2024) Recent advances in understanding of basophil function and differentiation. Folia Pharmacol Jpn 159:32–38

O'Dowd G, Bell S, Wright S (2023) Blood, haematopoiesis and bone marrow. In: Wheater's functional histology. Elsevier, pp 46–64

Pawlina W (2024) Blood. In histology a text and atlas with correlated cell and molecular biology, 5th edn. Wolters Kluwer, pp 298–343

Peng JY, Siracusa MC (2021) Basophils in antihelminth immunity. Semin Immunol 53:101529

Pfister H (2022) Neutrophil extracellular traps and neutrophil-derived extracellular vesicles: common players in neutrophil effector functions. Diagnostics 12:1715
Sollberger G (2022) Approaching neutrophil pyroptosis. J Mol Biol 434:167335
Tanaka S (2024) New developments in mast cell/basophil research. Folia Pharmacol Jpn 159:44–47
Weiss L (1988) In: Weiss L (ed) The blood. In cell and tissue biology – a textbook of histology, 6th edn. Urban & Schwarzenberg, pp 211–254

Haemopoiesis

9

What You Will Learn in This Chapter

In recent years, much progress has been made in understanding the processes of blood cell formation (haemopoiesis) and the latest developments are described in this chapter. Haemopoiesis begins in early embryonic stages. The sites of formation vary as the embryo develops, but when the first bones are produced, haemopoiesis establishes in the bone marrow. The bone marrow is a highly vascularised connective tissue that contains the haematopoietic cells. All adult blood cells are produced from the same lineage, originating from a haematopoietic stem cell. From this cell, derivatives are produced and differentiate to form the various blood cell types. This chapter describes in detail the processes of blood cell formation and blood cell precursors.

9.1 Introduction

A. The formation of blood cells is called *haemopoiesis* or *haematopoiesis* (from the Greek αἱμο- haimo- and αἱματο- haimato-, 'blood', and ποίησις poíēsis, 'action, creation').

B. Traditionally, two blood cell lines have been distinguished during haematopoiesis:
   1. *Lymphoid line*, which includes lymphocytes.
      a. The formation of lymphocytes is called *lymphopoiesis*.
   2. *Myeloid line* (from Greek μυελός myelós 'marrow' and -οειδές -oeidés 'similar to'), which includes the so-called *myeloid cells*, which are erythrocytes, granulocytes, monocytes and platelets.
      a. The formation of myeloid cells is called *myelopoiesis*. Myelopoiesis includes:
         (1) *Erythropoiesis* (erythrocyte formation)
         (2) *Granulopoiesis* (granulocyte formation)
         (3) *Monopoiesis* (monocyte formation)
         (4) *Thrombopoiesis* (platelet formation)

F. J. Sáez, I. Badiola, *Essential Oral Histology*, Learning Materials in Biosciences,
https://doi.org/10.1007/978-3-032-03165-5_9

C. When referring to the formation of leukocytes, the term *leukopoiesis* is used.
D. In the adult human being, all blood cells come from the same precursor, a *haemopoietic stem cell* (*HSC*).
   1. This is called *monophyletic theory*.
   2. In the adult, stem cells are found in the *bone marrow*. When they divide, the daughter cells they produce can follow two fates:
      a. Some remain as stem cells (self-renewal).
      b. The remaining ones are differentiated into blood cells.
E. Whether the daughter cells take one pathway or the other is controlled by regulatory agents in the environment in which the cell is found. This regulation can be:
   1. By *direct contact* with neighbouring cells.
   2. By the action of *soluble molecules* secreted by neighbouring cells, or hormones produced in other tissues.
F. The molecules that regulate haemopoiesis are called *haemopoietic growth factors*, and these include:
   1. *Interleukins* (*IL*).
   2. Other *cytokines* (such as *interferon*).
   3. *Colony-stimulating factors* (*CSF*).
G. Some of these factors act early, i.e. they regulate the early stages of haemopoiesis, but others are late and may be *lineage-specific*.
H. Many regulatory factors are produced in the bone marrow and other factors are produced in other organs and reach the bone marrow via the blood, i.e. they are *hormones*.
I. It should be noted that much of what is known about haemopoiesis has been studied in mice. Human haemopoiesis has been found to be very similar, but not exactly the same.

9.2 Haemopoiesis During Human Development

A. In the embryo, the production of the first blood cells occurs by complex and poorly understood mechanisms.
B. The first blood produced in extra-embryonic and embryonic organs is produced from erythroid and myeloid precursors.
C. In later embryonic stages and in the adult, haemopoiesis occurs from a *haemopoietic stem cell (HSC)*.
D. During embryonic development, blood formation goes through several waves.
   1. *Primitive haemopoiesis*: Not derived from HSCs. It is an extra-embryonic haemopoiesis. It occurs at two sites:
      a. In the *yolk sac*:
         (1) It begins between 16 and 18.5 days post-coitus (dpc).
         (2) Clusters of cells, called *blood islands*, are formed, whose surface cells form the endothelium of the first blood vessels and the inner cells form the first blood cells.

  (3) Several cell types are produced:
    (a) Large nucleated primitive erythrocytes (*megaloblasts*).
    (b) *Megakaryocytes*.
    (c) Primitive *monocytes/macrophages*. In later weeks, these cells give rise to the microglia cells that become permanently established in the central nervous system.
  (4) By 21–22 dpc, the heart begins to beat, intraembryonic blood circulation is established and megaloblasts are observed in the cardiac cavities.
 b. In the *placenta*, additional *macrophages (Hofbauer cells)* are produced. These macrophages assist in the enucleation of primitive erythrocytes.

2. *Transient definitive or prodefinitive haemopoiesis*: It begins with precursors installed in the yolk sac and probably in the embryo proper. It is produced from two different progenitors:
   a. *Yolk sac-derived myeloid progenitors (YSMPs)*: These cells colonise the liver between the third and fourth week. From here they produce:
      (1) Tissue-resident macrophages, such as microglia cells of the central nervous system, Langerhans cells of the epidermis and Kupffer cells of the liver.
      (2) Erythrocytes.
   b. *Lymphoid cell precursors (LMPs)*: Their origin is unknown. These cells colonise quickly the liver and give rise to B and T lymphocytes.
   c. Cells similar to YSMPs produced in the embryonic endothelium have also been identified.
3. *Definitive haemopoiesis*: It is produced from *haemopoietic stem cells (HSCs)*. These cells originate in the *aorta-gonad-mesonephros region (AGM region)* between week 4 and 5.
   a. Endothelial cells (of mesodermal origin) form *intra-aortic haemopoietic clusters (IAHCs)*. These cells form a *haemogenic endothelium* or *HE* on the ventral side of the dorsal aorta. From this HE, the first HSCs are produced.
   b. HSCs migrate to the yolk sac and placenta. During this migration they rarely differentiate but may give a few blood cells.
   c. Later, HSCs colonise the liver, where they settle around the sixth week. Here blood cells are already continuously formed.
   d. Finally, HSCs seed the bone marrow in the second trimester.

E. In summary, during the first and second trimester of human embryonic development, the primary site of blood cell formation is the liver, which acts as a confluence point for all embryonic haemopoietic pathways.

F. From the third trimester onwards, embryonic haemopoiesis occurs mostly in the bone marrow.

G. Initially, all bones produce blood. In the adult some bones lack haemopoietic bone marrow.

9.3 Models of Haemopoiesis

A. Until the turn of the century, haemopoiesis was considered to follow a *hierarchical model*, with the haemopoietic cell lineage following a pattern that resembles the branches of a tree.
B. The hierarchical model is based on the assumption that each time a cell divides, each of its daughter cells have two possible and completely independent differentiation paths.
   1. Once they have taken a differentiation pathway, the process is irreversible.
   2. Accordingly, the cells take a differentiation pathway that depends on the previous one.
C. This scheme was reached by *bone marrow transplantation* experiments:
   1. Mice were treated with *X-rays* to destroy their haemopoietic cells.
   2. These animals were injected with bone marrow extracts from animals with active hematopoiesis.
   3. *Colonies* with one or more blood cell types settled in the *spleen* of the animal (Fig. 9.1).
   4. The injected bone marrow contained cells with different chromosomal mutations.
      a. In this way, it was possible to verify that each colony was derived from a single transplanted cell.
      b. Consequently, it was assumed that there is a cell type that is a precursor of the cell types observed in the colony.
   5. The precursor cells that were hypothesised to have founded the colonies were called *colony-forming units* (*CFU*). Until recently, it was common to call the precursors of a haemopoietic lineage by this name, but it is no longer in use.
D. Recent studies have challenged the hierarchical model of haemopoiesis.
   1. Some differentiation pathways or lineages were found not to be independent:
      a. Some cells that should apparently produce one line or one lineage may produce two different lines or two different lineages.
      b. That is, there are several intertwined differentiation pathways and a cell can cross from one pathway to another.
   2. The differentiation pathway that a progenitor follows depends on the regulatory *environment* it encounters in the bone marrow, i.e. there may be *progenitor compartments* or *niches* in the bone marrow that condition the fate of the progenitor cell.
E. Figure 9.2 shows a diagram representing the current model of haemopoiesis.
F. Recent genomic and molecular analyses indicate that all HSCs are not the same. In mice, three types of HSCs have been differentiated:
   1. *LT-HSCs* (long-term cells): When transplanted into an irradiated animal, they can produce all blood cell types for a long time. At rest they do not divide, but under stress they reactivate and divide again.

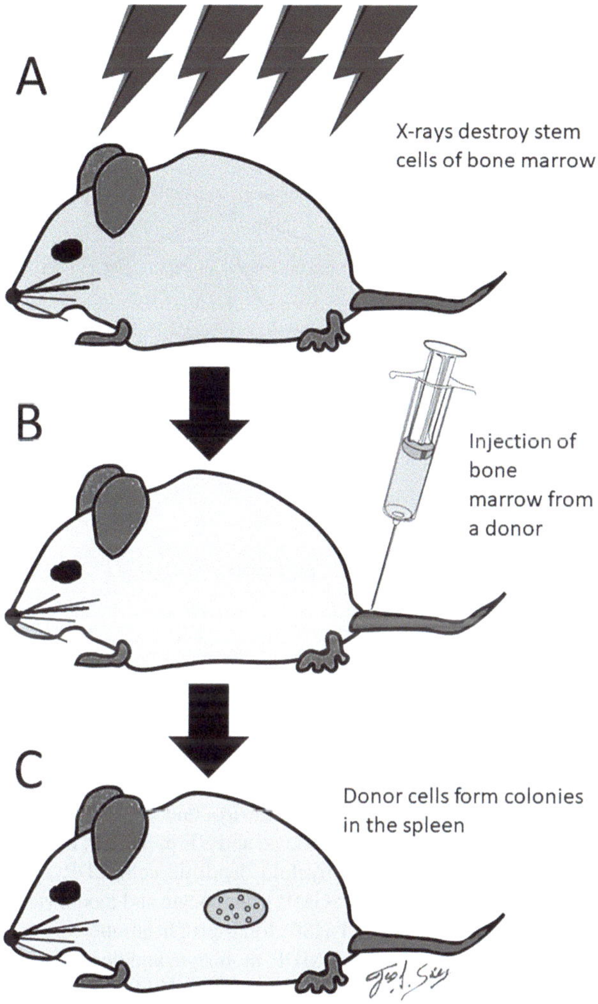

Fig. 9.1 Experiment demonstrating the existence of haematopoitic stem cells and progenitor cells. The experiment consists of destroying the haemopoietic stem cells of a mouse (**A**) with X-rays. After a while, the mouse will die from lack of blood cell production. However, if bone marrow from another mouse is injected (**B**), the mouse survives and it is observed that colonies of haemopoietic cells are produced in the spleen (**C**). If the injected marrow contains cells with different chromosomal mutations, each colony can be shown to derive from a single transplanted cell. Consequently, it can be assumed that there is a precursor cell type for the cell types observed in a colony. The precursor cells that are assumed to have founded the colonies were called colony-forming units (CFU), although this name is being replaced as new data are obtained

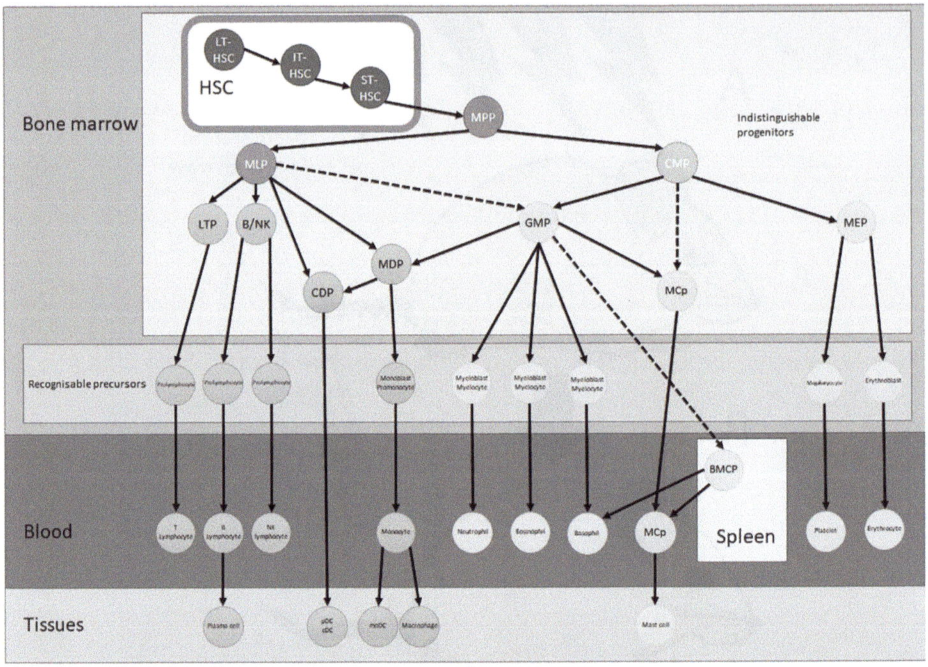

Fig. 9.2 Diagram representing the currently accepted model of human haematopoiesis. From top to bottom, indistinguishable progenitors, recognisable bone marrow precursors, circulating cells in the blood and, where they differ from circulating cells, differentiated cells that can be observed in tissues are shown. A box with precursor cells developing in the spleen is also included. The solid arrows represent the predominant differentiation pathways and the dashed arrows represent some alternative or minority pathways. B/NK, B lymphocyte and NK precursor; BMPC, bipotent basophil-mast cell progenitor; cDC, conventional or myeloid dendritic cell; CDP, common dendritic cell precursor; CMP, common myeloid progenitor; GMP, granulocyte and monocyte progenitor; IT-HSC, intermediate term haemopoietic stem cell; LT-HSC, long-term haemopoietic stem cell; LTP, T lymphocyte precursor; MCp, mast cell progenitor; MDP, monocyte and dendritic cell progenitor; MEP, megakaryocyte and erythrocyte progenitor; MLP, multilymphoid progenitor; moDC, monocytic dendritic cells; MPP, multipotent progenitor; pDC, plasmacytoid dendritic cells; ST-HSC, short-term haemopoietic stem cell

   2. *ST-HSCs* (short-term cells): When transplanted into an irradiated animal, they can produce blood cells for a short time (8–12 weeks).
   3. *IT-HSCs* (intermediate term cells), which have a limited self-renewal capacity and are therefore at an intermediate stage of differentiation.

G. It is not clear whether all three types of HSCs exist in humans, although it is possible.

H. HSCs can pass through the walls of blood vessels and circulate in the blood.

I. ST-HSCs can give rise to *multipotent progenitors* (*MPPs*):
   1. They divide rapidly and produce all blood cells, like a HSC.
   2. Unlike HSCs, they do not have the ability to self-perpetuate and must regenerate from ST-HSCs.

J. There are differences in murine and human haemopoiesis. Human haemopoiesis will be described here, so there may be differences with descriptions in other texts that may be based on mouse haemopoiesis.

K. From the MPP, two indistinguishable types of progenitor cells (*common progenitor cells*) are produced, producing two different lines:

1. *Common myeloid progenitor (CMP)*, the precursor of the myeloid line. It produces:
   a. *Granulocyte and monocyte progenitor (GMP)*: From this cell are formed:
      (1) *Mast cell progenitor (MCp)*: This cell circulates in the blood to give rise to mast cells in tissues.
      (2) Granulocytes.
      (3) *Monocyte and dendritic cell progenitor (MDP)*: From this cell can originate:
         (a) Monocytes.
         (b) The *common dendritic cell precursor (CDP)*.
      (4) It is possible that this GMP can circulate through the blood to the spleen, where it gives rise to a *bipotent basophil-mast cell progenitor (BMCP)*, from which two cell types are produced and released into the bloodstream:
         (a) Basophils.
         (b) MCp, which in tissues give rise to mast cells.
   b. *Megakaryocyte and erythrocyte progenitor (MEP)*, from which platelets and erythrocytes are produced.
   c. To a lesser extent, it is possible that CMP can directly produce MCp.
2. *Immature lymphoid progenitor (multilymphoid progenitor* or *MLP)*, precursor of the lymphoid line. Two routes can be followed from these cells:
   a. *Majority route*: This differentiation pathway produces the cells of the lymphoid lineage:
      (1) *T cell precursor (LTP)*.
      (2) *Precursor of B and NK lymphocytes*.
   b. *Minority route*: This differentiation pathway occurs in certain cases and produces cells of the myeloid line:
      (1) MDP, which in turn can give rise to monocytes and CDPs.
      (2) Directly CDPs.
      (3) Under certain conditions, they may give rise to GMP and produce granulocytes (at least neutrophils).

L. All these progenitors mentioned above are cells that are indistinguishable under the microscope with standard histological techniques, but can be identified by molecular and genetic markers.

M. Finally, recognisable precursor cells emerge from these progenitor cells:

1. Most of these cells are, as usual, named with the suffix *-blast*.
2. These precursor cells are *monopotent*, i.e. each is capable of producing only one blood cell type.

N. Stem cells and common progenitor cells divide relatively slowly to maintain their population, while the frequency of mitosis increases in the subsequent precursor forms.

9.4 Bone Marrow

A. Bone marrow is a variety of *reticular connective tissue* found in the cavities of bones:
   1. Medullary canal of the diaphysis of long bones.
   2. Spaces between the trabeculae of spongy bone.
B. It consists of (Fig. 9.3):
   1. *Reticular cells*: These are a variety of stellate-shaped fibroblasts, which are usually arranged close to blood vessels or form extensions surrounding the vessels. They are therefore called *adventitial reticular cells* or *adventitial cells*.
   2. *Reticular fibres* (*type III collagen*): They are arranged in a criss-cross pattern.
   3. *Haemopoietic cells*: They are very varied and will be described below. In general they are:
      a. *Erythrocyte precursors*
      b. *Granulocyte precursors*
      c. *Lymphocyte precursors*
      d. *Monocyte precursors*
      e. *Megakaryocytes*: They produce platelets.
   4. *Adipocytes*, which are formed from the accumulation of lipids in some reticular cells.

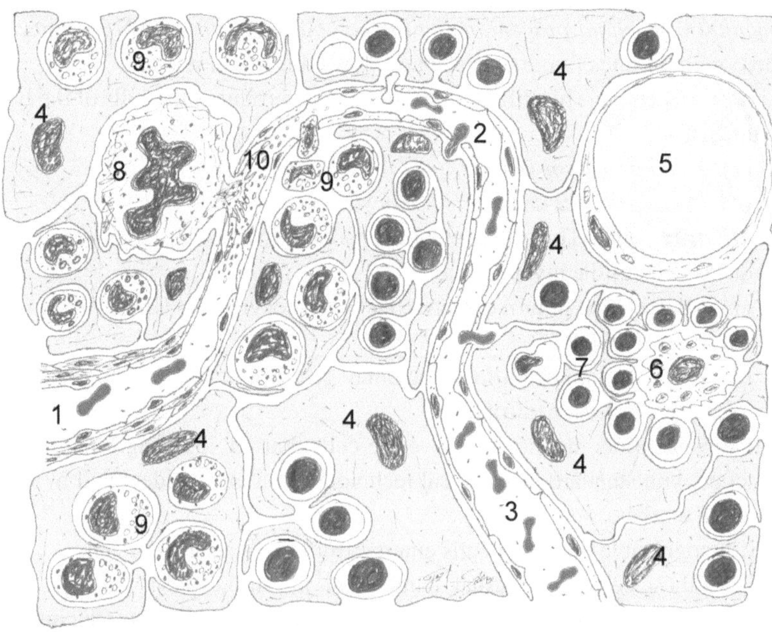

Fig. 9.3 Histological structure of the bone marrow. (1) Arteriole; (2) sinusoid; (3) postcapillary venule; (4) adventitial reticular cells; (5) adipocyte; (6) macrophage; (7) erythropoietic or erythroblastic islets; (8) megakaryocyte; (9) granulocyte precursors; (10) platelet release

5. *Macrophages*, which destroy non-useful or dead cells (old erythrocytes) and cellular debris, such as nuclei ejected from erythrocyte precursors.

C. It is called reticular tissue because the reticular cells emit cytoplasmic prolongations that extend over the reticular fibres and form a lattice. This meshwork of cells and fibres creates a favourable environment for haemopoietic cells to nest in the holes of the mesh.

D. The bone marrow is vascularised:
1. Small blood vessels, *arterioles*, enter, which branch out.
2. From the arterioles arise capillaries called *sinusoids* (from the Latin sinus 'sinus' and -oīdes, and this from the classical Greek -ειδής -eidḗs or -ειδές -eidés 'similar to'):
   a. Compared to most capillaries, sinusoids are very wide (30–75 μm).
   b. Its wall of endothelial cells, as well as the basal lamina, are perforated to allow blood cells to enter and exit.
   c. Some sinusoids are seen to be completely filled with blood cells, which is why it is believed that blood does not circulate in them, but rather accumulates.
      (1) These sinusoids act as reserve or storage areas.
      (2) When more blood cells are needed, the sinusoids mobilise the stored cells into the bloodstream and empty themselves, to be refilled later.
3. The blood arrives via arterioles, is distributed through sinusoids and finally passes into *postcapillary venules* that carry the blood to the venous circulation back to the heart.

E. Recently, some *haemopoietic niches* have been identified:
1. HSCs are located close to blood vessels, preferably near the surface of a bone trabecula:
   a. Most HSCs are attached to sinusoids, close to endothelial cells and adventitial cells.
   b. Others are attached to smaller arterioles, in contact with *pericytes*.
2. Erythrocyte precursor cells usually form clusters called *erythropoietic* or *erythroblastic islets*.
   a. These erythropoietic islets are often in contact with macrophages.
   b. They are also often close to sinusoids to facilitate the passage of newly formed erythrocytes into the blood.
3. *Megakaryocytes* are located next to the sinusoids and emit cytoplasmic extensions that penetrate into the lumen of the sinusoids, where they fragment and become platelets.
4. Lymphocyte precursors are controlled by *osteoblasts and are* therefore usually found close to the *endosteum* overlying the bone trabeculae.
5. Granulocyte precursors may be seen intermingled or in small groups away from the sinusoids. The more mature forms of these cells may move and travel towards the sinusoids.

F. Haemopoietic bone marrow is called *red bone marrow*, because of its colour when fresh.
   1. In the neonate, the entire bone marrow produces blood.
   2. In the adult, this function is restricted to the vertebrae, ribs, sternum, coccyx and proximal epiphyses of the humerus and femur.

G. The regions of the marrow that do not undergo haematopoiesis become adipose tissue and are called *yellow bone marrow*, due to their colour when fresh.

9.5 Erythropoiesis

A. Erythrocytes live for about 4 months, so they need to be constantly produced.

B. They are generated from an megakaryocyte and erythrocyte progenitor (*MEP*) belonging to the myeloid line, i.e. descended from *CMP*.

C. From these precursor cells, microscopically distinguishable cells are produced according to this sequence (Fig. 9.4):
   1. *Proerythroblasts*:
      a. Cells 14–20 μm in diameter.
      b. The cytoplasm is basophilic.
      c. The nucleus is large, euchromatic, with some evenly dispersed heterochromatin and two or more nucleoli.
      d. They divide by mitosis several times to produce the next cell type.
   2. *Basophilic erythroblasts*. Difficult to distinguish from proerythroblasts.
      a. They are somewhat smaller (15 μm).
      b. Its cytoplasm is very basophilic, as it has many free *ribosomes* to synthesise haemoglobin, and there is hardly any rough endoplasmic reticulum.
      c. They have a peripheral ring of microtubules that maintains cell shape.
      d. Its nucleus has clumped chromatin and no nucleoli, as ribosome synthesis has ceased.
      e. These cells also proliferate by mitosis.
   3. *Polychromatophilic erythroblasts*:
      a. Smaller (12–15 μm).
      b. The cytoplasm has a peculiar staining:
         (1) Basophilic regions with acidophilic patches, or areas with colour variations from bluish-green to grey-green of decreasing intensity.
         (2) This is because the number of ribosomes decreases as the *haemoglobin* concentration increases.
         (3) As haemoglobin binds eosin, while ribosomes bind alkaline blue dyes, the result is a variable staining mixture in each cell.
      c. Smaller nucleus, with fairly condensed chromatin showing a *checkerboard-like* grid pattern.
      d. These cells can also divide by mitosis.

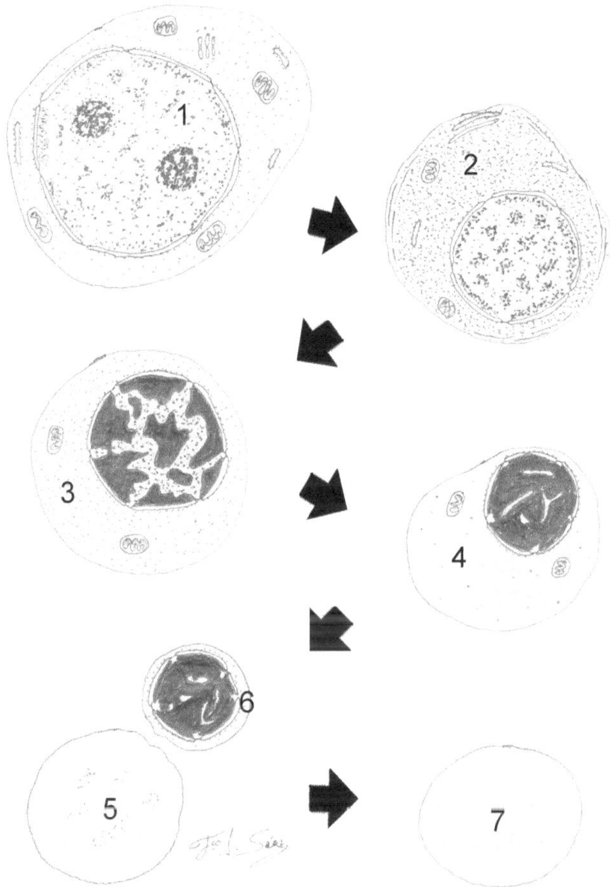

Fig. 9.4 Erythropoiesis. The arrows show the progress of differentiation. (1) Proerythroblast; (2) basophilic erythroblast; (3) polychromatophilic erythroblast; (4) orthochromatic erythroblast (normoblast); (5) polychromatophilic erythrocyte or reticulocyte; (6) cell fragment with erythroblast nucleus; (7) erythrocyte

4. *Orthochromatic* or *acidophilic erythroblasts (normoblasts)*:
   a. Cells smaller than above, around 10 μm.
   b. They have an eosinophilic cytoplasm due to an abundance of haemoglobin and a scarcity of ribosomes. The cytoplasm is devoid of organelles, except for some isolated mitochondria and a few free ribosomes.
   c. The small, off-centred nucleus is intensely stained. The nuclear chromatin is very compact, organised in thick masses.
   d. These cells do not divide.
   e. Finally, the nucleus is expelled surrounded by a thin film of cytoplasm with its plasma membrane and the cell becomes the reticulocyte. The ejected nucleus is destroyed by macrophages in the bone marrow.

5. *Polychromatophilic erythrocytes* or *reticulocytes*:
    a. They are the immature form of the erythrocyte.
    b. They are the size of an erythrocyte.
    c. They are anucleated cells originating from normoblasts that have expulsed their nuclei.
    d. They contain some ribosomes, which gives their cytoplasm a basophilic hue that is not seen in mature erythrocytes. When stained with *cresyl blue*, the ribosomes form a bluish mesh in the pink cytoplasm, hence the name *reticulocytes*.
    e. They pass through the walls of the sinusoids and enter the bloodstream.
    f. When they lose their ribosomes, they look like typical erythrocytes.

D. Only 1% of total red blood cells are found in the bone marrow as no red blood cells are stored in the bone marrow.
E. The process of erythrocyte formation from a basophilic erythroblast takes about a week.
F. Erythropoiesis is stimulated by *erythropoietin* (*EPO*), a hormone produced in the kidney.

9.6 Granulopoiesis

A. According to the traditional model of haemopoiesis, granulocytes (neutrophils, basophils and eosinophils) are produced from the *CMP* myeloid cell, which is common to other blood cells, and from which the monocyte and granulocyte progenitor (*GMP*) originates.
B. In certain circumstances, GMP can also originate from the *MLP* lymphoid cell.
C. In all cases, GMP produces precursors of each type of granulocyte identifiable in microscopic slides.
D. The early precursors of the different granulocyte types (neutrophils, eosinophils and basophils) are morphologically similar to each other, and even the first cells of the process are indistinguishable.
E. The sequence of differentiation that occurs in the bone marrow is (Fig. 9.5):
    1. *Myeloblasts*: Neutrophil-producing myeloblasts, basophil-producing myeloblasts and eosinophil-producing myeloblasts are morphologically identical and therefore indistinguishable. They exhibit these characteristics:
        a. Cells 15–20 μm in diameter.
        b. They have an oval nucleus with peripheral heterochromatin and three or more nucleoli.
        c. It has a basophilic cytoplasm and no granules.
        d. They divide by mitosis.
    2. *Promyelocytes (progranulocytes)*: As with myeloblasts, it is impossible to distinguish whether a promyelocyte is of the neutrophil, eosinophil or basophil lineage.
        a. They are larger cells (18–24 μm).

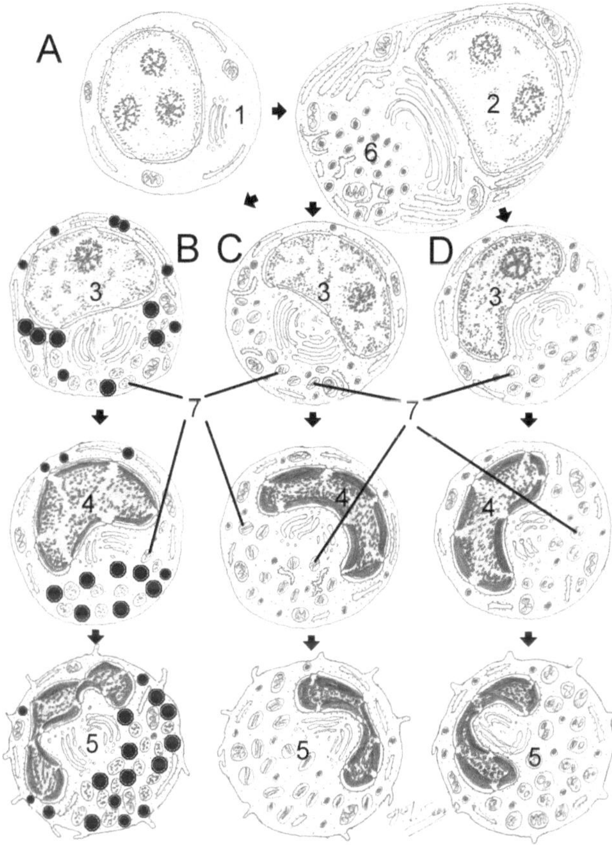

Fig. 9.5 Granulopoiesis. Arrows show the progress of differentiation. The common cell types for the three granulocyte types (**A**) and the precursor forms of neutrophils (**B**), eosinophils (**C**) and basophils (**D**) are shown. (1) Myeloblast; (2) promyelocyte (progranulocyte); (3) myelocytes; (4) metamyelocytes; (5) mature granulocyte (segmented stage); (6) azurophilic granules; (7) specific granules. The immature band cell is not shown

b. The nucleus is kidney-shaped, with peripheral heterochromatin and other dispersed heterochromatin, and has one or two nucleoli.
c. Characteristics of the cytoplasm:
   (1) They have *azurophilic granules*:
      (a) These granules are synthesised only at this stage.
      (b) Under the electron microscope, the azurophilic granules are dense, 0.1–0.25 μm.
   (2) Endoplasmic reticulum cisternae and free ribosomes are observed.
   (3) The Golgi complex is located next to the nuclear cleavage.
d. These cells grow from an initial 18 μm to a final 24 μm. Mitotic divisions then take place, again giving rise to smaller cells.

3. *Myelocytes*: Three types of myelocytes can be distinguished: neutrophil myelocytes, eosinophil myelocytes and basophil myelocytes.
   a. They measure between 10 and 16 μm in diameter.
   b. They have nuclei of varying shapes, usually indented, and with a single nucleolus.
   c. In the cell cytoplasm, *granules specific* to each granulocyte type are already visible.
   d. They divide by mitosis.
4. *Metamyelocytes*: Neutrophilic, eosinophilic and basophilic metamyelocytes can be distinguished.
   a. Their size is between 12 and 15 μm.
   b. The nucleus has a large indentation and may be kidney shaped.
   c. Specific granules are more abundant than azurophilic granules.
   d. They are cells incapable of dividing.
5. *Immature granulocyte (band stage)*: At least in the case of neutrophils, the first granulocytes that are produced have a rod-shaped nucleus, in which the characteristic lobulation is not yet discernible. This form can already circulate in the blood. This form is called *band (stab) cell.*
6. *Mature granulocyte*: When two lobes are clearly distinguishable in the nucleus, the granulocyte is considered to be mature (*segmented stage*). In the case of the neutrophil, it is called a *segmented neutrophil.*

F. The development of neutrophils is known in more detail:
   1. The mitotic phase of their development has a duration of 1 week.
   2. The post-mitotic phase lasts 1 week.
   3. Its half-life in the blood is 6–12 h.
   4. The lifespan of these cells in the connective tissue is about 2–3 days, so it is necessary to produce these cells continuously.

G. There is usually a large reserve of metamyelocytes, band cells and segmented neutrophils in the bone marrow.
   1. They can be released in case of need, such as when a bacterial infection occurs.
   2. It has been estimated that there are 30 band or segmented cells in the bone marrow for every circulating cell.

9.7 Lymphopoiesis

A. Lymphopoiesis occurs from the mature lymphoid progenitor (*MLP*). This progenitor gives rise to lymphocyte precursors, *prolymphocytes*, which are difficult to identify morphologically, as they are cells similar to mature lymphocytes.

B. Lymphocyte precursors are produced in the bone marrow but can travel through the blood and complete their differentiation in other organs.
   1. *T lymphocytes*:

   a. Prolymphocytes move from the bone marrow to the *thymus*. There, *naïve Th and Tc lymphocytes* (which have not yet contacted their specific antigen) are produced and pass into the blood.
   b. Naïve lymphocytes seed the *secondary lymphatic organs* (lymph nodes, Peyer's patches, spleen, tonsils) where they can contact their antigen and become *effector cells* and *memory T cells*.
2. *B lymphocytes*:
   a. *Naïve B lymphocytes* develop from prolymphocytes in the bone marrow.
   b. Naïve lymphocytes move to the *lymphatic nodules* or *lymphatic follicles* of secondary lymphatic organs, where they mature on contact with antigen and become *effector cells* (plasma cells) and *memory B cells*.
3. *NK lymphocytes*: They mature in both the bone marrow and thymus.

9.8 Monopoiesis (Monocytopoiesis)

A. Monocytes are produced from a common precursor with granulocytes, *GMP*. From this cell, the monocyte and dendritic cell progenitor, *MDP*, is produced.
B. *Monoblasts* originate from the MDP and develop into *promonocytes*.
   1. Both are difficult to differentiate, as they are rare and similar to other precursors of other blood cells.
   2. They divide by mitosis.
C. The promonocyte is about 20 μm in diameter, has fine chromatin, basophilic cytoplasm with a well-developed Golgi apparatus and endoplasmic reticulum and numerous small azurophilic granules. Finally, they give rise to monocytes.
D. Monocytes are derived from the myeloid line but some may be formed from the *immature lymphoid progenitor (multilymphoid progenitor* or *MLP)*, which can give rise to MDP.
E. Differentiation of a monocyte in the bone marrow can take 2–3 days.
F. Monocytes remain in the blood for 12–24 h and migrate to connective tissue and other tissues, where they become *macrophages* and other cells of the *mononuclear phagocytic system*.

9.9 Thrombopoiesis

A. *Thrombopoiesis* is the formation of platelets, which have a common origin with erythrocytes from megakaryocyte and erythrocyte progenitor (*MEP*).
B. The first identifiable precursor originating from the MEP is the *megakaryoblast*:
   1. It measures 15–25 μm in diameter.
   2. Its nucleus, like that of most human cells, is *diploid* (2n chromosomes).

3. Successive replications of chromosomes, without division of the nucleus or cytoplasm (*endomitosis*), give rise to *polyploid* cells, with chromosome number of *4n*, *8n*, *16n*, *32n* and *64n*.
4. Centriole divisions also occur in the cytoplasm, so that the number of centriole pairs corresponds to the degree of polyploidy.
5. The resulting cell is the megakaryocyte.

C. *Megakaryocytes* (from Ancient Greek μεγα- mega- 'large' and κάρυον káryon 'nut', 'nucleus') are polymorphonuclear giant cells (Fig. 9.6):
   1. They are very large, 50–100 μm in diameter.
   2. They are spherical, but with irregular and rounded *pseudopods* on their surface.
   3. The nucleus is large, multilobed, with a variable number of lobes.
   4. It has *azurophilic granules*.
   5. Cytoplasmic *demarcation channels* are produced in the mature megakaryocyte:
      a. It is unclear whether they are produced by fusion of intracytoplasmic vesicles or by invaginations of the plasma membrane.
      b. These channels divide the cell periphery into small regions of the cytoplasm containing granules and some organelles.
      c. These regions are the precursors of platelets.

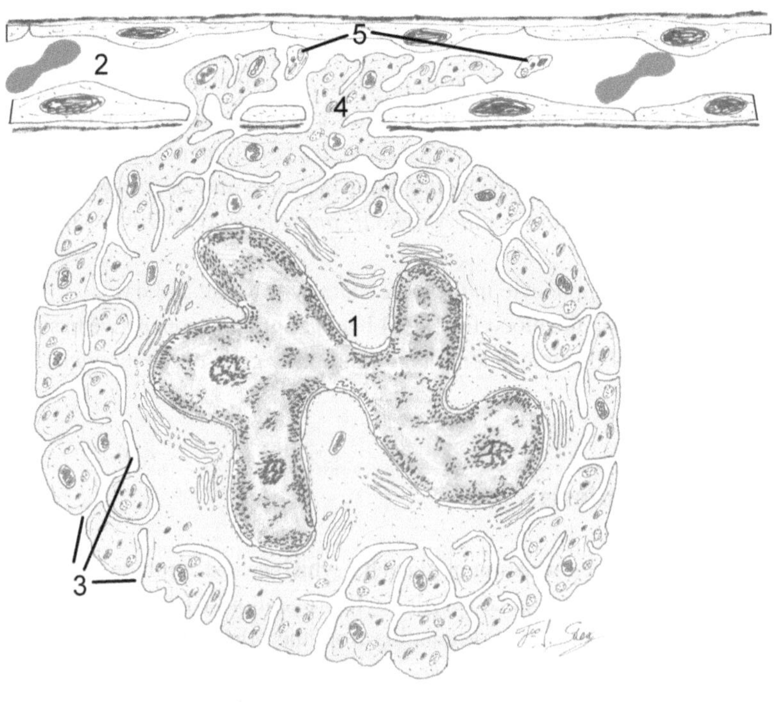

Fig. 9.6 Thrombopoiesis. (1) Megakaryocyte; (2) sinusoid; (3) demarcation channels; (4) pseudopod penetrating sinusoid; (5) platelets

6. The areas of cytoplasm with demarcation canals extend as extensions (*pseudopods*), which pass through the walls of the bone marrow sinusoids and extend into the lumen of the blood vessel.

D. The cellular extensions that penetrate the sinusoids fragment into platelets.
   1. A megakaryocyte can have up to six extensions.
   2. More than a thousand platelets can be produced from one of these extensions.

E. At the end of the process, the megakaryocyte dies and its remains are degraded by macrophages.

F. Thrombopoiesis is stimulated by *thrombopoietin*, a hormone produced in the *liver* and *kidney*.

9.10 Origin of Interdigitating Dendritic Cells (DC or IDC)

A. The origin of these antigen-presenting cells is becoming clearer in recent years.

B. There are several populations of these cells, which may have different origins:
   a. *Monocytic dendritic cells (moDC)* have recently been identified and are produced in tissues from monocytes.
   b. Conventional or myeloid dendritic cells (cDC) and plasmacytoid dendritic cells (pDC).
      (1) Some years ago it was thought that *conventional* or *myeloid dendritic cells (cDCs)* would be formed from the myeloid line and *plasmacytoid dendritic cells (pDCs)* from the lymphoid line.
      (2) It now seems clear that cDCs and pDCs are produced from the same precursor in the bone marrow. This precursor has been termed the *common dendritic cell precursor* (*CDP*). This precursor originates from:
         (a) *MDP*, which can be originated from the myeloid line (GMP), but can also be produced from the lymphoid line (MLP).
         (b) It can be produced directly from the MLP.

C. It is worth remembering that *Langerhans cells*, which are considered to be a variety of dendritic cells infiltrating some epithelia, are produced during embryonic development from yolk sac precursors.

9.11 Origin of Mast Cells

A. Mast cells are produced from bone marrow precursors, specifically the myeloid precursor, *CMP*.

B. It has been discovered that there may be several differentiation pathways that give rise to mast cells.

C. All pathways have in common that a *mast cell progenitor* (*MCp*) is produced, which enters the bloodstream and passes into the tissues, where it is converted into mast cells.

D. The routes discovered so far are:
  1. MCp is produced in the *bone marrow*. This route has two alternatives:
     a. MCp is produced directly from CMP.
     b. MCp is produced from GMP, which in turn descends from CMP.
  2. MCp is produced in the *spleen*:
     a. The GMP cell travels from the bone marrow through the blood to the spleen.
     b. In the spleen, GMP produces a *bipotent basophil-mast cell progenitor (BMCP)*. This cell gives rise to basophils and MCp, which move through the blood.

Take-Home Message

Haemopoiesis is the production of blood cells. During embryonic development, the first blood cells are produced in the yolk sac and placenta (primitive haemopoiesis). Later, myeloid and lymphoid progenitor cells are observed in the yolk sac and embryonic endothelium (transient definitive haemopoiesis). Finally, haemopoietic stem cells originate in the aorta-gonad-mesonephros region and migrate to the bone marrow (definitive haemopoiesis). In the adult, a common myeloid progenitor (CMP), which produces granulocyte, monocyte, platelet and erythrocyte progenitors, and an immature lymphoid progenitor (MLP), which produces mostly lymphocyte precursors and rarely monocyte precursors and dendritic cells, are formed from these stem cells. The bone marrow has cells and reticular fibres that provide a suitable niche for haemopoietic cells. There are blood sinusoids with perforated walls that facilitate the entry of formed blood cells. Erythrocyte precursors (erythroblasts) synthesise haemoglobin and lose the nucleus and organelles. Granulocyte precursors (myeloblasts and myelocytes) synthesise primary and secondary granules. Lymphocyte precursors (prolymphocytes) produced in the bone marrow mature here (B-lymphocytes) or move to the thymus to mature (T-lymphocytes). Monoblasts form promonocytes that give rise to monocytes. Platelets are produced as fragments of the cytoplasm of polyploid giant cells, the megakaryocytes. Interdigitating dendritic cells are produced from monocytes (monocytic dendritic cells) or from the common dendritic cell precursor (conventional dendritic cells and plasmacytoid dendritic cells). Langerhans cells originate during embryonic development from yolk sac precursors.

Further Reading

Bonnet D (2002) Haematopoietic stem cells. J Pathol 197:430–440
Calvanese V, Mikkola HKA (2023) The genesis of human hematopoietic stem cells. Blood 142:519–527
Dahlin JS, Hallgren J (2015) Mast cell progenitors: origin, development and migration to tissues. Mol Immunol 63:9–17
Doulatov S, Notta F, Laurenti E, Dick JE (2012) Hematopoiesis: a human perspective. Cell Stem Cell 10:120–136
Gremmel T, Frelinger ALIII, Michelson AD (2024) Platelet physiology. Semin Thromb Hemost 50:1173–1186
Ivanovs A, Rybtsov S, Ng ES et al (2017) Human haematopoietic stem cell development: from the embryo to the dish. Development 144:2323–2337
Jee WSS (1988) Bone marrow. In: Weiss L (ed) Cell and tissue biology—a textbook of histology, 6th edn. Urban & Schwarzenberg, pp 467–478
Kierszenbaum AL, Tres LL (2019) Blood and hematopoiesis. In: Histology and cell biology—an introduction to pathology, 5th edn. Elsevier, pp 199–236
Konieczny J, Arranz L (2018) Updates on old and weary haematopoiesis. Int J Mol Sci 19:2567
Krstić RV (1991) Bone marrow and Hematopoiesis. In: Human microscopic anatomy. An atlas for students of medicine and biology. Springer Verlag, pp 5–29
Laurenti E, Göttgens B (2018) From haematopoietic stem cells to complex differentiation landscapes. Nature 553:418–426
McGrath KE, Frame JM, Palis J (2015) Early hematopoiesis and macrophage development. Semin Immunol 27:379–387
Miyake K, Ito J, Karasuyama H (2024) Novel insights into the ontogeny of basophils. Front Allergy 5:1402841
Morrison SJ, Scadden DT (2014) The bone marrow niche for haematopoietic stem cells. Nature 505:327–334
Murphy TL, Grajales-Reyes GE, Wu X et al (2016) Transcriptional control of dendritic cell development. 34:93–119
Ng AP, Alexander WS (2017) Haematopoietic stem cells: past, present and future. Cell Death Discov 3:17002
O'Dowd G, Bell S, Wright S (2023) Blood, haematopoiesis and bone marrow. In: Wheater's functional histology. Elsevier, pp 46–64
Park J, Kang S-J (2024) The ontogenesis and heterogeneity of basophils. Discov Immunol 3:kyae003
Rogkoti T, Donnelly H, Dalby MJ et al (2025) Mechanical signatures and models of the bone marrow niche. Nat Rev Bioeng. https://doi.org/10.1038/s44222-025-00305-6
Weiss L (1988) The life cycle of blood cells. In: Weiss L (ed) Cell and tissue biology—a textbook of histology, 6th edn. Urban & Schwarzenberg, pp 445–466
Zhang Y, Gao S, Xia J, Liu F (2018) Hematopoietic hierarchy—an updated roadmap. Trends Cell Biol 28:976–986

Muscle Tissues

10

What You Will Learn in This Chapter

This chapter describes the three types of muscle tissue: smooth muscle, skeletal muscle and cardiac muscle. They are all made up of cells that can contract due to the special organisation of their cytoskeleton and the sliding of actin microfilaments between myosin filaments. It is explained that skeletal and cardiac muscle show transverse striations when viewed under the microscope, while smooth muscle does not. It is also shown how the cytoskeleton and other organelles are organised in the cells of each of these muscle types and how contraction occurs.

10.1 General Characteristics

A. The main characteristic of muscle tissue is its ability to *contract*.
   1. All muscle tissues are characterised by their cells being elongated and oriented in a direction that coincides with the direction in which contraction occurs.
   2. The contraction of muscle tissue enables the movement of some parts of the human body relative to others. This enables locomotion, constriction, pumping of fluids, erection of hair, gesticulation, etc.

B. *Muscle cells* or *myocytes* (from the ancient Greek μῦς, μυός mŷs, myós, 'muscle') are often observed to form elongated microscopic structures:
   1. These elongated structures are called *muscle fibres*.
   2. In many cases these fibres are the muscle cells themselves, and the terms muscle fibre and muscle cell are often used synonymously.[1]

C. Contractile organs made up of muscle tissue are called *muscles*.

D. Some components of muscle cells may have special names, for example:

[1] It is important to emphasise that the muscle fibre is a living cell, and therefore has not equivalence with the fibres of the extracellular matrix, collagen and elastic fibres, which are lifeless protein aggregates.

F. J. Sáez, I. Badiola, *Essential Oral Histology*, Learning Materials in Biosciences,
https://doi.org/10.1007/978-3-032-03165-5_10

   1. The plasma membrane is called the *sarcolemma* (from the Greek σαρκο- sarko-, 'flesh' and λέμμμα lémma, 'cortex').
   2. The cytoplasm is called the *sarcoplasm.*
   3. The smooth endoplasmic reticulum is called the *sarcoplasmic reticulum.*

E. Muscle tissue is of mesodermal origin and originates from embryonic *mesenchyme.* Some mesenchymal cells develop into muscle cells.

F. In the adult, conversion of *mesenchymal reserve cells* into muscle cells is possible.

10.2 Varieties of Muscle Tissue

A. Traditionally, three varieties of muscle tissue are distinguished, producing two categories of muscle (Fig. 10.1):
   1. *Smooth muscle* is so called because it has no transverse bands.
      a. It is made up of spindle cells.
      b. Its contraction is slow.
   2. *Striated muscles* are made up of muscle cells that show regularly spaced transverse bands. There are several distinct varieties of striated muscle:

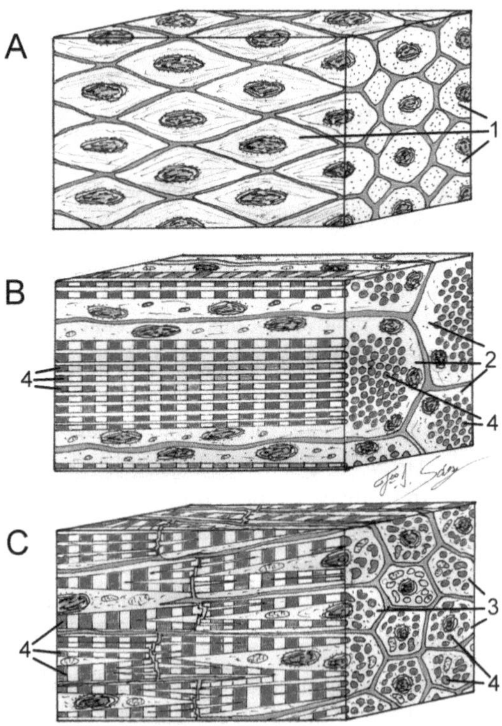

Fig. 10.1 The three main varieties of muscle tissue: smooth muscle (**A**), skeletal muscle (**B**) and cardiac muscle (**C**). (1) Smooth muscle cell; (2) skeletal muscle cell; (3) cardiac muscle cell; (4) myofibrils

a. *Skeletal muscle*, which constitutes the somatic musculature, i.e. the musculature of the body wall and limbs.
   (1) It consists of very large, multinucleated, cylindrical cells.
   (2) Its contraction is rapid and vigorous.
b. *Visceral striated muscle*. It is not associated with the skeleton, but is found in soft organs.
   (1) It is found in the tongue, the pharynx, the lumbar part of the diaphragm and the upper part of the oesophagus.
   (2) It has the same morphological characteristics as skeletal muscle.
c. The *cardiac muscle*, which forms most of the wall of the heart (the *myocardium*).
   (1) It is made up of long, but shorter than skeletal muscle, branched cells.
   (2) Its contraction is vigorous and rhythmic.

B. Skeletal muscle, with some exceptions, is of *voluntary contraction*, whereas visceral striated muscle, smooth muscle and cardiac muscle are characterised by *involuntary contraction*.

10.3 Smooth Muscle Tissue (Smooth Muscle)

A. The smooth muscle fibre or cell (*smooth myocyte*) is elongated and spindle-shaped.
   1. Their length varies from 20 to 200 μm, although in the gravid uterus they can reach 500 μm.
   2. Their diameter can be between 3 and 10 μm.

B. Smooth muscle cells may be present:
   1. Forming *smooth muscle*. Smooth muscle is made up of myocytes grouped either in *bundles* (from the Proto-Indo-European root bhendh- 'to bind') or *fascicles* (from Latin fascicŭlus 'little bundle'), or in *layers*.
      a. In bundles or layers, cells overlap with each other—with the thick portion of a fibre lying next to the thin ends of neighbouring cells—to form a compact mass.
         (1) In longitudinal sections, the fibres appear *spindle-shaped*.
         (2) In cross-sections, a mixture of *rounded profiles* of different sizes can be seen, depending on whether each cell has been cut in its central part or at its thinned end. Nuclei are visible only if the cut goes through the thick central part.
      b. The bundles or layers are separated by *loose connective tissue* with fibroblasts, collagen and elastic fibres, blood capillaries and nerves.
      c. Within each bundle or layer of muscle, between the cells, there is an *extracellular matrix* produced by the cells themselves, containing collagen, reticular and elastic fibres and proteoglycans.
   2. Isolated or in small groups within connective tissue (e.g. in the connective tissue of intestinal villi).

a. For this reason, smooth muscle cells are considered to be *resident cells* of connective tissue.
b. In some tissues there are *myofibroblasts*, cells with mixed characteristics of fibroblasts and smooth muscle cells.

C. Smooth muscle is found in:
   1. *Visceral musculature*, i.e. that found in the walls of the digestive tract and its associated glands, respiratory tract, urinary and genital tracts, blood vessels and lymphatic vessels.
   2. Small *contractile units*, such as the erector pili muscles in the skin, or the ciliary body muscles and iris muscles of the eye.
   3. The *areola of the breast*. Here it is responsible for the erection of the nipple.
   4. The *scrotum*, where it is responsible for its puckering.

D. Structure of the smooth myocyte. The main characteristic of smooth muscle cells is the absence of a repetitive organisation of cytoskeletal filaments, so that no striation is observed in their cytoplasm.
   1. Each cell is surrounded by a *basal lamina* called the *external lamina* (Fig. 10.2). Although they are very close to each other (between 40 and 80 nm), the cells are joined together not by junctional complexes typical of epithelial cells but by the intervening extracellular matrix, including the external lamina.
   2. The smooth muscle cell *nucleus* is elongated, relatively clear, with some peripheral heterochromatin and a nucleolus. When the cell contracts, the nucleus coils into a corkscrew.
   3. *Sarcoplasm* is eosinophilic under light microscopy (staining red with the usual haematoxylin and eosin stain) and shows no striations or other special features, hence the name smooth muscle.
   4. The Golgi apparatus, mitochondria, rough endoplasmic reticulum and lysosomes are found in two regions of the cytoplasm close to the nucleus located at opposite poles on the long axis of the cell.

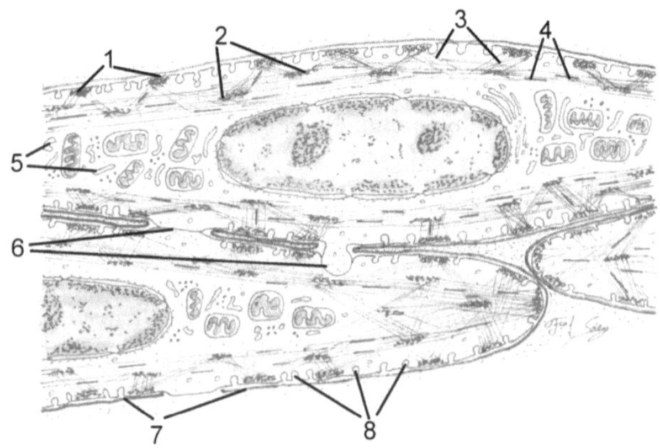

Fig. 10.2 Smooth muscle cells. (1) Dense plaques or adhesion plaques; (2) dense bodies; (3) actin microfilaments; (4) myosin filaments; (5) sarcoplasmic reticulum; (6) gap junctions; (7) external lamina; (8) caveolae

5. *Glycogen inclusions* are frequent.
6. Calcium-storing *sarcoplasmic reticulum* is observed in the peripheral regions of the cell.
7. The electron microscope shows the organisation of the cytoskeleton.
   a. The following structures have been described:
      (1) *Dense plaques* or *adhesion plaques*: These are dense areas attached to the inner side of the sarcolemma by binding proteins (*integrins*). Integrins are associated with *α-actinin*.
      (2) *Dense bodies*: Dense areas distributed throughout the cytosol. They also contain α-actinin.
      (3) *Contractile bundles of filaments*, consisting of:
         (a) *Actin microfilaments*: These filaments are organised in a little-known but different way from that of striated muscle.
            1. They form *bundles* arranged longitudinally or obliquely in the cell.
            2. Each bundle consists of 13–14 actin filaments.
            3. The ends of each bundle terminate in a dense plaque or dense body, where it binds to α-actinin.
         (b) *Myosin II thick filaments*: A filament of this type is associated with a bundle of microfilaments.
      (4) *Desmin intermediate filaments*: In the smooth muscle of blood vessels there are desmin and *vimentin intermediate filaments*.
         (a) These filaments may contact dense bodies and dense plaques.
         (b) Its function is to structurally support the web of contractile microfilaments.
   b. The filaments of the cytoskeleton are responsible for the contraction of the cell by the following mechanism:
      (1) Contractile bundles of filaments are anchored to the dense sarcolemmal plaques at one end and to dense cytosolic bodies at the other.
      (2) Other contractile bundles are anchored at both ends to the cytosolic dense bodies.
      (3) As a result, the cytoplasm is traversed by bundles of filaments connected to each other and to the sarcolemma, forming '*chains*' of bundles running in various directions, criss-crossing each other and connecting one point on the cell surface to a distant point on the cell surface.
      (4) The simultaneous contraction of all these filament chains produces a shortening of the smooth muscle cell. The simultaneous shortening of all the cells of the tissue produces shortening, i.e. contraction of the tissue.
8. The sarcolemma has depressions or invaginations in the shape of a flask or the letter 'Ω', called *caveolae*.
   a. The true significance of caveolae in these and other cells may be:
      (1) Perform *endocytosis*.

(2) Bringing the plasma membrane closer to the interior or creating special *microenvironments* with cell surface materials.
(3) Coupling the stimulus signals received by the cell to produce contraction of the internal cytoskeleton. This would be done by:
  (a) *Ca^{2+} pumps* abundant in the sarcolemma of the caveolae.
  (b) Maintain close proximity to the superficial smooth endoplasmic reticulum cisternae.

9. Smooth muscle cells are in close proximity to each other, although a specialised junction like the desmosomes of other cells is not clearly visible. However, the *dense plaques* of one cell face the dense plaques of neighbouring cells, so it is thought that the intervening extracellular matrix acts as a bridge, a cohesion, between these structures.
10. In many smooth muscles, neighbouring cells contact each other at small points where *gap junctions* occur, allowing the cells to couple and contract simultaneously.

E. Contraction mechanism:
1. Extracellular *calcium* and calcium accumulated in the smooth endoplasmic reticulum are transported to the cytosol via specific pumps and channels.
2. Calcium combines with the protein *calmodulin*, a cytosolic protein, and a *Ca^{2+}–calmodulin complex* is formed. The complex formed by calcium and calmodulin activates a *kinase* that phosphorylates myosin.
3. The phosphorylated myosin heads interact with the actin microfilaments. This causes displacement of the actin microfilaments over the myosin microfilaments.
4. Because the myosin heads have different orientations, some actin filaments move in one direction and others in the opposite direction, shortening the contractile bundles.
5. Shortening of the contractile bundles causes shortening of the cell.
6. When the concentration of Ca^{2+} decreases in the cytosol, a *phosphatase* is activated that dephosphorylates myosin, leading to the separation of actin and myosin and muscle relaxation.

F. Smooth muscle is *involuntary in contraction* and the increase of calcium in the cytosol is controlled by several mechanisms:
1. *Hormonal control*, via *hormones* and *local chemical mediators*: Some of the molecules that stimulate or inhibit contraction are noradrenaline, nitric oxide, oestrogen, oxytocin (in utero), histamine, vasopressin and serotonin.
2. *Nerve control*:
   a. Smooth muscle is innervated by the *autonomic nervous system.*
      (1) In most organs, smooth muscle is innervated by the *sympathetic* and *parasympathetic* nervous systems.
      (2) In the gastrointestinal tract, innervation comes from the *enteric division* of the autonomic nervous system.

b. The main neurotransmitters used by the autonomic nervous system to control smooth muscle contraction are *noradrenaline* and *acetylcholine*.
   (1) These neurotransmitters are antagonistic, i.e. one stimulates and the other inhibits contraction.
   (2) The effect of each neurotransmitter varies from organ to organ. In some, noradrenaline stimulates and acetylcholine inhibits, and in others it is the opposite.
   (3) The effect caused by the neurotransmitter depends on the *receptors* on each muscle cell.
c. Nerve endings do not form synapses with myocytes, but are somewhat separated (10–100 nm):
   (1) Neurotransmitters must diffuse through the matrix.
   (2) These neurotransmitters can be effective even at distances of 1 μm, although the effect is slower.

3. *Physical forces*: Tissue distension stimulates muscle contraction. This occurs because the stretching of the cell causes a change in the sarcolemmal potential.
4. *Intrinsic contractile activity*: In some cases, smooth muscle has some ability to contract without stimuli.

G. Depending on the innervation, two types of smooth muscle can be distinguished:
1. *Multi-unit smooth muscle*:
   a. Each smooth myocyte has its own innervation and can contract independently.
   b. The contraction is rapid and is followed by complete relaxation (*phasic contraction*).
   c. Spontaneous contraction never occurs.
   d. This muscle is found in the iris and vas deferens.
2. *Single-unit smooth muscle*:
   a. Only some myocytes are innervated, but as they are connected to the rest by *gap junctions*, the stimulus spreads to all cells, which contract simultaneously.
   b. These muscles always have a certain degree of contraction:
      (1) Stimuli can increase or decrease the degree of contraction.
      (2) The contraction is slow and contained (*tonic contraction* or *tone*).
   c. *Spontaneous contractions* can occur, for example, if the muscle is stretched.
   d. Some cells of these muscles act as *pacemakers*, i.e. they are spontaneous stimulus-generating centres that are transmitted to the rest of the cells via the gap junctions.
   e. This muscle is typical of the blood vessels, the digestive tract, the biliary and urinary tract and the uterus.

H. Smooth muscle cells can divide by *mitosis*, although this is rare.
1. Damaged or ageing cells are replaced in the blood vessels.
2. In the outer muscle layers of the stomach and colon there are regular divisions to slowly increase the thickness of these layers.

3. In the pregnant uterus, the muscular wall increases by two simultaneous mechanisms:
   a. *Hyperplasia* (from the ancient Greek ὑπερ- hyper- 'excess' and πλάσις plasis 'formation'), which is the increase in the number of cells due to cell divisions.
   b. *Hypertrophy* (Greek τροφία trophía 'nourishment, growth') or enlargement of cells.

I. Damaged muscle can be regenerated by the production of new smooth myocytes:
   1. From existing *myocytes* that divide by mitosis.
   2. By multiplication and differentiation of *reserve mesenchymal cells*, normally located in the adventitia of blood vessels.
   3. From *endothelial cells* and *pericytes* (cells of the blood vessel wall).

J. Smooth muscle cells can perform other activities unrelated to contraction:
   1. Synthesise *extracellular matrix* materials: collagen (type III and IV), elastin, laminin, glycosaminoglycans and proteoglycans.
   2. *Paracrine and endocrine regulation*:
      a. Production of *growth factors*.
      b. Some modified smooth muscle cells of the *juxtaglomerular apparatus* of the wall of the afferent arteriole of the renal glomerulus produce *renin*, which is an important component of the *renin–angiotensin system* that controls blood pressure.

10.4 Skeletal Muscle and Visceral Striated Muscle Tissues

A. The *muscles* that form the body wall and the muscles of the limbs are made up of *skeletal muscle tissue*.

B. *Visceral striated muscle* has the same morphological characteristics.

C. In these muscles, the muscle cells or fibres are long, more or less cylindrical, multinucleated structures.

D. Skeletal muscle is derived from mesenchymal cells that develop into *myoblasts*. The multinucleated cells of skeletal muscle are formed during development by the fusion of myoblasts.

E. Muscle fibres are associated in *bundles* or *fascicles* associated with *connective tissue* (Fig. 10.3):
   1. Within each fascicle, between the muscle fibres, there is a delicate reticulum of connective tissue, the *endomysium*, which has few fibroblasts.
   2. Surrounding each fascicle is a connective tissue sheath called the *perimysium*.
   3. All the fascicles of a muscle are held together by a layer of connective tissue called the *epimysium*.
   4. Epimysium, perimysium and endomysium are in continuity with each other, so this division is artificial, although it serves to facilitate the study and description of muscle organisation.

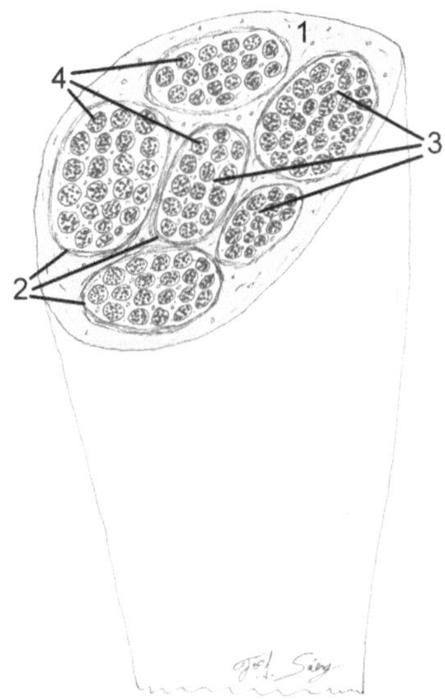

Fig. 10.3 Cross-section of skeletal muscle. (1) Epimysium; (2) perimysium; (3) endomysium; (4) skeletal muscle fibres, inside of which are myofibrils cut transversely

5. A rich network of *blood vessels*, *lymphatic vessels* and *nerves* is distributed throughout these connective tissue regions.
6. At the ends of the muscles, these connective sheaths form *tendons*, which attach the muscles to the bones.

F. Muscle fibre size:
   1. They can be as long as the muscle, but in most cases the fibres appear to be shorter. In some muscles they are less than 1 mm long, in others up to 30 cm.
   2. Muscle fibre thickness varies from 10 to 100 μm, or even more, depending on the muscle. Thinner and thicker fibres can be found within a muscle, but there is usually an average thickness.
   3. Physical activity can lead to increased muscle fibre thickness (*hypertrophy*). Lack of activity can lead to reduced thickness (*atrophy*).

G. A characteristic of skeletal muscle is that it is a *striated muscle*:
   1. A *transverse striation* is visible under the microscope. In addition, a finer *longitudinal striation* can be observed.
   2. Both striations are due to the same cellular structures, the *myofibrils*.
      a. Myofibrils are thin fibres, 1–2 μm thick, running longitudinally inside the cell.
      b. They are very numerous, arranged in parallel within the cytoplasm and each has a transverse striation.
   3. As the transverse striations of each myofibril are arranged at the same level, i.e. they coincide or are *in alignment*, the transverse striation is visible in the cell, hence the characteristic striated appearance of these muscles (Fig. 10.4).
   4. Myofibrils are visible in cross-sections as fine dots. They may be evenly distributed throughout the cell, or separated into polygonally contoured clusters called *Cohnheim fields*, which are artefacts produced by manipulation of the tissue for study.

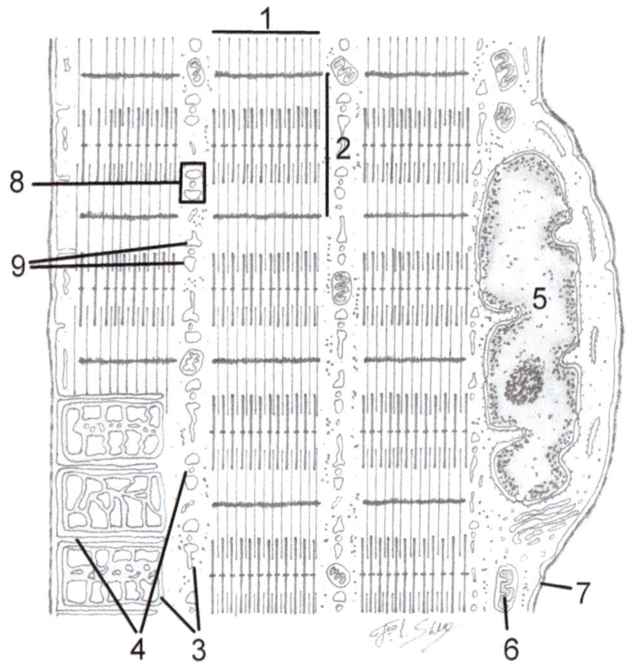

Fig. 10.4 Ultrastructure of a skeletal muscle cell. The skeletal muscle cell is cylindrical and very long, so a small part of the cell is shown in the illustration. (1) Myofibril; (2) sarcomere; (3) sarcoplasmic reticulum; (4) T tubules; (5) nucleus; (6) mitochondria; (7) external lamina; (8) triad; (9) terminal cisternae

H. When the transverse striation is observed in detail in longitudinal sections of muscle cells, there is an alternation of light and dark bands. This difference is best seen when the tissue is stained with *ferric haematoxylin*.
   1. *Light bands*: These are called *I-bands* because when viewed under a polarised light microscope, the light bands are *isotropic* and *monorefringent*.
      a. The I-bands are longer in the stretched muscle, somewhat shorter in the resting muscle and much shorter in the contracted muscle.
      b. Each I-band is crossed at the midpoint by a dark transverse line, the *Z line* (from the German zwischen, 'in-between'), which divides the I-band into two *half-bands* of identical size.
   2. *Dark bands*: They are called *A bands* because they are *anisotropic* when they are observed under polarised light. They are *birefringent*, because their refractive index changes with the plane of polarisation.
      a. The A bands have a constant length.
      b. In certain preparations of exceptional quality, a small lighter region can be seen in the centre of the A band, referred to as the *H band (Hensen band)*.
      c. In the centre of the H band there is a dark line called the *M line*.

3. Each segment of myofibril between two Z lines is called a *sarcomere*.
   a. The sarcomere is the repetitive *contractile unit* of the myofibril.
   b. It consists of two I half-bands between which is an A band, all located between two Z lines.

I. The organisation of the sarcomere is best understood when studied under the electron microscope. A simplified schematic is shown in Fig. 10.5.
   1. The length of the sarcomere at rest, i.e. the distance between two contiguous Z lines, is between 2.2 and 3 μm.
   2. *Z lines* are complex fibrillar networks that possess actin-binding proteins, such as *α-actinin* and *Cap-Z*.
   3. *Actin microfilaments* start from the Z lines and run parallel to each other towards the centre of the sarcomere. In resting muscle they do not reach the centre, as their length is 1 μm.
   4. At the centre of the sarcomere are *thick myosin filaments*, 1.5 μm long. These filaments are parallel to each other and are interspersed at their ends with actin filaments from the Z lines.
   5. The *A band* is the region occupied by the myosin filaments. Within the A band, actin filaments are sandwiched between the myosin filaments, but they do not reach the central region.
      a. *H band*: is the central region of the A band where only myosin filaments exist because actin filaments do not reach it.
      b. *M line*: formed by proteins that bind together the myosin filaments in the centre of the H-band.
   6. The *I Half-band* is the region occupied by actin filaments up to where they interca late with myosin filaments.
   7. The ends of the myosin filaments are attached to the Z lines by thin filaments of a protein called *titin*. This protein is distensible and elastic, so that lengthening and shortening of the sarcomere does not alter it.

J. The skeletal cell is multinucleated.
   1. The number of nuclei depends on the length of the cell, but in a long cell they can number up to several hundreds.
   2. They are located at the cell periphery, below the sarcolemma, which is easy to see in cross-sections.
   3. They are elongated nuclei possessing one or two nucleoli and peripheral chromatin.
   4. Sometimes, next to the muscle cell, nuclei belonging to elongated cells that are attached to the surface of the fibre are observed, occupying shallow depressions. These cells are called *satellite cells*.
      a. As the satellite cell has very little cytoplasm and is attached to the surface of the muscle fibre, it may appear under the light microscope to be another nucleus of the muscle fibre, although the appearance of these nuclei is usually somewhat different.

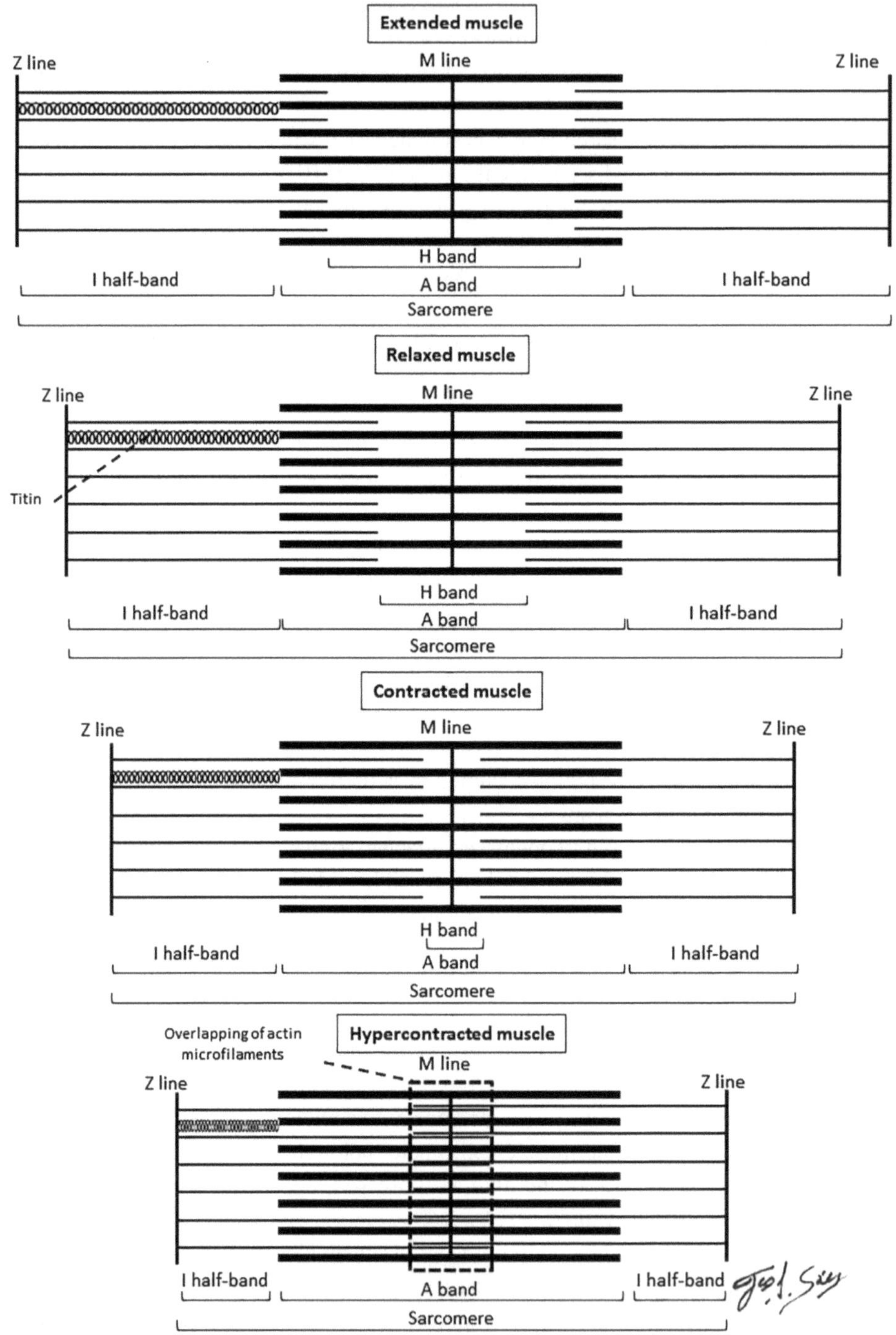

Fig. 10.5 Schematic representation of the structure of a sarcomere in different states: extension, relaxation, contraction and hypercontraction. The length of the sarcomere and of the I- and H-bands varies in each state, but the A-band always remains constant in length. Each myosin filament is attached to the Z-disc by a large extensible protein called titin, which is depicted here as a helical line. For simplicity, only one titin molecule has been shown in this diagram for each sarcomere

b. They are resting *myoblasts*, i.e. reserve cells that can be used to repair damaged muscle.

K. The skeletal muscle cell is surrounded by an *external lamina*. The satellite cells are within the same external lamina.

L. The cytoplasm or *sarcoplasm* appears brownish red in fresh tissue, due to the abundance of *myoglobin*, an oxygen-binding protein.
   1. When stained for microscopy, the cytoplasm is *eosinophilic*, due to myofibrils and abundant mitochondria.
   2. The *mitochondria* are arranged in rows between the myofibrils and, above all, in large clusters below the sarcolemma.
   3. Near the nuclei, in the cell periphery, other organelles accumulate, e.g. many small Golgi apparatuses, as well as ribosomes and some cisternae of rough endoplasmic reticulum.
   4. *Lysosomes* are present in varying numbers.
   5. There are *glycogen inclusions* throughout the cell and *lipid inclusions* between the myofibrils and next to the nuclei.

M. The *sarcoplasmic reticulum* has a complex organisation.
   1. The tubules of the sarcoplasmic reticulum are oriented longitudinally, which is why they have been called the *L system*.
   2. These tubules anastomose to each other, forming a network that inserts between the myofibrils and surrounds them in different ways depending on the region:
      a. H band: Here, the tubules form a very dense network.
      b. A band–I band boundary: Here, the tubules end in dilatations that surround the myofibrils like rings.
         (1) They are called *terminal cisternae* because they are the end of the longitudinal tubules.
         (2) Calcium accumulates in them.
      c. I band: In this zone, the tubules are arranged longitudinally, parallel to the myofibrils.

N. Special structures are observed in the plasma membrane:
   1. Ring-shaped regions at the level of the Z discs are called *costameres*. In the costameres there are many proteins that link the proteins of the cytoskeleton to those of the external lamina that envelops the cell.
   2. Tubular-shaped invaginations of the plasma membrane that extend into the cytoplasm and between the myofibrils. These structures are called *transverse tubules* or *T tubules* and are in close relation to the sarcoplasmic reticulum (Fig. 10.4):
      a. At the boundary between A band and I band, between two terminal cisternae, runs a T tubule. The two terminal cisternae and the associated T tubule form a structure called a *triad*.
      b. As the sarcoplasmic reticulum stores calcium, the junction with the T tubules allows the activation of the plasma membrane produced by the arrival of a stimulus to be transmitted very rapidly to the sarcoplasmic reticulum membrane, causing the release of calcium, which triggers cell contraction.

O. Each sarcomere is approximately 2.2–3 μm long at rest, but shortens with contraction (up to 1 μm in extreme contractions) and lengthens if the muscle is stretched (to more than 4 μm).
  1. The length of actin and myosin filaments is constant.
  2. Muscle *contraction* is caused by actin filaments *sliding* over myosin filaments. This causes shortening of the sarcomere because the actin filaments penetrate deeper between the myosin filaments.
     a. A shortening of the I half-bands and of the H band, which may become extinct, is therefore observed.
     b. The A band remains the same length.
  3. If *hypercontraction* of the muscle occurs, the actin microfilaments overlap the M line and overlap with those of the opposing I half-strand.
     a. I half-bands are reduced to a minimum.
     b. The H band is not visible.
     c. The A band remains constant.
  4. If muscle *extension occurs*, the cell and its myofibrils elongate.
     a. I half-bands and H bands increase in length.
     b. The A band remains constant in length.

P. The mechanism of contraction, as in smooth muscle, depends on the interaction of myosin heads with actin microfilaments:
  1. The process consumes *ATP*.
  2. During contraction, myosin heads perform several actions:
     a. They attach to the microfilament.
     b. They fold and 'push' the actin microfilament.
     c. They are released from the microfilament.
     d. They return to their original shape.
  3. This process is repeated for the duration of the initial stimulus.
  4. Myosin filaments are *bipolar*. The heads at one end are oriented in the opposite direction to the heads at the other end. The result is that the actin microfilaments of each I half-band slide towards the centre of the sarcomere, which shortens.
  5. The shortening of all the sarcomeres of a myofibril shortens the myofibril. The shortening of all the myofibrils of a muscle fibre shortens the fibre.

Q. Several types of skeletal muscle cells are distinguished:
  1. *Type I fibres* or *red fibres*. They are slow shrinking.
     a. They are small fibres.
     b. They have many mitochondria, oxidative enzymes and myoglobin. They usually contain lipid droplets.
     c. *Myoglobin* is an oxygen-binding protein. It is responsible for the reddish colour of fresh fibres.
     d. They are resistant, fatigue slowly and are involved in movements that do not develop great strength. They are abundant in muscles responsible for maintaining body posture.

   e. They use aerobic metabolism (*oxidative phosphorylation*).
   f. They are usually heavily irrigated.
2. *Type IIA fibres* or *intermediate fibres*. They are fast shrinking.
   a. They are medium-sized.
   b. They have many mitochondria and myoglobin.
   c. They have abundant glycogen.
   d. They contract rapidly and with resistance, generating a peak of muscle tension.
   e. They are resistant to fatigue.
   f. They can use anaerobic metabolism to produce ATP (*glycolysis* and *lactic fermentation*).
3. *Type IIB fibres* or *white fibres*. They are of rapid and discontinuous contraction.
   a. They are large fibres.
   b. They have few mitochondria and myoglobin.
   c. They have a lot of glycogen and abundant *glycolytic enzymes*.
   d. They can develop a very strong contraction for a shorter time.
   e. They fatigue quickly due to the accumulation of *lactic acid*.
   f. Aerobic metabolism cannot meet energy needs and they use *anaerobic metabolism*.
   g. They are less irrigated.

R. In each muscle there is a mixture of red, intermediate and white fibres in an adequate proportion to perform their function, but this proportion may vary with prolonged forced exercise.

S. Excitation of skeletal myocytes is produced by nerve stimuli.
1. Motor nerve fibre endings branch out and contact muscle cells in areas called *motor end plates (neuromuscular junctions)*.
   a. Each myocyte has a single motor plate in its medial region.
   b. These motor plates look like synapses[2] between fibres in the nervous system.
2. At the motor end plates, nerve fibres release a neurotransmitter, *acetylcholine*, which stimulates receptors in the sarcolemma of the muscle cell.
3. Neurons that innervate skeletal muscle are called *motor neurons*.
   a. A motor neuron can innervate a few muscle fibres or many muscle fibres. The motor neuron and the fibres it innervates form a *motor unit*.
   b. All muscle fibres of a motor unit are white, red or intermediate.
   c. The muscle fibres of a motor unit are distributed throughout the muscle.
4. Loss of innervation leads to *atrophy* of innervated muscle fibres.

T. Skeletal myocytes are produced during development:
1. Fusion of *myoblasts* occurs to form elongated multinucleated cells called *myotubes*, which eventually become myocytes.
2. Some myoblasts remain in adult tissue as *satellite cells*.
3. Differentiated myocytes cannot divide by mitosis, even during foetal life.

---

[2] Synapses will be discussed in the chapter on nerve tissue.

U. Muscle growth is produced by *hypertrophy*, i.e. an increase in the thickness of muscle fibres.

V. Damaged muscle tissue can regenerate:
   1. By differentiation of *satellite cells* into myoblasts. These divide, fuse together and produce new muscle fibres.
   2. If the loss of muscle fibres in an adult muscle is large, some of them may be replaced by connective tissue.

10.5 Cardiac Muscle Tissue (Cardiac Muscle)

A. *Cardiac muscle tissue* (from the Greek καρδιο- kardio-, 'heart') is another variety of *striated muscle* that differs from skeletal muscle in several respects:
   1. The cells are cylindrical and elongated but *uninucleate*, and are joined together at their ends by *intercalated discs*.
   2. The joining of cells at their ends produces multicellular *cardiac fibres*.
   3. The fibres branch out, connect at their ends to other fibres and form a three-dimensional network.
   4. The nuclei are not at the cell periphery but are located in the central area.
   5. The contraction is involuntary and rhythmic.

B. The muscle cells of the heart are organised in *layers* or *laminae*:
   1. Between the laminae are thin regions of connective tissue with *blood vessels* and *nerves*.
   2. Thin *capillaries* emerge from the blood vessels and penetrate the sheets of muscle tissue, forming a network that surrounds the muscle cells.

C. The muscle cell (Fig. 10.6) is called a *cardiomyocyte*.
   1. It is a large cell, 50–120 μm long and 15–20 μm in diameter, usually *mononucleated*, elongated, cylindrical in shape, but may be *branched*, bifurcated, and attached to other cardiac cells at their ends.
   2. The *nucleus*, rarely two nuclei, and many organelles are centred. The nucleus is elongated, euchromatic, with one or two nucleoli.
   3. In the cytoplasm near the nuclear tips there are *glycogen granules*, *lipid inclusions* and *lipofuscin* granules that increase in number with age.
   4. The rest of the cytoplasm has *myofibrils* similar to those of skeletal muscle and responsible for the transverse striation of these cells.
      a. These myofibrils are not uniform and cylindrical, but form thick, irregularly shaped structures that *anastomose* with adjacent myofibrils.
      b. The organisation of myofibril *sarcomeres* is identical to that of skeletal muscle, i.e. with Z and M lines and I, A and H bands.
   5. Interspersed between the myofibrils:

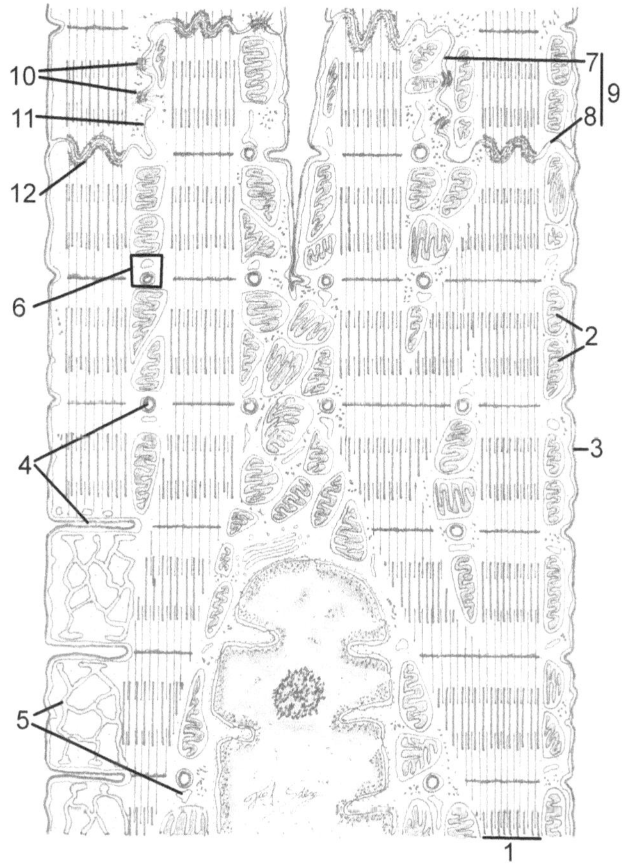

Fig. 10.6 Ultrastructure of a cardiac muscle cell. The illustration shows the central portion of the cell, which contains the nucleus, and shows that the cell branches. (1) Myofibril; (2) mitochondria; (3) external lamina; (4) T tubules with external lamina; (5) sarcoplasmic reticulum; (6) dyad; (7) longitudinal or lateral segment or component; (8) transverse segment or component; (9) intercalated disc; (10) desmosomes; (11) gap junction; (12) fascia adherens

   a. Numerous *mitochondria*, very large and with many ridges inside. They occupy almost half of the cell volume.[3]
   b. *Glycogen particles*: Glycogen is the main source of organic molecules for ATP synthesis.
   c. *Lipid inclusions*: Triglycerides also provide a lot of energy in the basal state.
6. *Myoglobin*, which is a protein that stores molecular oxygen, is very abundant.
7. The *T tubules* of these cells are very wide, are lined inside by the *external lamina* and are located at the level of the Z lines.
8. The *sarcoplasmic reticulum* is less developed than in skeletal cells.

---

[3] In skeletal muscle, mitochondria occupy only 2% of the cell volume.

a. It forms a system of somewhat anastomosed tubules that follow the sarcomere filaments.
b. Typical terminal cisternae do not form, but T tubules are associated with reticulum cisternae, usually one at a time, forming *dyads*.

D. The cells are joined at their ends by *intercalated discs*.
1. At these sites, the membranes of the two neighbouring cells come very close together, leaving a gap of just 15–20 nm.
2. In longitudinal sections of cardiac fibres, the intercalated discs are shown as straight or step-like lines, so that different segments can be distinguished:
a. *Transverse segments* or *components*: Arranged perpendicular to the major axis of the cardiac fibre and therefore perpendicular to the myofibrils. Two types of intercellular junctions can be distinguished:
(1) *Fascia*[4] *adherens*: They have the appearance of large desmosomes, as wide as the entire thickness of the myofibril that is anchored here.
(a) The cells are bound together by *cadherins*.
(b) To the inner side of the membrane of these fasciae are attached the thin filaments of the semi I bands of the terminal sarcomeres of the myofibrils, as if there were a Z disc here.
(2) Typical *desmosomes*, to the inner side of which *desmin filaments* are anchored. They increase the cohesion between cells.
b. *Longitudinal* or *lateral segments* or *components*: Arranged parallel to the major axis of the fibre and myofibrils. In these regions are observed:
(1) *Gap junctions*: They allow muscle cells to become electrically coupled, which facilitates the transmission of the electrical impulses responsible for contraction.
(2) *Desmosomes*.

E. The cardiac muscle cells of the walls of the atria contain some secretory vesicles, measuring 0.4 µm, called *atrial granules*.
1. They store and secrete *atrial natriuretic peptide (ANP)* and *brain natriuretic factor (ANF)*.
2. These peptides act on the kidneys to stimulate the elimination of sodium and water, which helps to lower blood pressure.
3. The cells in which these atrial granules are abundant are often referred to as *myoendocrine cells*.

F. Regulation of contraction in the cardiac muscle:
1. Cardiac muscle cells can contract spontaneously and rhythmically. To control and coordinate this ability, specialised cells in the muscle tissue itself produce and propagate an electrical impulse.

---

[4] A *fascia* (from the scientific Latin *fascia, 'band that holds', 'girdle', 'bandage'*) is an area of cell–cell junction with a relatively large surface area. When a junction between cells has a small, point-focused surface area, as in typical desmosomes, it is called a *macula* (Latin *macŭla, 'spot')*.

   2. The *autonomic nervous system* also plays a role in the control of rhythmic cardiac contraction. The action of the nerve endings modifies the rhythm of the contraction, but does not cause it.

G. Origin of the cardiac muscle:
   1. Cardiac muscle originates from a specific region of mesenchyme, the *myoepicardial mantle*.
   2. Cardiac muscle cells continue to divide until birth.
      a. After birth, cardiomyocytes stop dividing.
      b. Some retain some ability to divide by mitosis, so that a 50-year-old person has renewed 45% of his or her cardiac muscle cells since birth.
      c. Occasionally, the nuclei divide but the cells do not, and *binucleated cells* are formed, which may constitute 25–57% of the total cardiomyocytes.
      d. There is often chromosome multiplication without nuclear division, so that in the adult most cardiac muscle cells are *polyploid* (chromosome number 4n or 8n).
   3. *Cardiac muscle progenitor cells* have been identified in adults, which have a limited capacity to regenerate cardiac muscle cells and other cell types.
   4. After birth, the heart grows by increasing the size of its cells.

H. In case of death (*necrosis*) of cardiac tissue, e.g. due to lack of blood supply (*ischaemia*), the cardiac tissue rarely regenerates and the dead tissue is replaced by connective tissue.

Take-Home Message

Muscle tissues have cells (muscle fibres or myocytes) capable of contracting. In all muscle tissues, the cells are surrounded by a basal lamina (external lamina). Smooth muscle cells are spindle-shaped and uninucleated. Their cytoplasm is crossed by intertwined bundles of actin and myosin filaments, which slide over each other and contract the cell. This contraction is involuntary and is regulated by hormones and the nervous system. Skeletal muscle fibres are cylindrical, elongated, multinucleated and have light (I bands) and dark (A bands) transverse bands. Their cytoplasm has myofibrils arranged longitudinally and formed by successive sarcomeres, which are structures consisting of thin actin microfilaments intertwined with thick myosin filaments. The distribution of these filaments produces the I and A bands. The sliding of the filaments against each other shortens the sarcomere, and consequently the myofibril and the entire muscle cell. Contraction is voluntary in response to nerve signals. Cardiac muscle cells are elongated, sometimes branched, uninucleated, and contact each other at their ends in structures called intercalated discs. They have myofibrils with sarcomeres similar to those of skeletal muscle. Their contraction is involuntary and rhythmic.

Further Reading

Agarkova I, Ehler E, Lange S et al (2003) M-band: a safeguard for sarcomere stability? J Muscle Res Cell Motil 24:191–203
Au Y (2004) The muscle ultrastructure: a structural perspective of the sarcomere. Cell Mol Life Sci 61:3016–3033
Gartner LP (2017) Muscle. In: Textbook of histology, 4th edn, pp 179–210
Gauthier GF (1988) The muscular tissues. In: Weiss L (ed) Cell and tissue biology – a textbook of histology, 6th edn. Urban & Schwarzenberg, pp 255–276
Gómez de Ferraris ME, Campos Muñoz A (2009) Histología general humana. In: Histología, Embriología e Ingeniería Tisular Bucodental, 3rd edn. Médica Panamericana, pp 57–78
Kierszenbaum AL, Tres LL (2019) Muscle tissue. In: Histology and cell biology – an introduction to pathology, 5th edn. Elsevier, pp 237–260
Krstić RV (1985) Muscular tissue. In: General histology of the mammal. An atlas for students of medicine and biology. Springer Verlag, pp 237–291
Lin BL, Song T, Sadayappan S (2017) Myofilaments: movers and rulers of the Sarcomere. Compr Physiol 7:675–692
Lowe JS, Anderson PG (2015) Contractile cells. In: Stevens & Lowe's human histology, 4th edn, pp 71–83
Luther PK (2009) The vertebrate muscle Z-disc: sarcomere anchor for structure and signaling. J Muscle Res Cell Motil 30:171–185
O'Dowd G, Bell S, Wright S (2023a) Appendix 1: introduction to microscopy. In: Wheater's functional histology. Elsevier, pp 432–436
O'Dowd G, Bell S, Wright S (2023b) Appendix 2: notes on staining techniques. In: Wheater's functional histology. Elsevier, pp 437–444
O'Dowd G, Bell S, Wright S (2023c) Muscle. In: Wheater's functional histology. Elsevier, pp 101–121
Pawlina W (2024) Muscle tissue. In: Histology a text and atlas with correlated cell and molecular biology, 9th edn. Wolters Kluwer, pp 344–387
Pettinato AM, Ladha FA, Hinson JT (2022) The cardiac sarcomere and cell cycle. Curr Cardiol Rep 24:623–630
Rassier DE (2017) Sarcomere mechanics in striated muscles: from molecules to sarcomeres to cells. Am J Physiol Cell Physiol 313:C134–C145
Schultz E, McCormick KM (1994) Skeletal-muscle satellite cells. Rev Physiol Biochem Pharmacol 123:213–257
Soussiyanicostas N (1991) Muscular ontogeny – From mesodermal induction to sarcomere formation. Bulletin de l'institute Pasteur 89:255–295
Squirre JM (1997) Architecture and function in the muscle sarcomere. Curr Opin Struct Biol 7:247–257

Nerve Tissue

11

What You Will Learn In This Chapter

This chapter describes the major characteristics of nerve tissue and the nervous system and their embryonic origin. It explains the cytological organisation and properties of neurons, the excitable cells responsible for the conduct of nerve impulses and how this impulse is produced. The mechanisms of impulse transmission between neurons at synapses are described. In addition, the different cell types of neuroglia that accompany neurons in nerve tissues and the characteristics of nerve fibres are studied. Finally, structures of the nervous system that can be found in the walls of the oral cavity and in the tissues of the tooth and periodontium, i.e. peripheral nerves and motor and sensory nerve endings, free and encapsulated, are described.

11.1 General Characteristics

A. Nerve tissue forms the *nervous system*, which is responsible for:
   1. *Communication* between the different tissues and organs of the body.
   2. The reception of signals from the external environment that can be transformed into *sensations*.
   3. The body's *reaction to* external stimuli.
   4. *Consciousness*
   5. *Behaviour*
   6. *Personality*

B. The functioning of the nervous system is based on the ability of nerve cells to produce and transmit signals, which is achieved because these cells possess two properties:
   1. *Irritability*, which is the ability to react to physical and chemical stimuli.
   2. *Conductivity*, which is the ability to transmit this reaction to other areas of the cell or to other cells.

C. The signal produced by nerve cells is an electrical current that is transmitted across the surface of the cell. This current is called a *nerve impulse*.

F. J. Sáez, I. Badiola, *Essential Oral Histology*, Learning Materials in Biosciences,
https://doi.org/10.1007/978-3-032-03165-5_11

D. Nerve cells have long cytoplasmic extensions through which they conduct the nerve impulse.
   1. The impulse produced at one point in the organism can be transmitted by these extensions to other cells or tissues.
   2. The length of these extensions can be relatively short or very long, and as a result the nerve impulse can be transmitted within the nervous system or to other organs (*effector organs*).
E. The nerve tissue is made up of several cell types that can be classified into two categories:
   1. *Neurons* (from the Greek νεῦρον neûron, 'nerve'): They are the essential cells of the nerve tissue, as they are responsible for carrying out the functions of the nervous system.
      a. They have a *soma* or *cell body*, which contains the nucleus and most of the organelles.
      b. They have several types of cytoplasmic prolongations that start from the soma. The prolongations responsible for transmitting the nerve impulse are called *nerve fibres*.
      c. The extensions of one neuron contact those of other neurons and form *neural circuits* that can be very complex and combine with other circuits to perform a function.
   2. *Neuroglial cells*, *neuroglia* or *glia cells* (from the Byzantine Greek γλία glia, 'glue'): These are various types of cells associated with neurons that perform, depending on their characteristics, protective, insulating, nourishing, defensive or barrier functions.
F. Signal transmission between neurons takes place at specialised cell junctions called *synapses* (from the Greek σύναψις synapsis, 'junction, link').
G. Some nerve cells can secrete regulatory substances, hormones or local chemical mediators, and are called *neuroendocrine cells*.

11.2 Organisation of the Nervous System

A. Several parts of the nervous system can be distinguished:
   1. The *central nervous system (CNS)* receives nerve impulses from all over the body, integrates them and elaborates responses.
      a. It consists of several organs, including the *cerebrum*, *cerebellum*, *brainstem* and *spinal cord*.
      b. The cerebrum, cerebellum and brainstem form the *brain*.
      c. In the organs of the central nervous system, two regions with different organisation are usually distinguished:
         (1) The *grey matter*, where the neuronal bodies or somas are found, surrounded by their cytoplasmic prolongations and other glial cells. The set of inter-

twined structures surrounding the neuronal somas form the *neuropil* (from the Greek νεῦρον nêuron, 'nerve' and πῖλος pîlos, 'felt').

(2) The *white matter*, which lacks neuronal somas and consists only of nerve fibres and glial cells.

d. The central nervous system possesses connective tissue envelopes that protect its organs (*meninges*).

e. There are also small amounts of connective tissue that surround the *blood vessels* that enter these organs.

2. The *peripheral nervous system (PNS)* comprises all the nerve tissue that is not located in the central nervous system and brings it into contact with all regions of our body. It is made up of:

a. *Nerve plexuses*: networks of nerve fibres.

b. *Peripheral nerves*: bundles of nerve fibres.

(1) Nerves originate from the brain (*cranial nerves*) and the spinal cord (*spinal nerves*).

(2) Nerves branch away from the central nervous system. When they reach their destination, they present free endings or form specialised terminal structures.

(3) Nerves often carry a mixture of different types of fibres:

(a) *Afferent nerve fibres* (from Latin afferente, 'leading to'), which carry impulses from the periphery to the central nervous system.

(b) *Efferent nerve fibres* (from Latin efferente, 'leading outwards'), which carry impulses from the central nervous system to the peripheral organs.

c. *Ganglia*: contain nerve somas. Two types can be distinguished:

(1) *Sensory ganglia*: Located in some cranial nerves and in the dorsal root of all spinal nerves.

(a) Inside are the somas of *pseudounipolar neurons*.

(b) These neurons have a cytoplasmic extension bifurcated into two branches:

1. One reaches the peripheral organs, through which the neuron receives stimuli.

2. The other is directed to the central nervous system, to which the received signal is transmitted.

(2) *Autonomic ganglia*: There are several types located in different parts of the body.

(a) They contain neuronal somas of the *vegetative nervous system*.

(b) These neurons receive nerve fibre impulses from neurons in the central nervous system and emit fibres that transmit the nerve impulse to the effector organs.

B. The distinction between the central nervous system and the peripheral nervous system is anatomical, but both can be involved in two types of functions, allowing us to consider two functional parts of the nervous system:

1. *Somatic nervous system* (from Greek σωματικός *sōmatikós* 'pertaining to the body') or voluntary nervous system.
   a. It comprises all the parts of the nervous system responsible for controlling our body's relationships with the environment.
   b. The efferent fibres of the somatic system carry the nerve impulse directly from the central nervous system to the effector organ. That is, the neuronal soma that produces the axon is located in the grey matter of the brain or spinal cord.
   c. Afferent (sensory) fibres are pseudounipolar neurons.
2. *Autonomic* or *vegetative nervous system*[1] (from Latin *vegetāre* 'to vivify', 'to be alive'), which controls visceral activity.
   a. It has two types of neurons:
      (1) *Efferent neurons (visceral motor neurons)*: Conducts involuntary impulses to smooth muscle, cardiac muscle and glandular epithelia. They form the following circuit:
         (a) Neurons (*presynaptic neurons*) located in the brain or spinal cord, which emit efferent fibres (*preganglionic fibres*) leaving the central nervous system.
         (b) Autonomic ganglion neurons (*postsynaptic neurons)*. They receive the signal from preganglionic fibres.
         (c) *Postganglionic fibres*, which are the axons that arise from the postganglionic neurons and transmit the impulse to the effector organ (smooth muscle, cardiac muscle and glands).
      (2) *Afferent neurons (visceral sensory neurons)*: They transmit pain and reflexes from the viscera to the central nervous system. They are pseudounipolar neurons.
   b. The autonomic nervous system has three divisions:
      (1) *Sympathetic nervous system* (from the Greek σύν sýn 'with', 'union' and πάθος patho 'suffering', 'feeling'): Prepares the body for 'fight or escape'. It causes increased breathing, blood pressure, heart rate and blood flow to the muscles. It also dilates the pupils and slows down visceral functions.
      (2) *Parasympathetic nervous system* (from Greek παρά pará 'along', 'beside'): It carries out *homeostasis*. Generally produces antagonistic effects to the sympathetic system.
      (3) *Enteric system* (from the Greek ἔντερον *énteron* 'intestine'): Innervates the digestive tract. It communicates with the central nervous system via sympathetic and parasympathetic fibres, but also functions independently.

C. The somatic and vegetative nervous systems have neuronal somas and fibres in both the central and peripheral nervous systems.

[1] It is often mistakenly referred to as the autonomous nervous system.

11.3 Embryonic Origin

A. In embryogenesis, the nervous system originates during *neurulation* when the *neuroectoderm* is produced.
B. The neuroectoderm forms two embryonic structures:
    1. The *neural tube*, from which the components of the brain and spinal cord are produced, including neurons, neuroglia (except microglia), ependymus and choroid plexus.
    2. The *neural crests*, whose cells migrate and give rise to various nerve structures:
        a. Almost the entire sensory peripheral nervous system.
        b. Sensory neurons of cranial and spinal ganglia.
        c. Autonomic ganglia and postganglionic neurons.
        d. Cells of arachnoid and pia mater (meninges).
        e. Peripheral ganglion satellite cells.
        f. Schwann cells.

11.4 Neuron

A. The neuron is the fundamental cell of the nervous system (Fig. 11.1). It can be distinguished:
    1. A *soma* or *cell body* consisting of the nucleus and the surrounding cytoplasm. This cytoplasm is called a *perikaryon*.
    2. A variable number of short, usually branched extensions called *dendrites* (from the Greek δένδρον déndron, 'tree').
    3. A single and possibly very long extension, called a *axon* (from the Greek ἄξων áxōn, 'axis'). The axon may have some branching (*axon collaterals*), and at the end may show fine branching (*terminal branching*[2]).
B. A neuron can be thought of as having two different 'compartments':
    1. Dendrites and soma are functionally and structurally similar, so that they can be considered to form a single compartment.
    2. The axon, because of its function and organisation, can be considered a distinct unit.
C. Neurons do not divide, but in some areas of the brain there are *neural stem cells* that, if injury occurs, can divide, migrate to the site of damage and differentiate to replace damaged neurons.
D. Transmission of the nerve impulse:
    1. Dendrites pick up signals from other neurons or sensory cells that trigger an *action potential* in their membrane, which is transmitted to the soma as a *nerve impulse*.
    2. The soma can also receive stimuli.

---

[2] In long axons, these terminal branches have also been called telodendria (singular, telodendron).

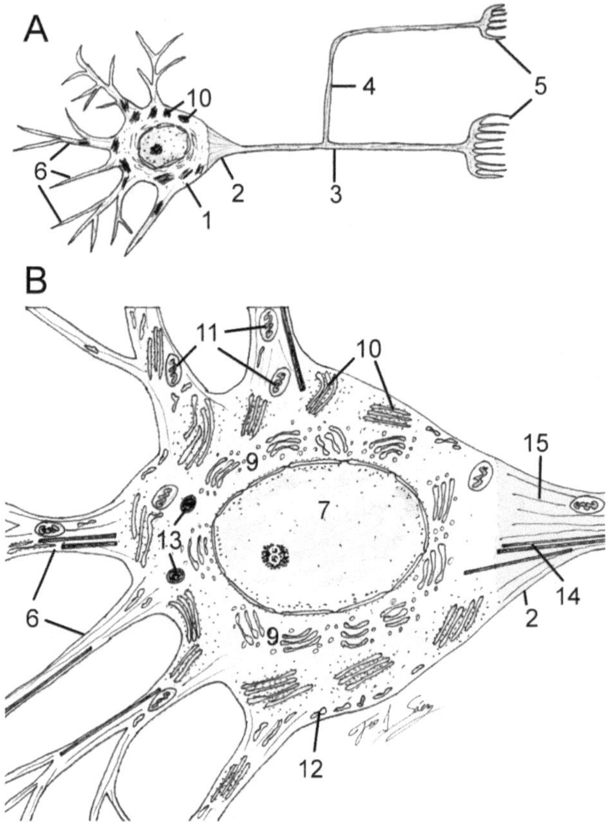

Fig. 11.1 Neurons can have many different shapes, but here is a hypothetical neuron (**A**), and details of the organisation of its soma (**B**). (1) Neuronal soma; (2) axon hillock; (3) axon; (4) axon collateral; (5) axon terminals; (6) dendrites; (7) nucleus; (8) nucleolus; (9) Golgi apparatus; (10) rough endoplasmic reticulum (Nissl bodies); (11) mitochondria; (12) smooth endoplasmic reticulum; (13) lipofuscin granules; (14) microtubules; (15) neurofilaments

3. All stimuli are integrated into the soma and, if a threshold is exceeded, a nerve impulse is emitted and transmitted to the end of the axon or *axon terminal.*

E. Neuronal soma:
   1. The neuronal soma can have a variable shape: rounded, ovoid, pyramidal, etc.
   2. The size varies greatly when comparing different neuronal types: there are 8 μm neurons (granule neurons in the cerebellum) and 100 μm neurons (giant pyramidal neurons in the cerebral cortex).
   3. The nucleus is usually large and *euchromatic* and usually shows a large *nucleolus*. It may be rounded or show indentations.
   4. In the perikaryon, the following are observed under the light microscope:
      a. Basophilic cytoplasmic regions, called *Nissl bodies* or *chromophilic substance*, which are regions with abundant cisternae of the rough endoplasmic reticulum.
         (1) This indicates that the neuron has intense synthetic activity.
         (2) Nissl bodies are also at the starting points of dendrites, but not at the point of origin of the axon, the so-called *axon hillock*.
      b. A lot of smooth endoplasmic reticulum, which is also found in dendrites and axons. This smooth endoplasmic reticulum often forms *hypolemmal cisternae*, i.e. located beneath the plasma membrane.

c. Many *Golgi apparatus*, also at the origin of dendrites.
d. Abundant *mitochondria*, also seen in dendrites and axons.
e. Abundant lysosomes, which turn into *lipofuscin granules*, increase with ageing and give some neurons a yellowish-brown colour.
f. *Neurofibrils*, consisting of *neurofilaments*, a variety of intermediate filaments, which are also visible in dendrites and axons.
g. *Microtubules*.
h. *Actin microfilaments* associated with the plasma membrane.

F. Dendrites:
1. They are extensions that can branch out into *dendritic trees*.
2. They increase the receptor surface area of neurons, i.e. they increase the possibility of establishing synapses.
3. A neuron can have many dendrites.
4. Most of the cytoplasmic organelles, including small Golgi outposts, are visible. The Golgi outposts are discrete structures not connected with the Golgi stacks of the soma.
5. The surface of the dendrites may be enlarged by the formation of 1 μm long, mushroom-shaped evaginations or outgrowths of the surface, which are called *dendritic spines*:
   a. They are usually sites of synapses.
   b. They have *actin microfilaments* that stabilise them, but also allow them to change shape.
6. An exception to the structure of dendrites is the 'dendrite' of *pseudounipolar neurons*, which is very long, its structure is typical of an axon, it has a myelin sheath, like axons, and behaves like an axon, and is therefore called a *dendritic axon* or *peripheral dendritic branch*.

G. Axon:
1. Each neuron can have only one axon.
2. The axon is specialised to transmit the action potential that is produced at its point of origin in the perikaryon, the *axon hillock*:
   a. The axon hillock has no Nissl bodies.
   b. The boundary between the cytoplasm of the perikaryon (with Nissl bodies) and the axon hillock (without Nissl bodies) can be abrupt.
3. The axon is covered by the plasma membrane (*axolemma*).
4. Within the axon, or *axoplasm*, thin tubules of smooth endoplasmic reticulum, thin mitochondria, *microtubules* and *neurofilaments* can be found.
5. Some axons produce right-angled *axon collaterals*.
6. In its final portion it may branch out and form many small branches (*terminal branching* or *telodendria*).
7. There are dilatations (*synaptic buttons* or *synaptic boutons*) that establish synapses with other cells. There are two types:

   a. *Bouton terminal* (French, terminal button), at the *axon terminals*. This occurs in most cases.
   b. *Boutons en passant* (French, button in passing), in the axon pathway, although this is less common.
8. Some axons are coated with a sheath produced by glia cells, called the *myelin sheath* (from the Greek μυελός *myelós* 'medulla' and *-ina* 'substance').
   a. Axons with a myelin sheath are *myelinated axons* or *fibres*.
   b. The remaining axons that lack this sheath are called *unmyelinated axons* or *fibres*.
9. The axon *initial segment* is the region between the axon hillock and the beginning of the myelin sheath:
   a. It has an actin filament filter that controls the passage of organelles and vesicles from the perikaryon to the axon.
   b. It is the point where the action potential is produced and transmitted down the axon.
10. Many substances are transported along the axon to the terminal (*anterograde transport*) or to the soma (*retrograde transport*). This transport is guided by microtubules.
11. In some neurons, signal molecules are transported and released at the axon terminal (*neurosecretion*).

H. Some differences between dendrites and axons are:
   1. A neuron may have many dendrites, but only one axon.
   2. Dendrites are shorter than axons.
   3. Dendrites are larger in diameter at the point of origin in the cell soma.
   4. Dendrites always lack a myelin sheath.
   5. Dendrites may be highly branched. Axons may have few axonal collaterals and some terminal branching.
   6. Axons lack Nissl bodies.

11.5 Types of Neurons

A. There are many different neuronal types, each in a particular region of the nervous system and specialised in a particular activity.

B. Despite the variety, neurons can be classified according to the branching of their cytoplasm (Fig. 11.2):
   1. *Anaxonic neuron*: A neuron without an axon. They are very rare. An example is the amacrine neuron in the retina, which has dendrites but no axon.
   2. *Unipolar* or *monopolar neuron*: It has a single extension, the axon. Not found in humans.
   3. *Pseudounipolar neuron*: These cells have a single extension which, a short distance from the soma, bifurcates into a T-shape.

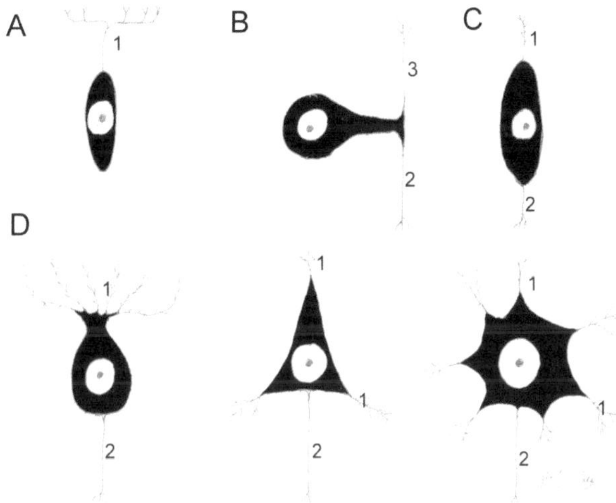

Fig. 11.2 Anaxonic (**A**), pseudounipolar (**B**), bipolar (**C**) and multipolar (**D**) neurons. The neuronal soma is shaded in black, with the nucleus in white. (1) Dendrites; (2) axons; (3) dendritic axon (peripheral dendritic branch)

a. Both branches have axon characteristics.
b. This neuron is formed from a neuron with an axon and a dendrite. During development, the two cytoplasmic extensions approach each other and fuse in their initial paths, until they acquire their final form (Fig. 11.3).
   (1) The dendritic extension then acquires the properties of an axon (*dendritic axon*). The distal end of this dendritic axon branches and shows typical small dendritic extensions that behave like dendrites: they receive signals and propagate the nerve impulse to the soma and axon.
   (2) The initial axon extension continues to behave like a typical axon.
c. Sensory ganglion cells are of this type. Their dendrite or dendritic axon is directed towards the periphery (*peripheral dendritic branch*). The axon is directed towards the central nervous system (*central axonal branch*).

4. *Bipolar neuron*: It has two extensions, the axon and a dendrite, usually originating from opposite poles of the perikaryon, which is usually fusiform. There are neurons of this type in the retina and the olfactory epithelium, for example.
5. *Multipolar neuron*: It has many dendrites, originating from various points in the perikaryon, and a single axon. It is the most frequent neuronal type in humans and there is a great variety of neurons of this type with very different organisation.

11.6 Membrane Potential

A. In the plasma membrane of neurons, there are channels and pumps responsible for transporting ions into or out of the cell (Fig. 11.4).
B. In an unstimulated neuron, there is a potential difference of −40 to −90 mV across the membrane, with the inside negative relative to the outside. This is called the *resting membrane potential*.

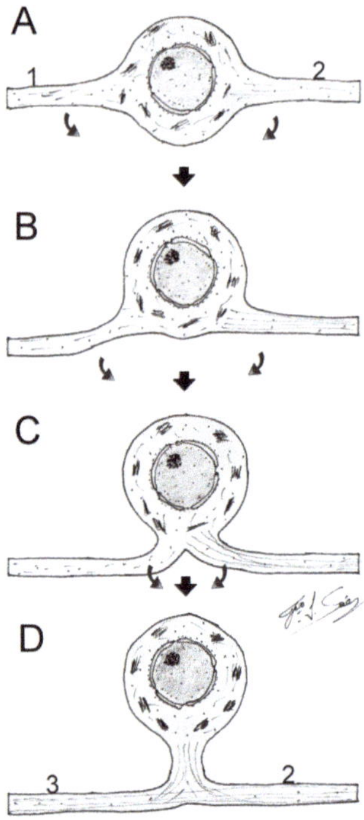

Fig. 11.3 Simplified representation of the development of a pseudounipolar neuron. The progression from one stage to another is indicated by the arrows. The point of origin of the dendrite (1) and axon (2) of a bipolar neuron (**A**) moves towards the same pole of the neuron in the direction shown by the curved arrows. The point of origin of the two extensions approaches each other (**B**, **C**) and finally converges, leading to the fusion of the two extensions at their initial part (**D**). The dendrite takes on axon-like characteristics and becomes a dendritic axon (3)

C. When a point on the surface of a neuron is stimulated, channels and pumps transport Na^+ into the cell and *depolarisation* occurs.
   1. The potential is reversed up to +30 mV, as the inside becomes positive with respect to the outside.
   2. This is called an *action potential.*
D. This situation is quickly reversed by the outflow of K^+ ions and returns to the resting situation in just a few milliseconds.
E. The depolarisation in the stimulated area propagates through the neuronal membrane as a *nerve impulse*.
F. All the signals received by a neuron are integrated into the soma, which ultimately transmits or fails to transmit the action potential to the axon.
G. When the action potential reaches the axon terminals, the stimulus is transmitted to other neurons or effector cells via *synapses*.

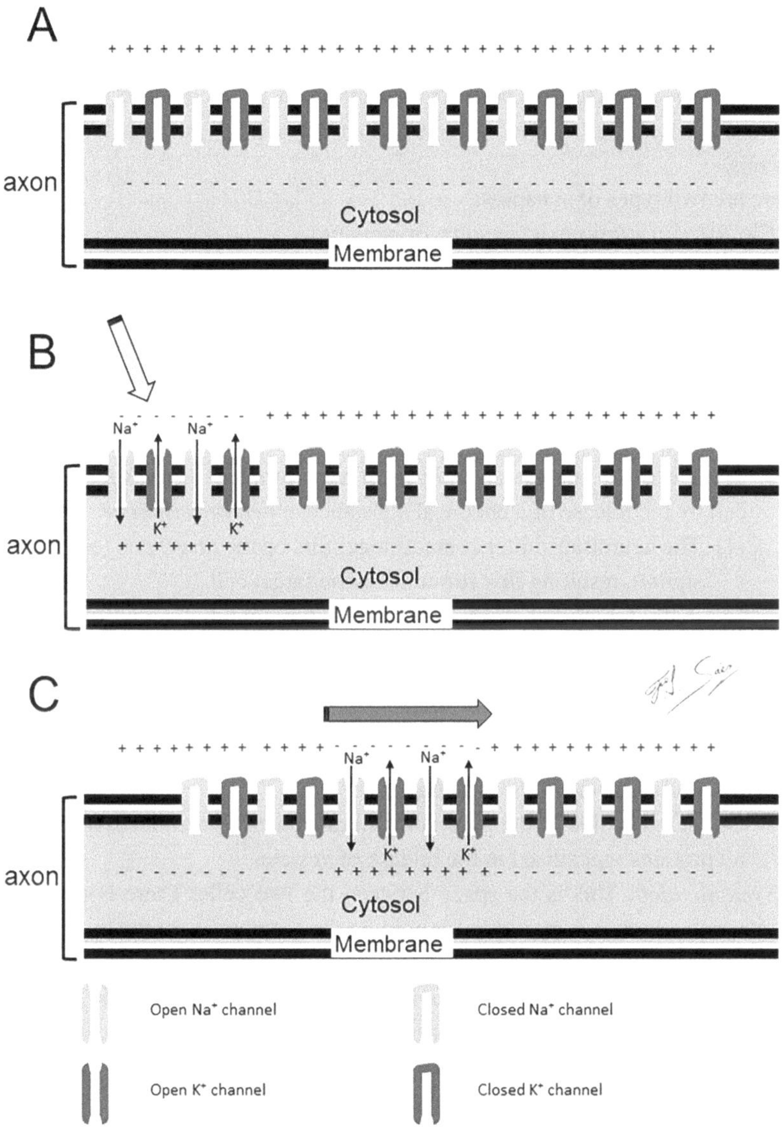

Fig. 11.4 Action potential propagation in the membrane of a neuron. In the resting state, the membrane has a negative membrane potential on the inside relative to the outside (**A**). When a stimulus (**B**), represented here as a light grey arrow, arrives, ion channels open, allowing sodium to enter and reversing the polarity. The situation recovers quickly due to the potassium outflow. This stimulus propagates as an action potential (**C**) in the direction shown by the dark grey arrow, due to the successive opening and closing of membrane channels

11.7 Synapses

A. A *synapse* is a specialised junction that allows a nerve impulse to be transmitted between two neurons or from one neuron to another cell type (muscle cells or glandular cells).
B. There are two types of synapses:
   1. *Electrical synapses* exist in some organisms:
      a. They are *gap junctions.*
      b. They allow *bidirectional transmission* of an electrical impulse.
      c. In humans, they appear to be infrequent and of relatively little importance.
   2. The most common type of synapse in humans is the *chemical synapse.*
      a. In these connections, the impulse is always transmitted in the same direction (*unidirectional transmission*).
      b. The electrical impulse arriving at an axon terminal is transmitted to the other cell by the release of a chemical substance, a *neurotransmitter.*
         (1) The neurotransmitter contacts receptors on the *target cell* (cell receiving the signal), resulting in a stimulus on the target cell.
         (2) If the target cell is another neuron, the stimulus is converted into a new electrical signal.
C. A typical synapse between two neurons consists of (Fig. 11.5A, B):
   1. *Presynaptic component*: This is the button region of the axon terminal.
      a. It has mitochondria and many vesicles 40–60 nm in diameter that carry neurotransmitters (*synaptic vesicles*).
      b. On the inner side of the plasma membrane, there is a *condensed region* formed by proteins specialised in the release of vesicles.
   2. *Synaptic cleft*: This is the space between the two cells. There is usually a dense material, which is responsible for holding the membranes of the two cells together.
   3. *Postsynaptic component*: This is the region of the neuron that receives the signal, in the dendrite, the perikaryon or sometimes in the axon. Beneath the membrane is a *condensed region.*
D. There are synapses that show some differences in their structure. Therefore, two types of synapses are usually distinguished (Fig. 11.5A, B):
   1. *Type I*: They have the following characteristics:
      a. Synaptic vesicles are spherical.
      b. Dense material is present in the synaptic cleft.
      c. The postsynaptic membrane is very thickened.
      d. In microscopic images the synapse is *asymmetrical.*
   2. *Type II*: Its characteristics are:
      a. Synaptic vesicles are oval.
      b. There is no dense material in the synaptic cleft.
      c. The postsynaptic membrane is not very thickened.
      d. Under the microscope they look *symmetrical.*

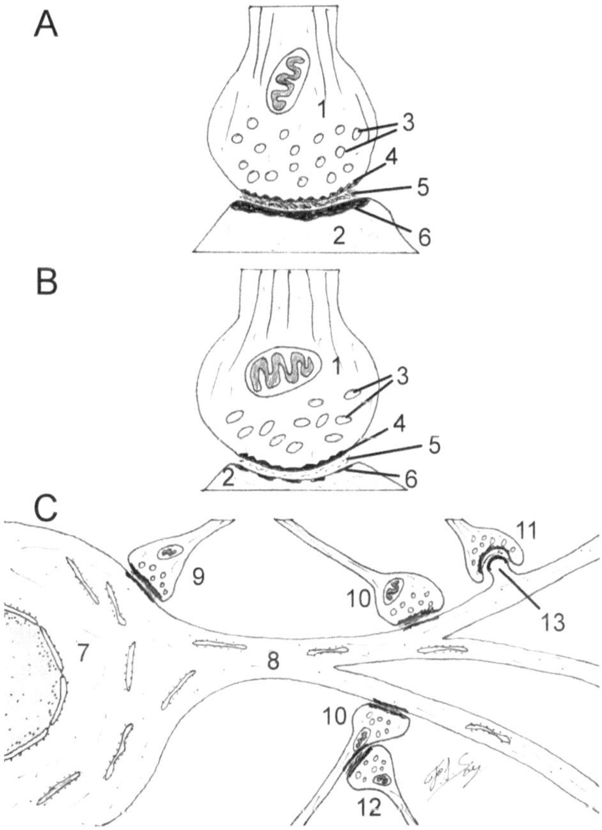

Fig. 11.5 Structure of type I (**A**) and type II synapses (**B**), and types of synapses according to their location (**C**). (1) Presynaptic component; (2) postsynaptic component; (3) synaptic vesicles; (4) presynaptic membrane condensation; (5) synaptic cleft; (6) postsynaptic membrane condensation; (7) neuronal soma; (8) dendrite; (9) axosomatic synapse; (10) axodendritic synapse; (11) axospinous synapse; (12) axoaxonic synapse; (13) dendritic spine

E. Synapses can also be classified according to the effect they have on the postsynaptic cell:
   1. *Excitatory synapses*, when the release of the neurotransmitter activates the postsynaptic cell. In these cases, the release of the neurotransmitter causes membrane *depolarisation* and the production of the action potential. As explained above, depolarisation is caused by the opening of Na^+ channels.
   2. *Inhibitory synapses*, where the release of the neurotransmitter causes *hyperpolarisation* of the postsynaptic membrane and no action potential is produced in the postsynaptic cell. Hyperpolarisation is caused by the opening of Cl^- channels that introduce this ion into the cell.
   3. In general, type I synapses are excitatory and type II inhibitory.
F. Depending on the parts of the neurons that establish synapses, several types of synapses can be distinguished (Fig. 11.5C):
   1. *Axodendritic*: Between the axon and a dendrite.
   2. *Axospinous*: A variant of the above that occurs between the axon and a dendritic spine.
   3. *Axosomatic*: Between the axon and the neuronal soma.

4. *Axoaxonic*: Between one axon and another axon.
5. *Dendrodendritic*: Between two dendrites. In the presynaptic dendrite, vesicles and membrane specialisations are similar to those of the presynaptic component of an axon.
6. *Neuromuscular synapse*: Between the axon and a muscle cell. They can be of two types:
   a. *Motor end plates*: these are the synaptic junctions between the axon terminals of motor neurons and *skeletal muscle cells*.
      (1) The outer lamina of the Schwann cell protecting the axon fuses with the external lamina of the muscle cell and the axon terminal bouton remains in close proximity to the plasma membrane of the muscle cell, with no outer lamina in between.
      (2) The postsynaptic membrane of the myocyte has numerous deep, narrow folds.
   b. *Autonomic synapses*: these are the junctions between the axon terminals of neurons of the *vegetative system* and *smooth muscle cells*.
      (1) The presynaptic membrane and the postsynaptic membrane are separated by the *external lamina* of the smooth myocyte.
      (2) The postsynaptic membrane of the smooth myocyte has *caveolae*.

G. Depending on the position of the synaptic button, two types of synapses can be distinguished:
   1. *En passant*: The synaptic button is in the path of the axon.
   2. *Terminal*: The synaptic button is at the end of the axon.

H. Each neuron usually produces a single type of neurotransmitter, which it releases at its axon terminals. Some of the most common neurotransmitters are:
   1. *Acetylcholine* (*cholinergic synapse*): Normally activates the target cell (postsynaptic), i.e. it is involved in excitatory synapses.
   2. *Amino acids and derivatives*: Among them, the most frequent are:
      a. *Glutamate* (*glutamatergic synapse*):. Found at excitatory synapses.
      b. *Gamma-aminobutyric acid, GABA* (*GABAergic synapse*): It is a very common inhibitory neurotransmitter.
      c. *Glycine* (*glycinergic synapse*): It is inhibitory, very frequent in the spinal cord.
      d. *Aspartate*.
   3. *Neuropeptides*: These include:
      a. *Substance P*: Excitatory undecapeptide.
      b. *Neuropeptide Y*: Made up of 36 amino acids. Excitatory.
      c. *Vasopressin*: Consists of nine amino acids.
      d. *Somatostatin*: Tetradecapeptide.
   4. *Monoamines*: For example:
      a. *Serotonin* (*serotonergic synapse*): Can be excitatory or inhibitory.
      b. *Dopamine* (*dopaminergic synapse*): Like the previous one, it can be excitatory or inhibitory.

c. *Adrenaline* or *epinephrine (adrenergic synapse).*
d. *Noradrenaline* (*noradrenergic synapse*): Also called *norepinephrine*, can be excitatory or inhibitory.
5. *Purines*: *ATP* may act as a neurotransmitter at some synapses, both in the central nervous system and in the peripheral nervous system. It appears to act as an excitatory.

I. Peptides are produced in the soma and transported along the axon. Most small neurotransmitters are produced at the synaptic terminal itself.

11.8 Neuroglia

A. Although neurons are the most important cells in the nervous system, the number of *neuroglia* cells is probably five or ten times the number of neurons, which gives us an idea of their importance.
B. We find different types of glial cells:
   1. In the central nervous system:
      a. *Macroglia*:
         (1) Ependymal cells
         (2) Astrocytes
         (3) Oligodendrocytes
      b. *Microglia*:
         (1) Microglial cells
   2. In the peripheral nervous system:
      a. Satellite cells or amphicytes
      b. Schwann cells
C. In addition to these frequent glial cells, there are others that are less abundant:
   1. *Radial neuroglia*: They are observed during embryonic development of the neural tube and guide the developing neurons. Only a few descendants are observed in the adult:
      a. *Müller cells* of the retina.
      b. *Bergmann glial cells* of the cerebellum.
   2. *Pituicytes*: These are cells of the neurohypophysis that contribute to the control of neurohormone production in this organ.
D. Origin of glial cells:
   1. Almost all glial cells, except for microglia, are derived from the *neuroectoderm.*
      a. The glial cells of the central nervous system originate from the neural tube.
      b. The glial cells of the peripheral nervous system are derived from neural crest cells.
   2. Microglial cells originate during embryonic development from *haemopoietic cells* of the yolk sac wall.

11.9 Ependymal Cells

A. *Ependymal cells* form an epithelium-like structure called *ependyma* (from modern Latin ependyma, derived from Greek ἐπί epí- 'on, over' and ἔνδυμα endyma 'garment').
B. The ependyma lines the internal cavities of the central nervous system (*ventricles of the brain* and *central canal of the spinal cord*), filled with *cerebrospinal fluid*, and lines the *choroid plexuses*, where it forms the *epithelium of the choroid plexus*.
C. Ependymal cells are cubic or cylindrical epithelial cells, linked by *desmosomes*, *banded desmosomes* (*zonula adherens*), *interdigitations* and *gap junctions*. In the choroid plexus, they have apical tight junctions.
D. Mitochondria and *intermediate filaments* are present in the cytoplasm.
E. At their apical pole, they have *microvilli* and *cilia*.
F. They lack a basal lamina (except in the choroid plexus).
G. Its functions are:
    1. They separate cerebrospinal fluid from nerve tissue.
    2. They reabsorb cerebrospinal fluid.
    3. They regulate the exchange between fluid and nerve tissue.
    4. Cilia cause movement of cerebrospinal fluid.
    5. The epithelium of the plexus produces cerebrospinal fluid.
H. In the *circumventricular organs*, the ependymal cells (called *tanycytes*) have apical tight junctions and lack cilia. On their basal surface, they emit processes that reach the blood vessels, where they form vascular feet similar to those of astrocytes.

11.10 Astrocytes

A. *Astrocytes* (from the Greek ἄστρον astron, 'star'), found in the *central nervous system*, are stellate-shaped cells connected by *gap junctions*, allowing them to dock and function synchronously.
B. They have *intermediate filaments (gliofilaments)*, formed by the *glial fibrillary acidic protein*, which give them a certain rigidity.
C. The end of the extensions expands and form end-feet or *pedicels*, which can be of various types:
    1. *Perivascular feet*, which cover the blood vessels and thus contribute to the *blood–brain barrier*.
    2. *Perineuronal feet*, which cover the surface of neurons.
    3. *Subpial feet*, which cover the outer surface of the organ and form the *glial limiting membrane*.
D. The pedicels help to separate the central nervous system from the rest of the body.
E. Astrocytes perform many functions:
    1. They support the neurons: They provide rigidity due to the gliofilaments.

2. Separation and isolation:
   a. They separate nerve tissue from the bloodstream. Their pedicels form part of the *blood–brain barrier* that isolates nerve tissue from blood vessels.
   b. The *glial limiting membrane* separates the nerve tissue from the surrounding tissues.
3. Chemical homeostasis:
   a. They are able to detect variations in *potassium* or *pH* and react by regulating the extracellular ionic medium so that the action potential can be produced in neurons.
   b. Astrocytes store excess *glucose* by synthesising *glycogen*.
      (1) Neurons need glucose but cannot store it.
      (2) When there is a lack of glucose, astrocytes degrade glycogen into glucose and glycogen into *lactate*, which is released for neurons to use in their energy metabolism.
4. Regulation of synapses:
   a. Its extensions surround the synapses, isolating them to prevent the diffusion of neurotransmitters.
   b. They provide *electrical insulation*.
   c. They regulate the concentration of neurotransmitters because they can destroy them.
5. They produce *cytokines* that regulate the production of myelin sheaths by oligodendrocytes.
6. When lesions are present, astrocytes contribute to the removal of molecules released by damaged cells and participate in tissue healing (*sclerosis*).

F. Four types of astrocytes are distinguished:
1. *Protoplasmic astrocytes* (Fig. 11.6):
   a. They are mainly found in the *grey matter*.
   b. The cell body is about 15–25 μm, and the nucleus is rounded and clear.
   c. They have many short, highly branched prolongations. The peripheral part of these extensions is thin and lamellar.
2. *Fibrous astrocytes* (Fig. 11.7):
   a. Present mainly in the *white matter*.
   b. They are smaller (10–12 μm).
   c. Its extensions are slender, long and not very branched.
3. *Interlaminar astrocytes*:
   a. They are located in the superficial layer of the grey matter of the brain.
   b. They emit extensions to the pia mater and deep layers of the cerebral cortex.
4. *Varicose-projection astroctyes*:
   a. They are found in the deep layers of the cerebral cortex.
   b. They have low branched extensions with numerous varicosities.

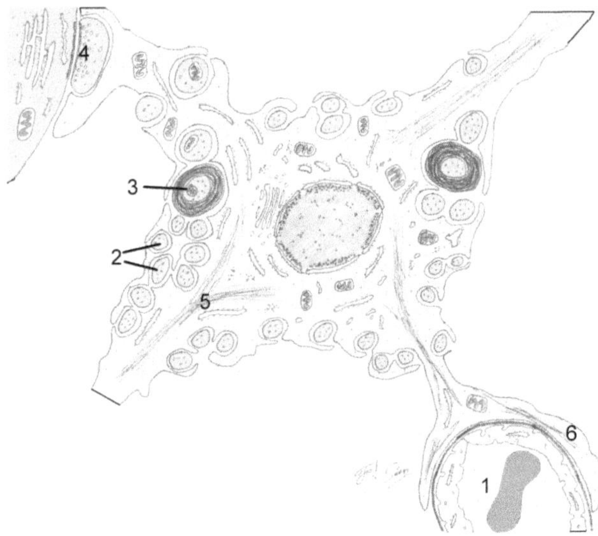

Fig. 11.6 Protoplasmic astrocyte. (1) Blood capillary; (2) amyelinated axons; (3) myelinated axon; (4) axosomatic synapse; (5) gliofilaments; (6) foot of astrocyte (pedicel)

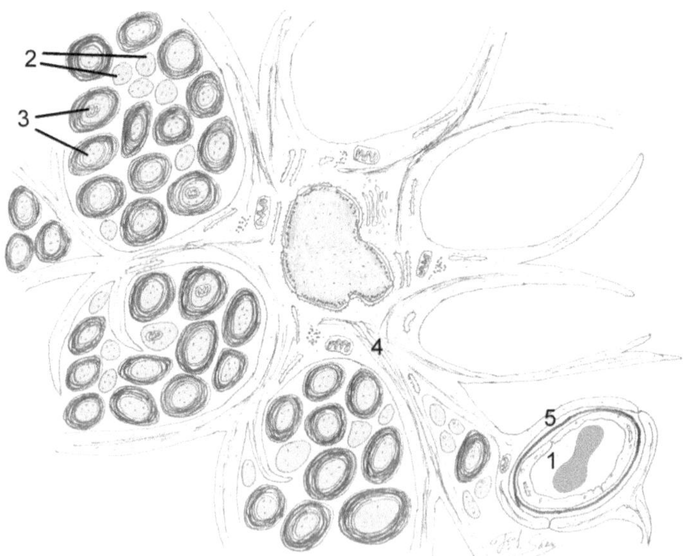

Fig. 11.7 Fibrous astrocyte. (1) Blood capillary; (2) amyelinated axons; (3) myelinated axons; (4) gliofilaments; (5) foot of astrocyte (pedicel)

11.11 Oligodendrocytes

A. *Oligodendrocytes* (from the ancient Greek ὀλιγο- oligo-, 'little' and δένδρον déndron, 'tree') are present in the white matter and grey matter.

B. As their name suggests, oligodendrocytes are cells with few and little-branched prolongations (Fig. 11.8).

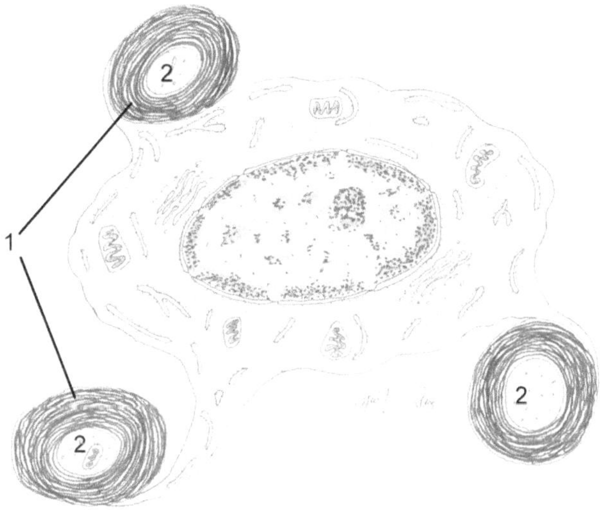

Fig. 11.8
Oligodendrocyte. (1) Myelin sheaths; (2) axons

C. The cell body is 6–8 μm. The nucleus is ovoid. The cytoplasm, dense under the electron microscope, contains rough endoplasmic reticulum, ribosomes, mitochondria, Golgi apparatus and microtubules.
D. They lack intermediate filaments.
E. Two types of oligodendrocytes have been described:
   1. *Perineuronal satellite oligodendrocytes*, attached to the neuronal soma in grey matter of the central nervous system.
   2. *Interfascicular oligodendrocytes*, abundant in the white matter, are located between the nerve fibres.
F. The cytoplasmic extensions of both types of oligodendrocytes form *myelin sheaths* around axons in the central nervous system.
   1. An oligodendrocyte can form 10–50 myelin sheath segments.
   2. An oligodendrocyte can form myelin segments on one or more axons.

11.12 Microglial Cells

A. Microglial cells are also called *Rio Hortega's cells.*
B. They have traditionally been considered to belong to the mononuclear phagocytic system.
C. They can appear in two states:
   1. When at rest, the cells have a small cell body and long, branching extensions that scour the tissue.
   2. When there is an injury or infection, they take on an *amoeboid* or branched shape, with thicker extensions than at rest and a large body that can divide, migrate and perform *phagocytosis*, as well as *present antigens* to T lymphocytes.

D. Unlike other glial cells, microglial cells are of mesodermal/hematopoietic origin:
   1. They are produced during embryonic development from *haemopoietic cells* of the *yolk sac wall* that produce macrophages.
      a. All microglia originate from the yolk sac.
      b. Embryonic liver haematopoiesis does not contribute to microglia.
   2. Precursors travel through the blood and nest in the embryonic central nervous system, where they proliferate rapidly and an increase in cells occurs.
   3. In the adult, in a resting state, the proliferation of these cells ceases.
   4. In case of cell loss, new cells can be produced by two mechanisms:
      a. Proliferation of the surviving microglial cells, which regain the ability to divide.
      b. Formation of new cells from *bone marrow* precursors.

E. They perform two functions:
   1. *Homeostatic maintenance*:
      a. During development:
         (1) They are involved in neuronal pattern formation by regulating neuronal survival and neuronal death.
         (2) They regulate the pattern of synapse formation.
      b. They regulate synaptic activity in the adult.
      c. In case of nerve tissue damage, they produce *cytokines* that protect and stimulate surviving neurons (*neuroprotection*).
   2. *Immune defence*: Its function is to perform phagocytosis, release cytotoxic substances to other cells and present antigens to T lymphocytes.
      a. They remove bacteria, injured cells and dead cell debris.
      b. When activated for immune defence, they can produce substances that are toxic to the neurons themselves (*neurotoxicity*), but they counteract this effect with their neuroprotective action.

11.13 Satellite Cells or Amphicytes

A. *Satellite cells* or *amphicytes* are cells present in the nerve ganglia of the peripheral nervous system. In the ganglia of the enteric system, there are *enteric neuroglial cells* with morphology and function similar to astrocytes of the central nervous system.

B. They are organised as flat or cubic cells joined together to form a sheath that surrounds the somas of the neurons in these ganglia.

C. In paravertebral and peripheral ganglia, the prolongations of neurons must penetrate between these cells to establish synapses (in sensory ganglia there are no synapses).

D. They are derived from *neural crest* cells.

E. These cells are equivalent to *Schwann cells* and are in continuation with the Schwann cells of the axons entering or leaving the ganglia.

F. Its functions are:
   1. Maintain a regulated microenvironment for the neuron.
   2. Provide electrical insulation.

11.14 Schwann Cells or Neurolemmocytes

A. *Schwann cells* are present in peripheral nerves, where they accompany nerve fibres and protect them, either by forming a *simple axonal sheath* or by forming a *myelin sheath*.
B. The cell is surrounded by a basal lamina or *external lamina*.
C. These cells are originated from the *neural crests*.
D. When the simple sheath is formed, the cell can protect several axons (*amyelinated axons* or *amyelinated fibres*), but when a myelin sheath is formed, the Schwann cell forms a single sheath segment around a single axon (*myelinated axon* or *myelinated fibre*).
E. The appearance of these cells varies according to whether or not they form a myelin sheath. In fact, three variants have been distinguished:
    1. *Nonmyelinating Schwann* or *Remak Schwann cells*: These are the ones that form simple sheaths (Fig. 11.9).
        a. They are elongated cells, parallel to the axon.
        b. Deep grooves are produced on its surface in which axons are located and may be completely enveloped by the Schwann cell.
    2. *Myelinating Schwann cells*: They form the *myelin sheath* around the axon. Surrounding the sheath is a peripheral portion of cytoplasm containing a flattened oval nucleus, a small Golgi apparatus, some cisternae of rough endoplasmic reticulum, some mitochondria and some lysosomes (Fig. 11.10). It also possesses *vimentin intermediate filaments*.
    3. *Repair Schwann cells*: They are responsible for repairing lesions in the peripheral nervous system. They are produced from either of the two previous types (myelinating and nonmyelinating) in response to injury.
F. The functions of Schwann cells are:
    1. Support and protection of nerve fibres.
    2. Debris removal.
    3. Guiding peripheral axon proliferation during axon regeneration after injury.

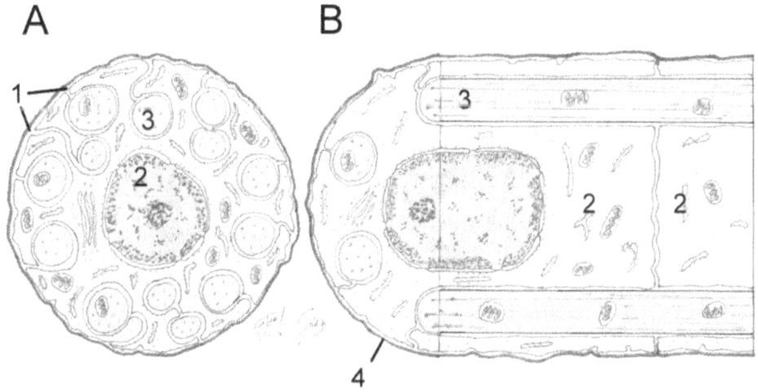

Fig. 11.9 Amyelinated nerve fibres in the peripheral nervous system. Cross section (**A**) and squared section (**B**). (1) Mesaxon; (2) Schwann cells; (3) axons; (4) external lamina

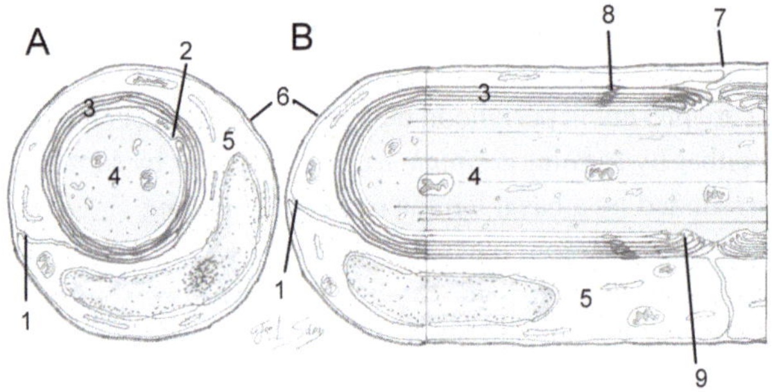

Fig. 11.10 Myelinated nerve fibres in the peripheral nervous system. Cross section (**A**) and squared section (**B**). (1) External mesaxon; (2) internal mesaxon; (3) myelin sheath; (4) axon; (5) Schwann cell; (6) external lamina; (7) node of Ranvier; (8) Schmidt–Lantermann incisures; (9) paranodal bulbs

11.15 Nerve Fibres

A. The nerve fibre is made up of an *axon*.
   1. In the peripheral nervous system, the fibre is always, from almost its beginning to almost its end, surrounded by *Schwann cells*, which may or may not form myelin sheaths, producing *myelinated fibres* and *amyelinated fibres*, respectively.
   2. In the central nervous system, there are also myelinated and amyelinated fibres.

B. *Amyelinated fibres*:
   1. In the peripheral nervous system, amyelinated fibres, also called *Remak fibres* or *grey fibres*, are enveloped by the cytoplasm of Schwann cells (Fig. 11.9), which are located in depressions on the surface of these cells.
      a. The place where the Schwann cell membrane forms the depression that envelops the axon is called the *mesaxon*.
      b. A Schwann cell may envelop several nerve fibres, each in a depression with its own mesaxon. Sometimes, if the nerve fibres are very thin, several of them may be in the same depression of the Schwann cell with a single mesaxon.
      c. In a mesaxon, a distinction can be made between the area next to the outside of the cell (*outer mesaxon*) and the area next to the axon (*inner mesaxon*).
   2. In the central nervous system, amyelinated axons are not covered by any cells:
      a. They are free axons that circulate, together with myelinated fibres, held (but not wrapped) between the extensions of astrocytes.
      b. There is no neuroglial cell in the central nervous system that is equivalent to the nonmyelinating Schwann cell.

C. *Myelinated fibres*:
   1. In the peripheral nervous system, the myelin sheath is formed by *myelinating Schwann cells* (Fig. 11.10). In the central nervous system, myelin sheaths are formed by *oligodendrocytes*. Both types of myelin sheaths are very similar.

2. *Myelin* is an envelope formed by the apposition of membranes of the myelinating cell (oligodendrocyte or Schwann cell, as appropriate).
3. The myelin sheath is produced by the following mechanism:
   a. The axon is enclosed in a depression or longitudinal invagination of the cytoplasm of the glial cell.
   b. Subsequently, the cytoplasm of the glial cell coils around the axon and several layers of membrane pile up around the axon.
   c. These layers exclude the cytoplasm they contain, so the membranes are intimately attached.
4. Due to its membranous origin, myelin is primarily made up of *lipids*, but there are also some characteristic proteins.
5. Each glial cell forms a sheath *segment* of 200 μm to 1.5 mm, which is in contact with other segments produced by other cells.
6. The points of contact between segments are called *nodes of Ranvier*. Here, adjoining cells emit loosely intertwining prolongations, while the myelin sheath membranes form terminal enlargements called *paranodal bulbs*.
7. In the peripheral nervous system, but rarely in the central nervous system, the myelin sheath has oblique grooves or clefts. These grooves are called *Schmidt–Lantermann incisures* or *clefts*.
8. The place where the sheath membranes connect to the cell soma membrane is called the *mesaxon*. There is an *outer mesaxon*, on the outside of the sheath, and an *inner mesaxon*, between the myelin and the axon.
9. The role of the myelin sheath is to allow faster propagation of the nerve impulse along the axon.
   a. The speed of propagation is faster in thicker axons, but the electrical insulation caused by the myelin sheath causes the impulse to be transmitted from one node of Ranvier to another, according to the so-called *saltatory* or *discontinuous conduction* of impulses.
   b. This allows for faster conduction, and is facilitated because more Na^+ channels are concentrated in the axonal membranes at the nodes of Ranvier.

D. Comparison of the myelin sheath in the peripheral nervous system and in the central nervous system:
   1. *Myelinating glial cell*:
      a. PNS: Schwann cell.
      b. CNS: oligodendrocyte.
   2. *Segments* that the glial cell can form:
      a. PNS: a segment of a single axon.
      b. CNS: several segments of the same or several axons.
   3. Fibre *insulation*:
      a. PNS: each fibre is surrounded by the Schwann cell, its external lamina and connective tissue of the endoneurium.

b. CNS: there may be several fibres in close proximity, partially surrounded by astrocyte extensions, but without any insulation, basal lamina or connective tissue.
4. *Mesaxons*:
a. PNS: there is an inner and an outer mesaxon.
b. CNS: there may not be an outer mesaxon because the outer cytoplasm layer does not form a complete circumference.
5. *Schmidt–Lanterman incisures*: these are less abundant in the CNS.

E. Nerve fibres in peripheral nerves are classified into several types:
1. *Type A fibres*: These are myelinated fibres from 3 to 20 μm in diameter. They are the most frequent, the largest and the fastest.
2. *Type B fibres*: These are thin myelinated fibres from 2 to 3 μm in diameter, slower than type A fibres.
3. *Type C fibres*: These are amyelinated fibres from 0.5 to 1.5 μm in diameter, slower than type A and type B fibres.

11.16 Peripheral Nerves

A. It is important to understand the organisation of peripheral nerves as they can be found in the tissues of the oral cavity, including the tooth and periodontium.

B. The peripheral nerves are composed of (Fig. 11.11):
1. *Bundles* or *fascicles* (from Latin fascicŭlus, 'bundle') of *myelinated and amyelinated nerve fibres*:
a. Each fascicle is made up of several fibres, from 10 to hundreds of fibres.
b. The fibres in the fascicle follow a wavy course, which allows the nerve to stretch without the fibres breaking.
c. The somas from which these nerve fibres arise are located in the organs of the central nervous system or in the ganglia.
2. *Schwann cells*, which form the simple sheaths and myelin sheaths that cover the axons.
3. *Connective tissue*:
a. May contain blood vessels (*vasa nervorum*).
b. It forms a series of sheaths which, although they are in continuity, are divided into three types for ease of study.
(1) *Endoneurium* (from the Greek ἐνδο- endo-, 'inside'):
(a) It is the small portion of loose connective tissue, with a high content of *reticular fibres*, which forms a felt around the external lamina of the Schwann cells (*reticuloendothelial lamina of Key Retzius*).
(b) May contain fibroblasts, macrophages, mast cells and blood capillaries.
(c) Reticular fibres can be synthesised by fibroblasts and Schwann cells.

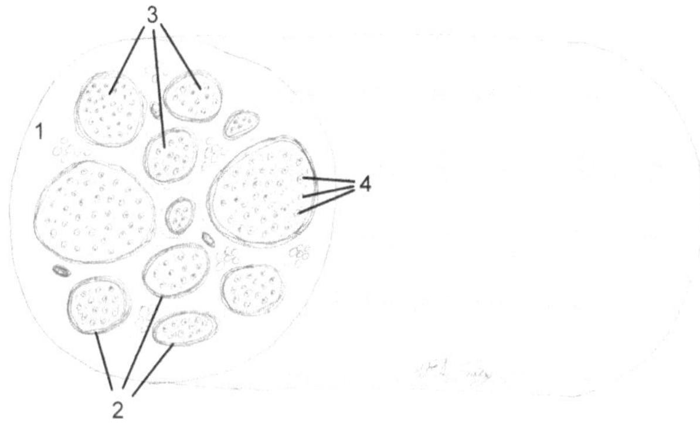

Fig. 11.11 Cross section of a peripheral nerve. (1) Epineurium; (2) perineurium; (3) endoneurium; (4) nerve fibres

(2) *Perineurium* (from the Greek περι- peri-, 'around'): It is the connective tissue surrounding a bundle of nerve fibres. It is made up of two layers:
   (a) *Fibrous portion*: external, of *dense connective tissue*, with collagen fibres oriented in a smooth spiral and with elastic fibres.
   (b) *Epithelial portion*: internal, consisting of several layers of flattened *fibroblast-like* cells, but with epithelial organisation. Each layer has an *inner basal lamina* and an *outer basal lamina*. Also known as *perineural epithelium* or *perineural sheath*.
(3) *Epineurium* (from the Greek ἐπί- epí-, 'over'):
   (a) The connective tissue that forms a *capsule* that surrounds and encloses several nerve bundles to form the peripheral nerve.
   (b) It is *dense irregular connective tissue* with collagenous fibres and thick elastic fibres.
   (c) *Adipocytes* and blood vessels of various calibres are frequently present, depending on the size of the nerve.

C. As they approach their distal end, i.e. the region furthest from the central nervous system, the nerves branch out, so that the fascicles separate. The epineurium becomes progressively thinner.
D. The smallest nerves, now consisting of only one fascicle, lack the epineurium and have their perineurium, which becomes thinner until it consists only of isolated, flattened cells.
E. At the end of the nerve ending, the perineurium disappears and only the Schwann cell persists, which may eventually disappear and, in the synaptic region, only its basal lamina persists.

11.17 Motor Nerve Endings

A. *Motor nerve endings* are the nerve endings that carry stimuli from the nervous system to other organs or tissues.
B. These are *efferent nerve endings*.

C. They can be:
   1. *Somatic endings*: These fibres originate from *motor neurons* in the spinal cord or brainstem and innervate skeletal muscles. There they produce the *motor plates* that stimulate the *skeletal myocytes.*
   2. *Visceral endings*: These fibres originate from neurons of the *ganglia of the autonomic system.* They are *amyelinated* and terminate in cardiac muscle, smooth muscle and glandular epithelia.
      a. In cardiac and smooth muscles, the fibres end close to or relatively far from the myocytes.
      b. In glandular epithelia, axons contact epithelial cells in two ways:
         (1) Hypolemmal termination: The axon passes through the basal lamina of the epithelium and remains at a distance of 10–20 nm from the epithelial cell.
         (2) Epilemmal termination: The axon is protected by the Schwann cell, except in an area approaching the epithelial cell, from which it is separated by 100–200 nm. Between the axon and the epithelial cell are the basal laminae of the Schwann cell and the epithelium.
      c. In no case are true synapses formed.

11.18 Sensory Nerve Endings

A. *Sensory nerve endings* function as sensory receptors that pick up peripheral stimuli and transmit them to the central nervous system.
   1. They are *afferent nerve endings* (*afferent receptors*).
   2. They are able to initiate an impulse in response to a stimulus.
   3. These are the axon terminals of *sensory neurons.*
B. Sensory receptors can pick up different stimuli, which makes it possible to distinguish between different types of sensory receptors:
   1. *Exteroceptors*: They perceive external stimuli.
   2. *Interoceptors*: They perceive stimuli from internal organs.
   3. *Proprioceptors*: They perceive stimuli related to the position of the body, produced by the musculoskeletal system or the vestibular apparatus of the ear.
C. There are receptors that pick up chemical stimuli, others pick up mechanical or physical stimuli. Two types of sensory receptors can be distinguished:
   1. *Somatosensory* receptors, i.e. receptors for heat, cold, pain, pressure, touch, vibration, position and visceral sensitivity.
   2. Receptors attached to the *sense organs*, responsible for the reception of special stimuli, such as *vision*, *hearing*, *balance*, *taste* and *smell.*
D. Stimulus uptake:
   1. Sometimes the stimulus is picked up by the nerve ending itself.
   2. Other times, the stimulus is picked up by specialised cells and transmitted to the nerve ending, usually at a *synapse.*

E. Two types of sensory endings can be distinguished according to their histological organisation:
   1. *Free nerve endings* (Fig. 11.12).
      a. These are the free terminal portions of sensory nerve fibres that may be partially covered by Schwann cells.
      b. Free nerve endings are distributed throughout the body. They are the most common nerve endings and are present in epithelia, connective tissues, muscles and serous membranes.
      c. There are free nerve endings for temperature (*thermoreceptors*), pain (*nociceptors*), chemicals (*chemoreceptors*) and movement (*mechanoreceptors*).
      d. Despite this diversity, they all have a similar appearance.
      e. Nociceptors react to potentially damaging stimuli, which can be intense mechanical stimuli, intense heat or cold and chemicals.
      f. Two types can be found in the epithelium of the mouth and skin:
         (1) Free nerve endings that enter the epithelium and are distributed between the cells.
         (2) Nerve endings contacting special sensory cells.
   2. *Specialised sensory cells*:
      a. These receive a stimulus and transmit it to a free nerve ending.
      b. They have been located in several places: taste buds, carotid bodies, epidermis (*Merkel cells*, also found in the epithelium of the oral mucosa), retina and inner ear.
   3. *Sensory corpuscles* (encapsulated endings). In addition to nerve endings, they possess neuroglial cells (*Schwann cells*) and connective tissue cells that form a peripheral *capsule* derived from the endoneurium, perineurium and epineurium. Because of their interest to the dentist, we will study the followings:

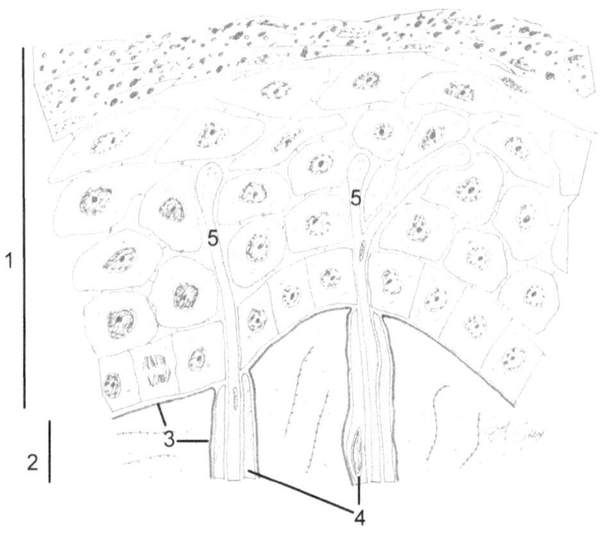

Fig. 11.12 Free nerve endings. (1) Skin epithelium; (2) connective tissue; (3) basal lamina; (4) Schwann cell; (5) axon terminal

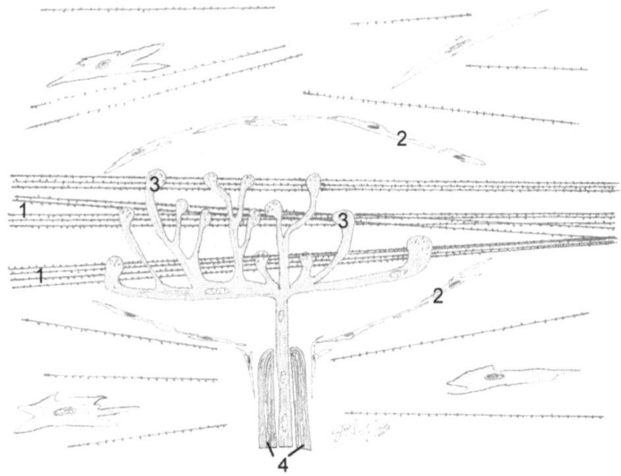

Fig. 11.13 Ruffini corpuscle. (1) Collagen fibres; (2) capsule; (3) nerve endings; (4) myelin sheath

a. *Ruffini corpuscles* (Fig. 11.13):
   (1) They are found in the skin, periodontium and joint capsules.
   (2) They are cylindrical in shape, 40–90 μm long × 7–15 μm wide.
   (3) They have a very thin capsule of connective tissue.
   (4) The interior is traversed by collagen fibres.
   (5) They have a sensory myelinated axon that enters from the side or end of the cylindrical structure, loses the myelin sheath and branches into many fine endings that end in a small knob-shaped bulb in contact with collagenous fibres.
b. *Golgi tendon organs* (Fig. 11.14):
   (1) They are encapsulated and are located at the transition between skeletal muscle and tendons or aponeurosis.
   (2) They are fusiform, up to 1.5 mm long and 120 μm in diameter.
   (3) Its sheath is perineural, with flattened cells.
   (4) Inside are collagen fibres that form longitudinal bundles.
   (5) Laterally, myelinated axons enter, lose their sheaths and branch out between the collagen fibres, through which they climb.
   (6) The branches are covered by Schwann cells and have thickened mitochondria which are the receptor regions.
c. *Vater–Pacini corpuscles* or *Pacinian corpuscles* (Fig. 11.15):
   (1) They are large lamellar corpuscles present in various locations, such as in the subcutaneous tissues and the periosteum of bones. They can be 1–4 mm long × 200–300 μm wide.
   (2) It has several sheaths:
      (a) A *connective tissue sheath*, the outermost.
      (b) A *perineural sheath*, inside the connective sheath, with flattened cells that may form several peripheral lamellae (up to 50).

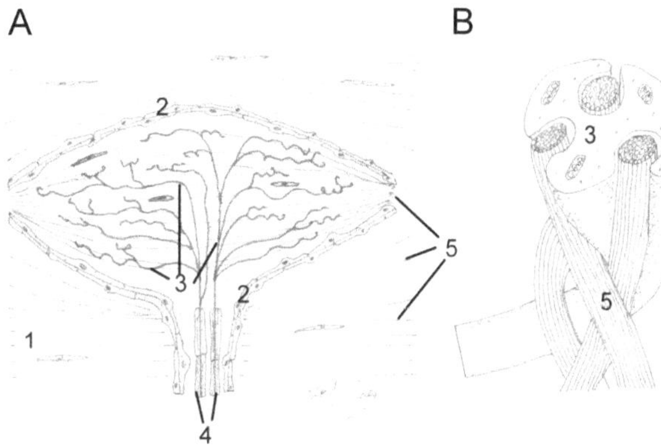

Fig. 11.14 Section of a Golgi tendon organ (**A**), and magnified image of a nerve ending encompassing the collagen fibres (**B**). (1) Dense connective tissue of the tendon; (2) capsule (perineural sheath); (3) nerve fibres; (4) Schwann cells; (5) collagen fibres

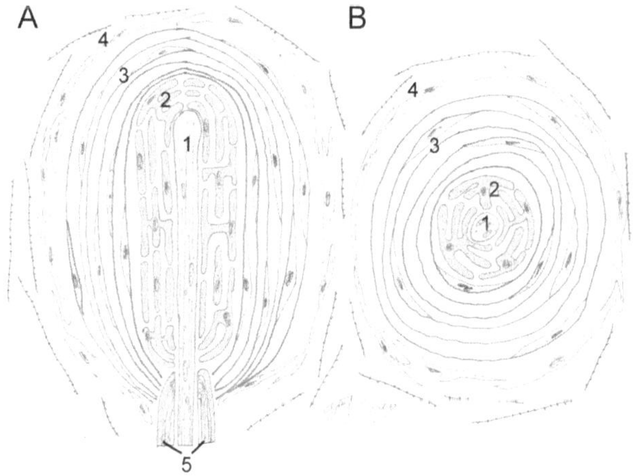

Fig. 11.15 Longitudinal section (**A**) and transverse section (**B**) of a Vater–Pacini corpuscle. (1) Sensory axon terminal; (2) inner sheath (Schwann cell sheath); (3) perineural sheath; (4) connective sheath; (5) myelin sheath

(c) An *inner core*, internal, formed by many lamellae (which may also be up to 50). These lamellae consist of the cytoplasm of Schwann cells.

(3) A thick sensory nerve ending penetrates the corpuscle and does not branch. It has mitochondria.

d. *Meissner's corpuscles* (Fig. 11.16):

(1) They are encapsulated receptors present in the skin and oral mucosa.
(2) They are ovoid or cylindrical, 120–180 μm long and 30–40 μm wide.
(3) They are covered by a poorly developed capsule.
(4) They contain the endings of several axons that are surrounded by stacks of lamellae formed by modified Schwann cells (*lamellar cells*).
(5) The lower part of the corpuscle is surrounded by one or two flattened cells originating from the perineurium. Between the perineural cells and the lamellae are thin collagenous fibres.

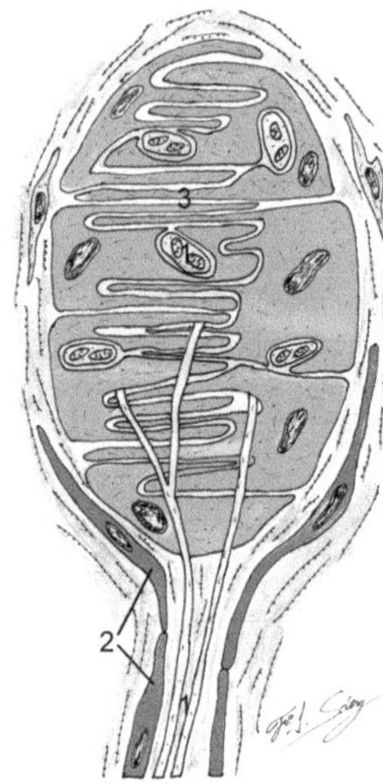

Fig. 11.16 Section of a Meissner corpuscle. (1) Sensory axon terminal; (2) perineurium cells; (3) laminar cells (modified Schwann cells); (4) capsule

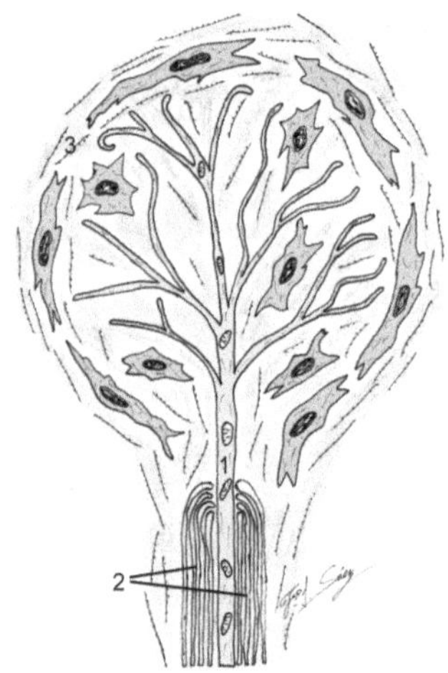

Fig. 11.17 Section of a Krause's end bulb (Krause's corpuscle). (1) Sensory nerve ending; (2) myelin sheath; (3) connective sheath

e. *Mucocutaneous terminal organs*: They are similar to Meissner's corpuscles, present in transitional zones between the skin and mucous membranes (anus, genitals and lips).
f. *Krause's end bulbs* or *Krause's corpuscles* (Fig. 11.17):
   (1) They are spherical nerve endings, about 50 μm in diameter.
   (2) They are present in the skin, joints, ocular conjunctiva, peritoneum, genitals, mouth and nose.
   (3) It consists of a myelinated nerve ending that loses its myelin and penetrates a conjunctival sheath, where the fibre usually branches and fills the corpuscle.

Take-Home Message

The nerve tissue forms the nervous system, which is responsible for communication, sensations, reactions, consciousness, behaviour and personality. Nerve tissues originate from the neuroectoderm: the neural tube forms the central nervous system and the neural crests form most of the peripheral nervous system. The neuron is the fundamental cell of nerve tissue. It has a soma or cell body that processes impulses received by a variable number of cytoplasmic prolongations called dendrites and emits a nerve impulse to other neurons or effector organs by a single prolongation called an axon. The nerve impulse is produced by variations in the neuron's membrane potential. Transmission of the nerve impulse between neurons occurs at contact sites called synapses, where substances (neurotransmitters) are released that excite the neighbouring neuron. Neurons are accompanied by neuroglial cells: astrocytes, oligodendrocytes, ependymal cells and microglial cells in the central nervous system, and satellite cells and Schwann cells in the peripheral nervous system. Axons form nerve fibres, which may be protected by a myelin sheath, produced by oligodendrocytes in the central nervous system and by Schwann cells in the peripheral nervous system. In the peripheral nervous system, nerve fibres are protected by several connective tissue sheaths (endoneuro, perineuro and epineuro) to form the peripheral nerves. Nerve fibre endings may be motor (somatic and visceral) or sensory (free or encapsulated).

Further Reading

Aten S, Kiyoshi CM, Arzola EP et al (2022) Ultrastructural view of astrocyte arborization, astrocyte-astrocyte and astrocyte-synapse contacts, intracellular vesicle-like structure, and mitochondrial network. Prog Neurobiol 213:102264

Baldwin KT, Murai KK, Khakn BS (2024) Astrocyte morphology. Trends Cell Biol 34:547–565

Cuadros MA, Sepulveda MR, Martin-Oliva D et al (2022) Microglia and microglia-like cells: Similar but different. Front Cell Neurosci 16:816439

Garaschuk O, Verkharatsky A (2019) Physiology of Microglia: Methods in Molecular Biology. In: Microglia: Methods and Protocols, vol 2034, pp 27–40

Garaschuk O, Verkhratsky A (2019) Microglia: the neural cells of nonneural origin. In: Methods in molecular biology – microglia: methods and protocols, vol 2034, pp 3–11

Gartner LP (2017) Nervous tissue. In: Textbook of histology, 4th edn, pp 211–250

Gómez de Ferraris ME, Campos Muñoz A (2009) Histología general humana. In: Histología, Embriología e Ingeniería Tisular Bucodental, 3rd edn. Médica Panamericana, pp 57–78

Hasel P, Liddelow SA (2021) Astrocytes. Curr Biol 31:R326–R327

Jessen KR, Mirsky R (1997) Embryonic Schwann cell development: the biology of Schwann cell precursors and early Schwann cells. J Anat 191:501–505

Jones EG (1988) The nervous tissue. In: Weiss L (ed) Cell and tissue biology – a textbook of histology, 6th edn. Urban & Schwarzenberg, pp 277–352

Kagawa T, Wada T, Ikenaka K (2001) Regulation of oligodendrocyte development. Microsc Res Tech 52:740–745
Kaur C, Hao AJ, Wu CH, Ling EA (2001) Origin of microglia. Microsc Res Tech 54:2–9
Kierszenbaum AL, Tres LL (2019) Nervous tissue. In: Histology and cell biology – an introduction to pathology, 5th edn. Elsevier, pp 261–298
Kimelberg HK, Norenberg MD (1989) Astrocytes. Sci Am 260:66
Krstić RV (1985) Nervous tissue. In: General histology of the mammal. An atlas for students of medicine and biology. Springer Verlag, pp 293–389
Lowe JS, Anderson PG (2015) Nervous tissue. In: Stevens & Lowe's human histology, 4th edn, pp 84–104
O'Dowd G, Bell S, Wright S (2023) Nervous tissues. In: Wheater's functional histology. Elsevier, pp 122–142
Osanai Y, Yamazaki R, Shinohara Y, Ohno N (2022) Heterogeneity and regulation of oligodendrocyte morphology. Front Cell Dev Biol 10:1030486
Pawlina W (2024) Nerve tissue. In: Histology a text and atlas with correlated cell and molecular biology, 9th edn. Wolters Kluwer, pp 388–441
Richardson WD, Kessaris N, Pringle N (2006) Oligodendrocyte wars. Nat Rev Neurosci 7:11–18
Salzer JL (2015) Schwann cell myelination. Cold Spring Harb Perspect Biol 7:a020529
Zhang LJ, Cao Y, Zhang X et al (2021) The origin and repopulation of microglia. Dev Neurobiol 82:112–124
Zorick TS, Lemke G (1996) Schwann cell differentiation. Curr Opin Cell Biol 8:870–876

Circulatory System

12

What You Will Learn in This Chapter

The circulatory system consists of the cardiovascular system and the lymphatic vascular system. In this chapter you will learn about the histological organisation of the heart, the larger blood vessels (arteries and veins) and the smaller blood vessels (arterioles, capillaries, sinusoids, venules, small veins). You will also learn how lymphatic vessels (lymphatic capillaries, collecting vessel and lymphatic channels or ducts) are organized.

12.1 General Characteristics of the Circulatory System

A. The *circulatory system* is made up of two related components:
   1. *Cardiovascular system*, through which *blood* circulates.
   2. *Lymphatic vascular system*, which collects *lymph* produced in the tissues and carries it to the cardiovascular system.

B. The *cardiovascular system* distributes blood throughout the body.
   1. Blood is a liquid tissue that distributes oxygen, nutrients and hormones throughout the tissues, while removing carbon dioxide and other waste products.
   2. To do this work, blood circulates through the cardiovascular system, which functions like the pipes in a city's water distribution system:
      a. There is a pumping station that drives the blood, the *heart*.
      b. From the pumping station, ducts branch off into smaller ducts that distribute the fluid to the tissues. These ducts are the *arteries*.
      c. In the tissues, the blood is distributed through a network of small-calibre ducts, which form the *microcirculation*.
      d. After passing through the tissues, the blood is collected in new tubes that merge into larger tubes until the blood returns to its starting point. These tubes are called *veins*.

F. J. Sáez, I. Badiola, *Essential Oral Histology*, Learning Materials in Biosciences,
https://doi.org/10.1007/978-3-032-03165-5_12

3. The heart propels blood into two continuous systems of tubular vessels:
   a. *Pulmonary circulation*: distributes blood to the lungs for oxygenation and removal of carbon dioxide, after which the blood is returned to the heart.
   b. *Systemic circulation*: distributes blood to all the body's tissues, where it supplies oxygen and collects carbon dioxide, and carries the blood back to the heart.
4. In both circulations, we can distinguish the following blood vessels:
   a. *Elastic* or *conducting arteries*: These are large diameter vessels that branch out from the heart and produce smaller arteries as they branch out.
   b. *Muscular* or *distributing arteries*: They originate from the smaller elastic arteries.
   c. *Arterioles* or *resistance vessels*: These are the small end branches of the small muscular arteries. They are produced in the tissues to be irrigated.
   d. *Capillaries*: These are the smallest vessels that arise from the terminal arterioles and in which most of the exchange of substances with the tissues takes place.
   e. *Postcapillary venules*: Small vessels that collect blood from the capillaries.
   f. *Venules*: They are produced by fusion of postcapillary venules. Larger in calibre than postcapillary venules.
   g. *Veins*: Produced by the fusion of venules. Small veins merge to form increasingly larger veins (medium veins) that eventually merge into large veins that carry blood back to the heart.

C. The *lymphatic vascular system* drains fluid from the tissues. The process is as follows:
   1. In the microcirculation there is an exchange of fluid between the blood and the tissues (Fig. 12.1):
      a. Within the blood capillaries, the *hydrostatic pressure* is higher than in the tissues and the *osmotic pressure* is lower, so both pressure differences drive the fluid that forms the blood plasma out of the capillaries into the tissues, and constitutes the *tissue fluid*.
      b. At the venous end of the capillaries, the trend is reversed, as the hydrostatic pressure in the capillary is lower and the osmotic pressure is higher than in the tissue, so that tissue fluid is pushed into the capillary.
      c. However, the amount of fluid that returns to the blood is less than the amount that has previously left it, so the excess tissue fluid enters *lymphatic capillaries* and forms *lymph*.
   2. Lymphatic capillaries originate in the tissues as tubes with their initial end blinded.
   3. The lymphatic capillaries merge into larger *lymphatic vessels* that eventually carry the lymph to veins near the heart, thus returning the fluid to the blood.
   4. As it travels through the lymphatic vessels, lymph passes through the *lymph nodes*, organs of the immune system where it is filtered for infectious agents.

D. This chapter will briefly describe the histological organisation of the heart and major blood vessels. The structure and organisation of the smaller blood and lymphatic vessels that can be found when studying the tissues of the mouth and the tooth will be described in more detail.

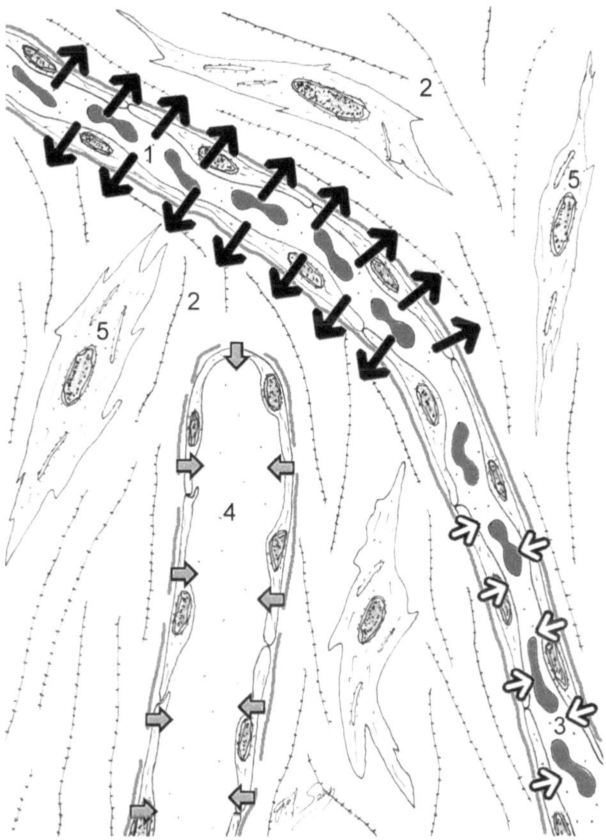

Fig. 12.1 Filtration of blood plasma to produce tissue fluid. At the arterial side of the capillary, (1) the outflow of fluid (black arrows) from the plasma into the extracellular matrix of the connective tissue (2) occurs. The outflow is due to the difference in hydrostatic pressure (higher in the capillary) and osmotic pressure (higher in the tissue) between the blood and the connective tissue. At the venous side of the capillary (3), where the hydrostatic pressure is lower and the osmotic pressure higher than that of the tissue, the opposite effect occurs, i.e. the passage of fluid from the tissue into the blood (white arrows). The volume of fluid returning to the blood at the venous side is less than that entering the tissue at the arterial side. To prevent excess fluid from accumulating in the tissues, it passes into blind-end capillaries (grey arrows), the lymphatic capillaries (4), which carry this fluid back to the heart. (5) Fibroblasts

12.2 General Characteristics of Blood Vessels

A. The vessel wall is made up of three layers or tunicae:
   1. *Tunica intima*: This is the innermost layer, in contact with the circulating blood. It is made up of:
      a. *Endothelium*: It is a simple squamous epithelium. Exceptionally, in some places the epithelium is cubic. It is characterised by:

(1) *Endothelial cells* are usually elongated in the direction of blood flow.

(2) It has a *basal lamina*. However, in some small vessels the basal lamina may be discontinuous or absent.

(3) Endothelial cells secrete:

(a) *Von Willebrand factor*, which is a multimeric plasma protein that is also synthesised in *megakaryocytes* in the bone marrow.

1. This factor transports factor VIII in plasma and plays an important role in platelet aggregation and sealing a site of injury.
2. In most blood vessels the secretion is continuous, but in the endothelium of arteries this secretion is stored in *Weibel–Palade bodies*.

(b) *Endothelin I*, which is a hypertensive agent.

b. *Subendothelial connective tissue*: This is a layer of connective tissue beneath the epithelium. Smooth muscle cells (*myointimal cells*) may be present and appear to have phagocytic capacity.

c. *Internal elastic lamina*, which is the boundary between the tunica intima and the tunica media.

(1) It is characteristic of *muscular arteries*, is less obvious in elastic arteries and is not seen in veins.

(2) In muscular arteries it usually forms a wavy hyaline layer.

(3) It may be double and in such cases two elastic laminae are seen in close proximity, anastomosing at some points.

2. *Tunica media*: It is well distinguished in arteries and worse in veins.

a. In arteries, it forms a layer in which connective tissue—with collagen and elastic fibres in varying proportions depending on the vessel—is mixed with smooth muscle tissue.

(1) It is the thickest layer of the arterial wall.

(2) *Elastic fibres* are more abundant in elastic arteries than in muscular arteries.

(3) Muscle cells are oriented in circular direction.

(a) The contraction of this muscle layer causes the decrease of vessel's calibre (*vasoconstriction*). Vasoconstriction reduces the lumen of the vessel, which decreases blood flow and increases *blood pressure*.

(b) Relaxation of this circular muscle increases vascular calibre (vasodilation) and thus blood flow, and lowers blood pressure.

(c) Muscle contraction and relaxation are controlled by hormones, nerve endings and local chemical factors.

b. In some arteries there is an *external elastic lamina* at the boundary with the most external tunica.

c. In veins it is usually a thin layer, sometimes imperceptible, consisting of connective tissue and smooth muscle cells.

d. In the smaller vessels, capillaries and postcapillary venules, the tunica media is replaced by cells called *pericytes*.

3. *Tunica adventitia*: The outermost layer, formed by connective tissue, which may have collagen and elastic fibres, depending on the case, and sometimes smooth muscle cells arranged longitudinally, especially in veins.
   a. In veins, it is the thickest and most distinguishable layer of the vein wall.
   b. In arteries it is usually thin and poorly distinguishable from the perivascular connective tissue.
   c. Other smaller blood vessels (*vasa vasorum*) and nerves (*nervi vascularis*), whose function is to supply and innervate the vascular wall, respectively, may be very evident in the large vessels. The endings of these vessels and nerves may reach the tunica media.

B. The smaller vessels have smaller and smaller tunica media and adventitia, until in the smallest vessels (capillaries) they practically disappear.

C. Blood vessels are organs whose tissues need to obtain *nutrients*. These nutrients can be obtained:
   1. Of the blood circulating in the vessel. This is feasible in thinner vessels and in tissues close to the lumen of larger vessels.
   2. Of the blood vessels supplying the tunica adventitia (*vasa vasorum*).
      a. This is necessary in the tissues away from the lumen (tunica media and adventitia) of the thicker blood vessels (arteries and veins).
      b. As blood circulating in the veins is poorer in nutrients than blood circulating in the arteries, *vasa vasorum* are more abundant in the veins.

D. Arteries and, to a lesser extent, veins receive *motor innervation* (*nervi vascularis*) in their tunica adventitia:
   1. From the *sympathetic nervous system*. These are nerve endings that do not penetrate the tunica media.
      a. They release *noradrenaline*, which diffuses to the tunica media and causes smooth muscle contraction and subsequent *vasoconstriction* of the vascular wall.
      b. In the veins, the motor endings also act on the tunica adventitia muscle.
   2. From the *parasympathetic nervous system*. This occurs in the arteries supplying the skeletal muscles, which receive *cholinergic innervation* that produces *vasodilation*.

E. Arteries also receive *sensory innervation*, which can be pressure receptors (*baroreceptors*) and *chemoreceptors*.

F. The wall of the heart has a peculiar organisation, but like the rest of the vessels it is made up of three layers, which are equivalent to the three tunicae described above.

12.3 Heart

A. The heart consists of a wall surrounding four chambers:
   1. Two *atria* receiving blood:
      a. *Right atrium*: Receives blood from the systemic circulation.
      b. *Left atrium*: It receives oxygenated blood from the pulmonary veins.

2. Two *ventricles*, which propel blood into the circulation:
   a. *Right ventricle*: Receives blood from the right atrium and propels it to the pulmonary arteries.
   b. *Left ventricle*: Receives blood from the left atrium and propels it into the systemic circulation via the aorta artery.

B. The heart wall is made up of three layers, which are equivalent to the three layers described for blood vessels:
   1. *Endocardium*: It is the innermost layer.
      a. It consists of an *endothelium* (simple squamous epithelium) and a *subendothelial connective tissue*, which in its deepest layer contains cardiac cells of the *conducting system.*
      b. Equivalent to the tunica intima of the vessels.
   2. *Myocardium*: It is the intermediate layer.
      a. It is the thickest layer, made up of *cardiac muscle.*
      b. It is the contractile layer of the heart.
      c. Equivalent to the tunica media.
   3. *Epicardium*: This is the outer layer.
      a. It consists of a *loose connective tissue* with abundant *adipose tissue*, *coronary vessels* and nerves, lined by *mesothelium*, a simple squamous epithelium.
      b. The epicardium is the *visceral layer of the pericardium*, the cavity in which the heart is located, and is continued by the serous layer of the parietal pericardium.
      c. Equivalent to the tunica adventitia of the vessels.

12.4 Arteries

A. Three types of arteries are usually distinguished, which follow one another as the circulation moves away from the heart:
   1. *Elastic* or *conducting arteries* (Fig. 12.2): These are the arteries leaving the heart and its first branches. They are not present in the tissues of the oral cavity, so their characteristics are briefly outlined.
      a. These are large-calibre arteries.
      b. Its tunica media is characterised by two intermingled components:
         (1) *Smooth muscle cells*, arranged circularly around the vascular lumen.
         (2) Thick *perforated elastic sheets*, formed by the fusion of elastic fibres.
   2. *Muscular* or *distributing arteries* (Fig. 12.3): Most of the arteries in our body are muscular arteries. Some have a relatively large calibre but most are of medium or small calibre.
      a. They have a prominent *internal elastic lamina*, 2–3 μm thick, wavy and hyaline in appearance, which separates the tunica intima from the media.

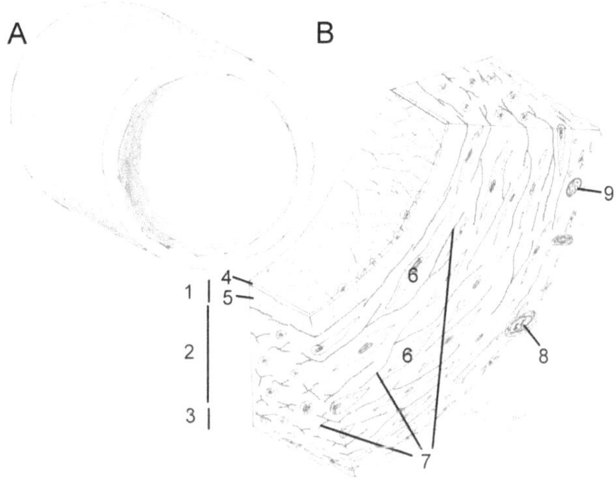

Fig. 12.2 Cross-section of an elastic artery (**A**), showing in detail the organisation of its wall (**B**). (1) Tunica intima; (2) tunica media; (3) tunica adventitia; (4) endothelium; (5) connective tissue; (6) smooth muscle cells; (7) elastic fibres; (8) *vasa vasorum*; (9) *nervi vascularis*

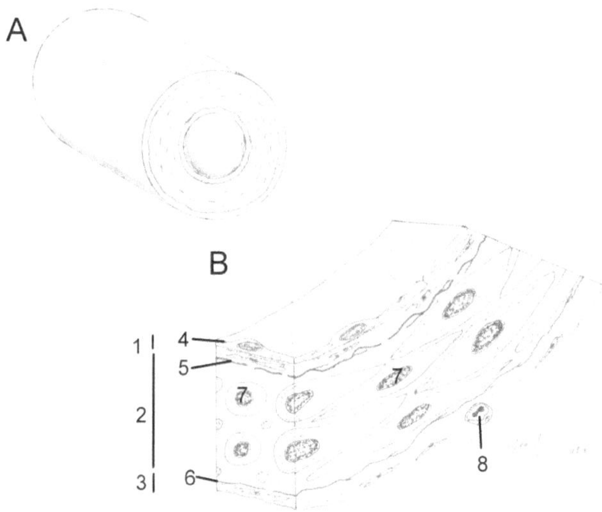

Fig. 12.3 Cross-section of a muscular artery (**A**), showing in detail the organisation of its wall (**B**). For simplicity, in (**B**) only a few layers of smooth muscle cells have been drawn in its tunica media, but normally the tunica media is thicker and has more layers of smooth muscle cells. (1) Tunica intima; (2) tunica media; (3) tunica adventitia; (4) endothelium; (5) internal elastic lamina; (6) external elastic lamina; (7) smooth muscle cells; (8) *vasa vasorum*

b. They have a tunica media where the *smooth muscle* is very visible. Depending on the calibre of the vessel, we can find between 4 and 30 concentric layers of muscle cells.
c. Between the smooth muscle cells there are *collagen* and *elastic fibres* that are barely visible, unless specific stains are used. This extracellular material is produced by the muscle cells themselves.

d. Sometimes, but not always, an *external elastic lamina* is observed between the tunica media and the adventitia, which may be multilamellar.

3. *Arterioles* or *resistance vessels* (Fig. 12.4): Arterioles are the smallest arterial vessels, with diameters ranging from 200–400 μm (the largest) to 50 μm for the smallest. Their characteristics are as follows:
   a. They are mainly responsible for the normal *blood pressure* in the arterial circulation. This is because the large number of arterioles present in the tissues produces a fairly large surface area, which leads to a very high resistance to flow.
   b. Arterioles reduce blood pressure to levels that are bearable for the thin walls of the capillaries.
   c. Histological organisation:
      (1) Its *tunica intima* is very thin and consists of the endothelium with very little underlying connective tissue. In some major arterioles, the *internal elastic lamina* can be seen, but in most others it is not visible.
      (2) The *tunica media* consists of only one to three layers of smooth muscle cells.
      (3) There is no external elastic lamina.
      (4) The *tunica adventitia* is connective tissue that blends and merges with the surrounding connective tissue, with no clear boundaries.

B. Arteries decrease in calibre as they branch and the transition from one type of artery to another is gradual. Thus, there are arteries with intermediate characteristics.

C. Endothelial cells in arteries possess *Weibel–Palade bodies* containing *von Willebrand factor*.

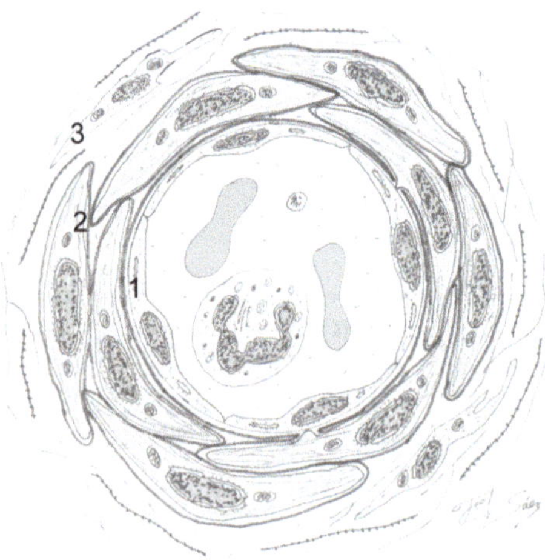

Fig. 12.4 Cross-section of an arteriole. (1) Endothelium; (2) smooth muscle cells; (3) adventitial connective tissue

12.5 Blood Pressure Regulation

A. Nerve regulation: The brain controls blood pressure.
   1. Vasomotor nerves of the *sympathetic system* release *noradrenaline* and cause *vasoconstriction*.
   2. Nerves of the *parasympathetic system* release *acetylcholine*, which causes endothelial cells to release *nitric oxide (NO)*. NO causes relaxation of the smooth muscle of the vessel and subsequent *vasodilation*.

B. Hormonal control: Muscle cells are sensitive to chemical mediators and hormones, for example:
   1. *Angiotensin II* is produced in the blood from *angiotensinogen* synthesised by the liver through the action of *renin*, secreted by the kidneys when blood pressure is low. It produces *vasoconstriction*.
   2. *Antidiuretic hormone (ADH)* or *vasopressin*, produced by the *pituitary gland*, produces *vasoconstriction*.

12.6 Microcirculation or Terminal Circulation

A. The *microcirculation* or *terminal circulation* is where the exchange of substances between the blood and the tissues takes place.

B. The blood vessels forming the microcirculation can be (Fig. 12.5):
   1. *Terminal* or *precapillary arterioles*.
   2. *Metarterioles*: Present in some tissues.
   3. *Capillaries*.
   4. *Postcapillary venules*.

C. Blood flow through the microcirculation is not always continuous. Therefore, two types of vessels can be distinguished:
   1. *Preferential channels*: These are blood vessels that are always open.
      a. Blood flow is continuous, although the amount of blood passing through at any one time may be greater or lesser and is regulated by the degree of contraction of the terminal arteriole or the metarteriole.
      b. There are two types of blood vessels that behave as preferential channels:
         (1) *Preferential capillaries*.
         (2) *Metarterioles*.
   2. *True capillaries*: These are capillaries that may be open or closed to circulation.
      a. Blood flow is discontinuous and depends on the needs of the tissue.
      b. Normally, blood is only allowed to flow through 25% of the capillary network in a tissue and the remaining capillaries only open if the need for oxygen and nutrients increases.

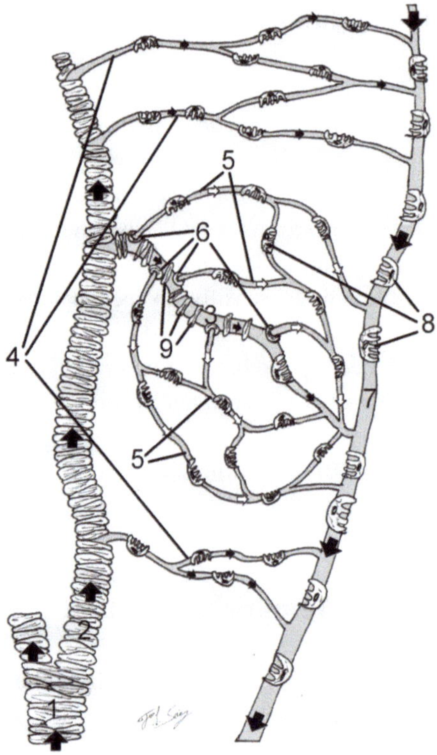

Fig. 12.5 Illustration representing a hypothetical external view of blood vessels of the microcirculation. (1) Smooth muscle-lined arteriole; (2) terminal arteriole; (3) metarteriole; (4) preferential capillaries; (5) true capillaries; (6) precapillary sphincters; (7) postcapillary venule; (8) pericyte; (9) smooth muscle cells. Black arrows indicate continuous blood flow, and empty arrows indicate optional blood flow through true capillaries

D. *Terminal* or *precapillary arterioles* are the smallest arterioles, with only one or two layers of smooth muscle in their tunica media. From their terminal wall, capillaries and metarterioles are produced.
E. *Metarterioles* arise in some tissues as branches of terminal arterioles.
   1. Its proximal portion is covered by smooth muscle cells arranged in a discontinuous pattern, while the distal part lacks smooth muscle and is similar to that of capillaries.
   2. In tissues where metarterioles are present, capillaries arise from them. Ultimately, the metarterioles flow into a postcapillary venule.
F. *Capillaries* (from the Latin capillus, 'hair') form an anastomosed network of very small calibre tubules leading to a postcapillary venule.
   1. Characteristics:
      a. They arise as very fine branches from terminal arterioles and metarterioles.
      b. Their diameter is 7–9 μm.

c. Each capillary is usually relatively short, about 50 μm in length, but the total number of capillaries in the human body is 96,000 km.
d. In true capillaries, the point of origin of the capillary at the wall of the terminal arteriole or metarteriole may have a ring of smooth muscle that acts as a *precapillary sphincter*, which when closed or opened prevents or allows, respectively, the passage of blood through the capillary. In some tissues there are no such sphincters.
e. Capillaries are the *exchange vessels* where most of the diffusion of substances between blood and tissues takes place.
f. The concentration of capillaries reflects the metabolic needs of a tissue. In the myocardium there are 2000 capillaries per mm^2. In the dermis 50.

2. Histological organisation. Its wall is very thin and consists of:
   a. *Endothelium*, which has a single layer of very flattened cells (simple squamous epithelium).
      (1) In small capillaries the lumen is surrounded by a single cell, in larger capillaries it may take two to three cells to surround the entire perimeter.
      (2) The edges of the cells overlap in marginal folds that project into the lumina with *occluding fasciae*.
      (3) The endothelial cells are very thin, 0.2–0.4 μm. The region containing the nucleus is somewhat thicker and protrudes into the vascular lumen.
      (4) Most of the few organelles are located near the nucleus.
      (5) *Caveolae* (up to 1000 caveolae per μm^2) are very abundant in the luminal and basal side membranes.
         (a) The caveolae may form cluster structures, where several caveolae detached from the membrane are in contact with each other.
         (b) Connections may be established between caveolae of the luminal and basal membranes, forming temporary channels.
      (6) In the cytoplasm there are vimentin or desmin *intermediate filaments* and *actin microfilaments*.
      (7) Cells are joined together by *tight junctions (zonula occludens)*, which can be discontinuous (to facilitate *paracellular transport*), *gap junctions* and *cadherin-mediated* adhesion junctions.
   b. *Basal lamina*.
   c. *A network of reticular and collagen fibres* that surround the basal lamina and intermingle with the extracellular matrix of the surrounding tissues.
   d. *Pericytes*: These are cells found surrounding terminal arterioles, capillaries and postcapillary venules.
      (1) Organisation:
         (a) They are cells with branched cytoplasm, with main branches that follow the main direction of the blood vessel, and transverse secondary branches that wrap around the capillary.
         (b) They are surrounded by a *basal lamina* that fuses with that of the endothelium.

(c) There are *gap junctions* with endothelial cells.
(d) They have *actin microfilaments* and *myosin filaments* and are contractile cells.
(e) Some pericytes may emit very fine extensions that contact the cytoplasm of pericytes in other nearby capillaries. These *interpericyte tunnelling nanotubes* form gap junctions with the pericytes they contact and serve to couple the contractile activity of both cells.

(2) Types of pericytes:
(a) *Helical pericyte*, the extensions of which coil around the capillary forming a helix. They are common in the central sections of capillaries.
(b) *Mesh pericyte*, whose prolongations are arranged in a mesh around the capillary. They are common in terminal arterioles and at the starting points of capillaries. They have a greater contractile capacity.

(3) Functions of pericytes:
(a) Pericyte contraction influences the perfusion capacity of the capillary (i.e. the volume of blood flow).
(b) When injury occurs, pericytes are activated and become endothelial and muscle cells of the new vessels, and are therefore considered to be equivalent to *mesenchymal reserve cells*.

3. Types of capillaries:
   a. *Continuous capillaries*: The endothelium is uninterrupted and the basal lamina is continuous (Fig. 12.6). Typical of muscle tissue, exocrine glands and nerve tissue.
   b. *Fenestrated capillaries* (from Latin fenestra, 'window'): Their endothelium is perforated by numerous pores (Fig. 12.7). These perforations are in discrete regions that form *sieve plates*, separated from each other by other non-perforated areas. The perforations are 60–100 nm in diameter. Two types of fenestrated capillaries are distinguished:

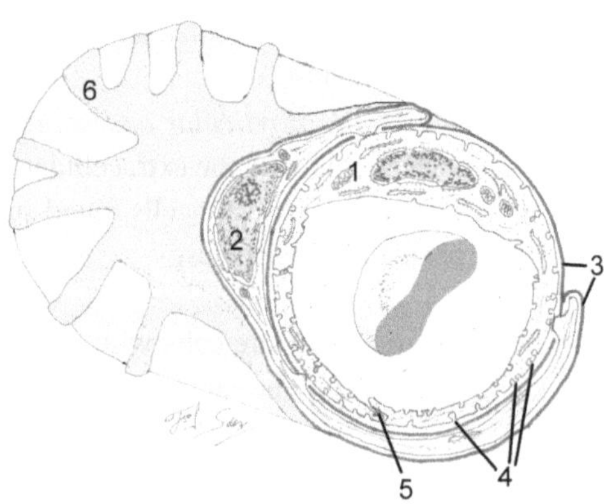

Fig. 12.6 External longitudinal view and cross-section of a continuous capillary. (1) Endothelium; (2) pericyte; (3) basal lamina; (4) caveolae; (5) occluding fascia; (6) pericyte extensions

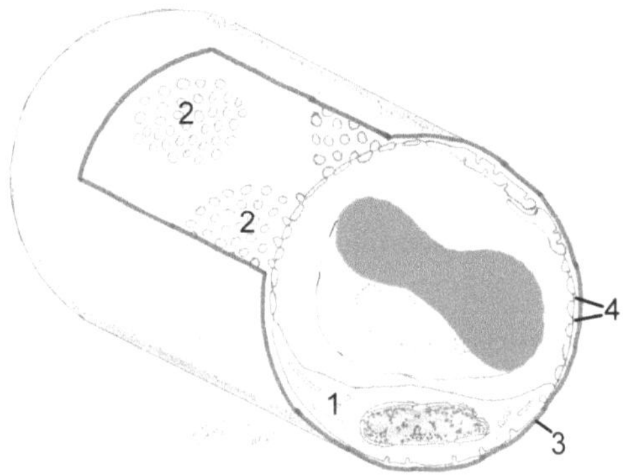

Fig. 12.7 External longitudinal view and cross-section of a fenestrated capillary. An area of the basal lamina has been removed to better observe the perforations. (1) Endothelium; (2) sieve plates with perforations; (3) basal lamina; (4) diaphragm

(1) *Fenestrated capillaries with diaphragm*: In these capillaries, the pores are closed by an ultra-thin, 4 nm thick, lamellar material called a *diaphragm*.
   (a) The diaphragm has a fibrillar appearance and in its centre there is a thickening called the *central nodule*.
   (b) These diaphragms may be artefacts produced by tissue processing.
   (c) They are found in kidney, intestine and endocrine glands.

(2) *Discontinuous capillaries*: These are fenestrated capillaries whose pores do not have a diaphragm. They are characteristic of the renal glomeruli.

c. *Sinusoids* (from Latin sinus 'sinus' and Greek -ειδής -eidḗs 'like'): They are capillaries with a wide span, 30–40 μm, and irregular contour, as they tend to adapt to the spaces between the components of the tissues they pass through.
   (1) They are found in liver, bone marrow and spleen.
   (2) The basal lamina may be missing or discontinuous.
   (3) There are several types of sinusoids:
      (a) In some organs, sinusoids are *discontinuous* because the cells contact each other at certain points, but in others they are separated by large gaps.
      (b) In others there are large *perforations* through the endothelial cell.
      (c) There are large *fenestrated capillaries* in the pituitary and adrenal glands that are considered sinusoids.

G. *Postcapillary venules* collect blood from the capillaries.
   1. They have a diameter of 15–30 μm.
   2. Its wall is capillary-like, usually with *fenestrations* and highly branched *pericytes*.
   3. The junction between endothelial cells is not as strong as in arteries and capillaries. These vessels are highly permeable and are the preferred site for *leukocytes* to leave the bloodstream and pass into the tissues, by a process called *diapedesis*. Leukocytes pass between or through endothelial cells.

4. They are very sensitive to vasoactive agents such as *histamine*, which causes *inflammation*. When inflammation occurs, gaps of up to 500 nm open up through which blood fluid escapes and accumulates as tissue fluid, forming *oedema*.
5. In some lymphatic organs, the endothelial cells in these venules are cubic or cylindrical and are specialised in the identification and segregation of lymphocytes that must pass into these organs (*high endothelial venules, HEV*).

12.7 Arteriovenous Anastomoses or Arteriovenous Shunts

A. In some organs, the ends of arterioles communicate with veins, in addition to capillaries, by larger connections called *arteriovenous anastomoses* (from the Greek ἀναστόμωσις anastómōsis, 'opening, mouth').

B. Histological organisation:
   1. They have thick vessel lumen walls, with smooth muscle and a lot of vasomotor innervation, especially from the *sympathetic system*.
   2. In the central region, the smooth muscle cells are thick and longitudinally arranged, and have therefore been described as *epithelioid cells*.
   3. There is no internal elastic lamina.

C. Function: Arteriovenous anastomoses act as *short circuits*:
   1. If the musculature of the anastomosis contracts, blood passes through the capillary network.
   2. If the muscle wall relaxes, blood passes directly from the arteriole to the veins without passing through the capillary network.

D. There are two types of arteriovenous anastomoses:
   1. *Simple bridge-like anastomoses (direct arteriovenous shunts)*: These are relatively simple straight ducts.
   2. *Glomus organs* or *glomus* (from Latin glomus, 'ball'): They are twisting ducts that form tangles surrounded by a connective tissue *capsule* and supplied with many adrenergic and cholinergic nerve fibres.

E. Location:
   1. *Erectile tissue* of the penis and clitoris. Its role is to control *erection*:
      a. Smooth muscle contraction closes the vessel and sends blood to the erectile tissues, causing the erection.
      b. Smooth muscle relaxation opens the anastomosis and has the opposite effect.
   2. Glomus organs abound in certain areas of the *skin*, where they play an important role in the body's *thermoregulation*:
      a. Contraction of the anastomosis increases blood flow through the capillary network of the skin and allows heat loss.
      b. Relaxation prevents blood flow through the capillaries and helps to conserve heat.

12.8 Veins

A. *Veins* carry blood from the tissues back to the heart. They usually run parallel to the arteries, although they carry blood in the opposite direction.
B. In the pulmonary circulation, the veins do not run parallel to the arteries.
C. A vein can be distinguished from an artery that runs parallel to it by several criteria:
   1. In general, the venous lumen is larger and its walls are proportionally thinner.
   2. Smooth muscle forms bundles, rather than layers, which sometimes run longitudinally.
D. The blood circulating in the veins circulates at very low pressure, which slows down circulation.
   1. In some cases, the movements of the skeletal muscle of the body press on the veins and drive the blood.
   2. This impulse is important in the lower limbs.
E. Veins are arbitrarily classified according to their size.
   1. *Venules*: These vessels are slightly larger than the postcapillary venules.
      a. The venules run parallel to the arterioles, but carry blood in the opposite direction, back to the heart,
      b. Several types can be distinguished:
         (1) *Collecting venules*, 30–50 μm in diameter.
            (a) Completely surrounded by pericytes. They carry blood to other venules of larger diameter.
            (b) They are sometimes considered postcapillary venules.
         (2) *Muscular venules*, 50–100 μm in diameter.
            (a) Structure:
               1. *Tunica intima*: formed by a continuous endothelium and subendothelial connective tissue.
               2. *Tunica media*: has one or two layers of circularly oriented smooth muscle cells.
               3. *Tunica adventitia*: usually thick with considerable collagen and elastic fibres.
         (3) In some organs, exceptionally, there are venules without muscle.
   2. *Veins* (*capacitance vessels*): The venules empty the blood into small and medium-sized veins, also parallel to the arteries, but with much greater capacity, so they contain a large volume of blood, larger than that of the arteries.
      a. Classification:
         (1) *Small veins*: These are veins with a diameter of 0.1–1 mm.
         (2) *Medium-sized veins*: They are between 1 mm and 1 cm in diameter.
      b. Structure:

(1) *Tunica intima*: it has the endothelium and a thin subendothelial layer of reticular fibres and, rarely, elastic fibres that do not form an elastic lamina. It contains some smooth muscle cells.

(2) *Tunica media*: there are several layers of smooth muscle cells. The muscle cells are loosely organised and intermingled with collagen fibres and fibroblasts.

(3) *Tunica adventitia*: this is the thickest layer, especially in medium-sized veins, and has longitudinally arranged collagenous and elastic fibres and scattered smooth muscle cells.

c. Blood flow:

(1) After passing through the capillary network, the blood has lost the pressure with which it was propelled from the heart and may have difficulty circulating, especially in the lower limbs, where it must rise against gravity.

(2) *Skeletal muscle* movements can push on the veins and propel blood. Now, these movements can push the blood forward or backward.

d. *Valves*:

(1) To prevent blood from flowing in the opposite direction, some medium-sized veins with a diameter of more than 2 mm, especially in the lower limbs, have *valves* that prevent the return of blood.

(2) The valves are made up of two semilunar *flaps* formed by folds of the tunica intima:

(a) If the blood moves towards the heart, it pushes the flaps apart, allowing it to move forward.

(b) If the blood moves in the opposite direction, it pushes the flaps backwards and the flaps close the passage, preventing backward movement.

(3) The wall of the veins is thinner in the area of the valves. When the blood flows backwards and closes the valve, this region bulges and forms the *valvar sinus*.

3. *Large veins*: These are veins more than 1 cm in diameter, some of which are very large. They have the following layers:

a. *Tunica intima*: it has endothelium, a thick subendothelial layer with fibroblasts, a network of elastic fibres and some smooth muscle cells.

b. *Tunica media*: it is relatively thin. The boundary between this layer and the tunica intima is diffuse. It has few layers of circularly arranged smooth muscle and some fibroblasts.

c. *Tunica adventitia*: this is very thick, with collagen and elastic fibres and smooth muscle cells arranged longitudinally, and is therefore often called *adventitia muscularis*. The cava and pulmonary veins may have cardiac muscle cells close to the heart (*myocardial sleeves*).

12.9 Portal Systems

A. A portal system carries blood from one capillary network to another capillary network.
B. Two types of portal system can be found:
   1. *Arterial portal system*: An artery or arteriole connects the two capillary networks. It is found in the *portal system of the kidney*, where it connects the renal corpuscles and the tubules of some nephrons.
   2. *Venous portal system*: A vein connects the two capillary networks. These are of two types:
      a. *Hepatic portal vein*, which connects the intestine to the liver.
      b. *Hypothalamo-pituitary portal system*, which connects the hypothalamus with the *pars distalis* of the pituitary gland.

12.10 Lymphatic Vessels

A. *Lymphatic circulation* is a drainage system:
   1. Blind-end vessels in the tissues collect excess tissue fluid and transport it in the form of *lymph* to larger lymphatic vessels.
   2. The last lymphatic vessels, the lymphatic ducts (*thoracic duct* and *right lymphatic trunk*), flow into the veins located in the neck region.
B. It is estimated that the lymphatic circulation returns 2–3 L of fluid to the blood every 24 h.
C. If the outflow of lymph is blocked in an organ, the lack of drainage in the tissue leads to *oedema*.
D. Functions of the lymphatic vessels:
   1. Circulatory functions:
      a. Drain excess tissue fluid and return it to the bloodstream.
      b. Transporting large molecules absorbed in the intestine that cannot penetrate into the blood capillaries.
   2. Immune functions:
      a. Transporting *lymphocytes* from one organ to another.
      b. Transporting *antigens* to the lymph nodes to trigger an immune response:
         (1) The lymphatic vessels reach the lymph nodes (*afferent lymphatic vessels*).
         (2) There, the lymph is scanned for pathogens and activates the immune response.
         (3) Lymph leaves the lymph nodes through new lymphatic vessels (*efferent lymphatic vessels*).
E. Composition of lymph: Lymph is a fluid derived from the tissue fluid and is therefore ultimately derived from blood plasma. It contains:
   1. Water, electrolytes and some protein (2–5%).

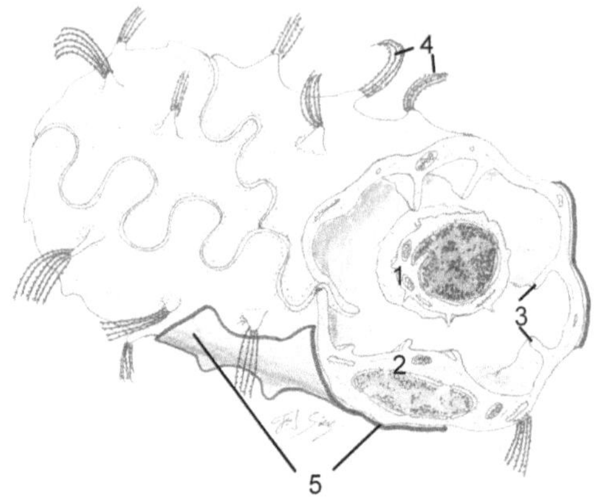

Fig. 12.8 External longitudinal view and cross-section of a lymphatic capillary. (1) Lymphocyte; (2) endothelium; (3) endothelial cell folds; (4) collagen fibres; (5) fragment of basal lamina

2. *Interdigitating dendritic cells* that have contacted a pathogen (*antigen*) in the tissue of origin, can penetrate the lymphatic vessels and travel through the lymph to reach the lymph nodes.
3. In the lymph nodes, many *lymphocytes* from the blood pass into the lymph, so the lymph often contains lymphocytes.
4. Lymphatic vessels in the intestine also carry *cholesterol* and *fatty acids* absorbed in the intestine.

F. The main types of lymphatic vessels are described as follows:
   1. *Lymphatic capillaries*: Lymphatic capillaries are the smallest vessels that collect lymph from the tissues (Fig. 12.8). The main characteristics are as follows:
      a. Their diameter is larger than that of blood capillaries (up to 100 μm).
      b. One end is blinded, so the lymph produced travels to the opposite end and into the larger lymphatic vessels.
      c. They have a very thin *endothelium*, except in the region where the cell nuclei are located.
         (1) Sometimes it has microvilli or folds that project into the capillary lumen.
         (2) Sometimes these folds occur at the cell periphery, at contacts with other endothelial cells, and extend to cover neighbouring cells.
      d. The junction between cells is loose, there are no continuous tight junctions, so there are gaps through which certain substances can easily pass.
      e. The *basal lamina* is discontinuous or absent and there are no pericytes.
      f. *Anchoring filaments* attach to the outer surface of the endothelium.
         (1) These are fibres from the extracellular matrix of the surrounding tissue that attach either directly to the endothelial plasma membrane or to the remnants of the basal lamina.
         (2) It appears that there is always a basal lamina remnant at the point of attachment.
         (3) The anchoring filaments ensure that the capillary does not close when there is a lot of pressure from the surrounding tissue.

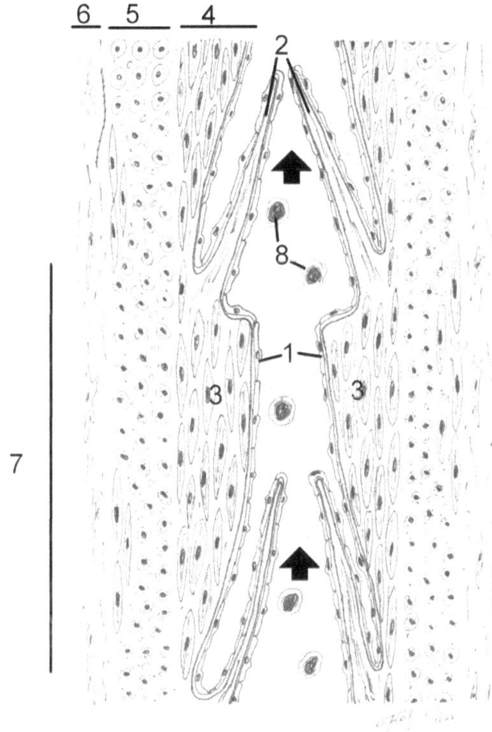

Fig. 12.9 Longitudinal section of a collecting lymphatic vessel. (1) Endothelium; (2) valves; (3) subendothelial pads; (4) tunica intima; (5) tunica media; (6) tunica adventitia; (7) lymphangion; (8) lymphocyte. The arrows indicate the direction of lymph flow

2. *Collecting vessels*:
   a. The lymphatic capillaries join together to form larger vessels, with a very thin endothelial wall, thin connective tissue and some smooth muscle cells.
   b. These vessels gradually form thicker vessels, in which several layers with vague boundaries can be distinguished (Fig. 12.9).
      (1) The *tunica intima* has endothelium with its basal lamina. Sometimes an incomplete internal elastic lamina separating the tunica intima from the tunica media.
         (a) The tunica intima usually has *valves* similar to those in the veins and with the same function of preventing backflow. They consist of two layers of endothelium with a central zone of collagenous and reticular fibres.
         (b) In the tunica intima there may be *subendothelial pads* projecting into the vessel lumen. These pads consist of *smooth muscle cells oriented in* the same direction as the vessel (longitudinally).
      (2) The *tunica media* has one or at most three layers of smooth muscle cells arranged circularly with respect to the vessel lumen. Occasionally, there may be some longitudinally oriented smooth muscle cells.
      (3) The *tunica adventitia* is thin.

c. The presence of valves and subendothelial pads divides the lymphatic vessel into bulb-like segments, which are separated by the valves. Each segment is called a *lymphangion*.

3. *Lymphatic channels* or *ducts*:
   a. The major lymphatic vessels eventually converge into two major channels, the *right lymphatic trunk* and the *thoracic duct*, which drain the lymph into the venous stream.
   b. They have a tunica intima, a tunica media and an adventitia with longitudinal bundles of smooth muscle and *vasa vasorum*.
   c. Their size may make them look like veins, but the greater thickness of their tunica media differentiates them.

G. Lymphatic vessels are usually seen close to a small artery or arteriole and a small vein or venule. The criteria for differentiating lymphatic vessels from veins are:
1. The lumen of the lymphatic vessels is devoid of visible cells, except for lymphocytes. Veins may have many erythrocytes and some leukocytes.
2. Lymph usually stains more intensely than blood plasma.
3. Valves are more numerous in lymphatic vessels than in veins.

H. *Lymph flow*: As the lymphatic vascular system lacks a driving pump, lymph flow is slow and takes place by two mechanisms:
a. *Intrinsic mechanism*: by contraction of the smooth muscles of the *lymphangion*. In large vessels they are slow: 2–3 per minute.
b. *Extrinsic mechanism*: due to compression caused by the contraction of adjacent skeletal muscles.

Take-Home Message

The cardiovascular system distributes blood throughout the body tissues. The large blood vessels have a wall formed by three layers of tissues, which from the inside out are the tunica intima, tunica media and tunica adventitia. The heart pumps blood into arteries, which branch into smaller arteries until they form arterioles. The arterioles carry the blood to the microcirculation, formed by terminal arterioles, metarterioles, capillaries and postcapillary venules; this is where the exchange of gases and other molecules between the blood and the tissues takes place. The blood returns to the heart by circulating through venules and veins that converge into increasingly larger veins. The lymphatic vascular system is responsible for draining excess fluid from the tissues and returning it to the bloodstream. It consists of lymphatic capillaries that collect excess tissue fluid and carry it through collecting vessels to lymphatic channels or ducts that converge in veins near the heart.

Further Reading

Breslin JW, Yang Y, Scallan JP et al (2019) Lymphatic vessel network structure and physiology. Compr Physiol 9:207–299
Friederici HH (1968) Tridimensional ultrastructure of fenestrated capillaries. J Ultrastruct Res 23:444
Frokjaerjensen J (1985) The continuous capillary – structure and function. Matematisk-Fysiske Meddelelser Kongelige Danske Videnskabernes Selskab 41:11–16
Kierszenbaum AL, Tres LL (2019) Cardiovascular system. In: Histology and cell biology – an introduction to pathology, 5th edn. Elsevier, pp 421–446
Krstić RV (1991) Cardiovascular system. In: Human microscopic anatomy. An atlas for students of medicine and biology. Springer Verlag, pp 39–67
Martinez-Lemus LA (2012) The dynamic structure of Arterioles. Basic Clin Pharmacol Toxicol 110:5–11
Maul GG (1969) Ultrastructuc of pores of fenestrated capillaries. J Cell Biol 43:A87
Maul GG (1971) Structure and formation of pores in fenestrated capillaries. J Ultrastruct Res 36:768
O'Dowd G, Bell S, Wright S (2023) Circulatory system. In: Wheater's functional histology. Elsevier, pp 144–158
Ota T (2023) Functional vascular anatomy and arteriovenous shunts of the spine. Semin Neurol 43:356–369
Paaske WP, Sejrsen P (1989) Permeability of continuous capillaries. Dan Med Bull 36:570–590
Pawlina W (2024) Cardiovascular system. In: Histology a text and atlas with correlated cell and molecular biology, 9th edn. Wolters Kluwer, pp 442–481
Petrova TV, Koh GY (2020) Biological functions of lymphatic vessels. Science 369:157
Rapp R, Ellaban NB, Kramer IRH (1976) Fenestrated capillaries in human dental pulps. J Dent Res 55:B85–B85
Rovenská E, Rovensky J (2012) Lymphatic vessels: structure and function. Isr Med Assoc J 13:762–768
Ryan TJ, Curri SB (1989) Blood-vessels and lymphatics. Clin Dermatol 7:25–36
Sherman JL (1963) Normal Arteriovenous anastomoses. Medicine 42:247
Shore RC, Moxham BJ, Berkovitz BKB (1984) Fenestrated capillaries in the periodontal-ligament. J Dent Res 63:513–513
Simionescu N, Simionescu M (1988) The cardiovascular system. In: Weiss L (ed) Cell and tissue biology – a textbook of histology, 6th edn. Urban & Schwarzenberg, pp 211–254
Smaje LH (1984) Permeability of fenestrated capillaries. Int J Microcirc Clin Exp 3:324–324
Sumagin R, Sarelius H (2013) Emerging understanding of roles for arteriole in inflammation. Microcirculation 20:679–692

Odontogenesis 13

What You Will Learn in This Chapter

This chapter introduces the morphology and structure of teeth, the basic terms for understanding dental organisation and histology, the different types of human dentition and basic notions about the chronology of their development. The embryonic origin of teeth from the oral epithelium and underlying ectomesenchyma and the processes of tooth crown and root formation are then described in detail. Finally, the molecular mechanisms that occur during odontogenesis and hypotheses on the formation of the dentition pattern are discussed.

13.1 Tooth Morphology and Structure

A. The teeth are made up of:
   1. A skeleton of mineralised tissue, the *dentine*.
   2. An internal cavity, the *pulp cavity*, surrounded by dentine and filled with connective tissue that constitutes the *pulp*.
   3. Other mineralised tissues lining the dentine:
      a. *Enamel*. Coats the dentine in the part of the tooth that protrudes into the oral cavity.
      b. *Cementum*. Coats the dentine in the part of the tooth that is included in the gingiva and is not visible.

B. The teeth consist of two parts (Fig. 13.1):
   1. *Crown*. This is the part that protrudes from the gingiva and is visible in the oral cavity.
      a. Here, the dentine is covered by enamel.
      b. The boundary between dentine and enamel is called the *dentinoenamel junction (DEJ)*.
      c. In reality, not all of the enamel-covered area protrudes into the oral cavity, which is why two types of crowns are usually distinguished:

F. J. Sáez, I. Badiola, *Essential Oral Histology*, Learning Materials in Biosciences,
https://doi.org/10.1007/978-3-032-03165-5_13

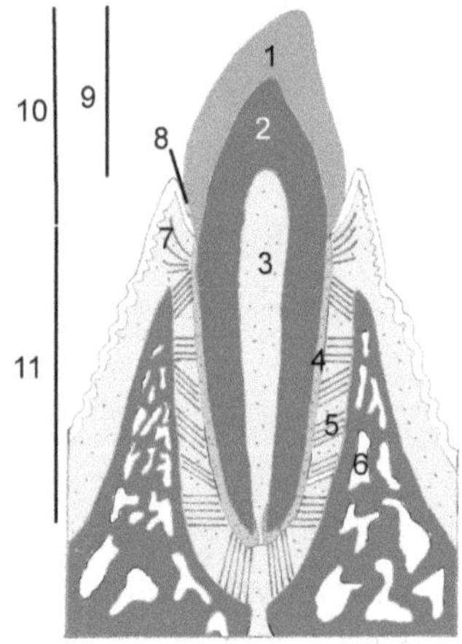

Fig. 13.1 Inciso-apical section of a tooth. (1) enamel; (2) dentine; (3) dental pulp; (4) cementum; (5) periodontal ligament; (6) alveolar bone; (7) gingiva; (8) dentogingival junction; (9) clinical crown; (10) anatomical crown; (11) root

(1) *Clinical crown*: the part that is visible because it is not covered by the gingiva.

(2) *Anatomical crown*: this is the entire part of the tooth covered with enamel. It includes the clinical crown but also a small area that is not visible in the oral cavity because it is covered by the gingiva.

2. *Root*. This is the part of the tooth that is not visible, as it is inside the gingiva, and is not covered with enamel. Here, the dentine is covered by:
   a. The cementum. The area where dentine and cementum meet is called the *cementum-dentine junction (CDJ)*.
   b. The *periodontium*, made up of all the tissues that support and protect the tooth. Two components can be distinguished:
      (1) *Protective periodontium*, consisting of the *gingiva* and the *dentogingival junction*.
      (2) *Insertion periodontium*, consisting of:
         (a) *Cementum* (covering the dentine). Although strictly a part of the tooth, it is considered part of the periodontium because of its close relationship with the other components of the periodontium.
         (b) *Periodontal ligament* (fibrous connective tissue).
         (c) *Alveolar bone*, which forms the alveolus (teeth socket), i.e. the cavity in which the root is housed.

C. The crown and root contact in the *neck* or *cervical region*.

D. In order to describe a tooth, different anatomical areas are distinguished (Fig. 13.2):
   1. The side of the tooth facing the oral cavity is called the *lingual side* (in the lower jaw) or the *palatal side* (in the upper jaw).
   2. The side of the tooth that faces the *vestibule* (the space between the teeth and the cheeks or lips) is called the *free side*.

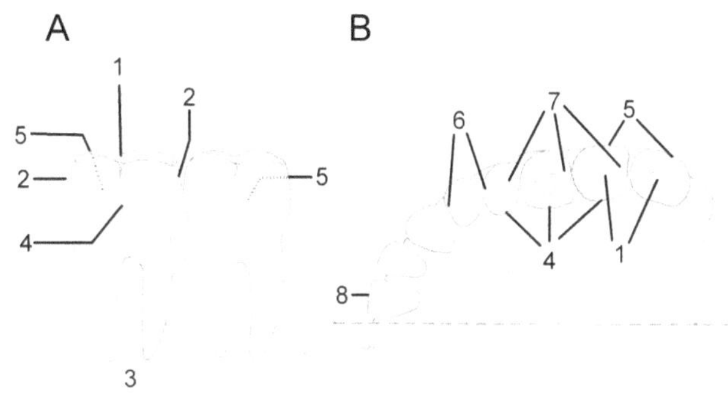

Fig. 13.2 Representation of the relationship between neighbouring teeth (**A**) and a quadrant or hemiarch of the upper jaw (**B**) to show the planes and surfaces of a tooth. (1) occlusal side; (2) proximal side; (3) apical side; (4) palatal (upper jaw) or lingual (lower jaw) side; (5) free side (buccal side in posterior teeth, labial side in the anterior teeth); (6) mesial (proximal) side; (7) distal (proximal) side; (8) incisal edge

   a. In anterior teeth (incisors, canines and first molars), it is often referred to as the *labial side.*
   b. In posterior teeth (second premolars and molars), it is usually called the *buccal side.*
3. The faces of the tooth that face a neighbouring tooth are called *proximal sides.*
   a. Normally, teeth have two proximal sides:
      (1) The *mesial side* is closest to the midline (the one that divides the human being into two nearly symmetrical halves).
      (2) The *distal side* is furthest from the midline of the body.
   b. The mesial side of each tooth contacts the distal side of the anterior tooth and vice versa.
   c. There are exceptions to the latter:
      (1) The mesial sides of the central incisors contact each other.
      (2) The distal aspect of the last molar does not contact any teeth.
4. The side of the tooth that contacts the tooth of the other arch is called the *occlusal side*, although in incisors and canines it is usually called the *incisal side* or *incisal edge.*
5. The end of the root, i.e. the part furthest away from the crown, is called the *apical side.*

E. Description of the histological components of a tooth. There are two ways to describe the relative position of the components of the tooth (Fig. 13.3):
   1. According to the *occlusal-apical axis*, when two microscopic dental structures are put in relation to each other, they are called:
      a. *Coronal*, closest to the crown or occlusal face.

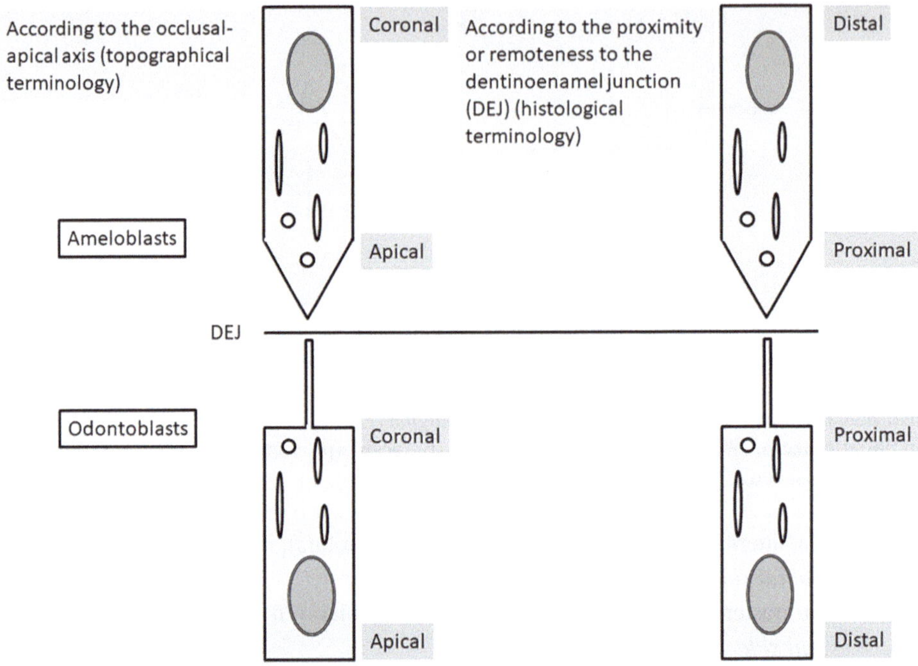

Fig. 13.3 Histological terminology of tooth structures

   b. *Apical*, that which is closest to the apical side of the tooth. This can be confusing because the term apical is used in epithelia to describe a pole of the cell opposite to the basal.

2. Taking the *DEJ* (dentinoenamel junction) at the crown or the *CDJ* (cementum-dentine junction) at the root as a reference, the terms can be used:
   a. *Proximal*, closest to DEJ or CDJ.
   b. *Distal*, furthest from DEJ or CDJ.

F. In humans, we have two dentitions, which makes us *diphyodontic* beings:
   1. *Primary, deciduous* or *milk dentition*.
      a. These teeth erupt into the oral cavity during the first years of life and begin to be replaced from the age of six.
      b. It consists of 20 teeth, 10 in the maxilla and 10 in the mandible, arranged in an arch. On each side of the arch (*quadrant* or *hemiarch*), there are:
         (1) Two *incisors*: one central and one lateral.
         (2) A *canine*.
         (3) Two *molars*, first and second.
   2. *Secondary* or *permanent dentition*.
      a. It is made up of teeth that erupt from the age of six.
      b. There are 32 pieces distributed as follows per quadrant:
         (1) Two *incisors*: one central and one lateral.

(2) A *canine*.
(3) Two *premolars*, first and second.
(4) Three *molars*, first, second and third.

13.2 General Characteristics of the Formation of the Human Dentition

A. The basic process of tooth formation is the same for all teeth.
B. The formation of teeth is called *odontogenesis* (from the Greek ὀδούς, ὀδόντος odoús, odontos, 'tooth' and γένεσις genesis, 'origin').
   1. Two phases can be distinguished in this process:
      a. *Morphogenesis*. The process during which a structure is formed and acquires the shape and organisation of the definitive tooth.
      b. *Histogenesis*. Process linked to the previous one, during which each constituent tissue of the tooth acquires its definitive cytological and histological characteristics, i.e. *differentiation* takes place.
   2. In this chapter, dental morphogenesis will be described. The processes of histogenesis will be explained in the chapters dedicated to the individual tooth tissues.
C. Origin of human dentitions:
   1. *Primary* or *deciduous dentition*. It is made up of teeth whose development begins in the embryo and is completed after birth.
   2. *Secondary* or *permanent dentition*. It gradually replaces the deciduous dentition from the age of six. It comprises two types of teeth:
      a. *Successional teeth*. They replace the deciduous teeth. They are the permanent *incisors, canines and premolars*. They originate from a different structure than primary teeth and accessional teeth.
      b. *Accessional teeth*. These are teeth that form at the back of the dental arches, in places where there were no milk teeth. They are the *molars* of the permanent dentition. They have the same origin as deciduous teeth, but erupt much later.
D. Depending on the structure from which the teeth are formed, two generations of teeth can be distinguished:
   1. *First generation*: deciduous and accessional teeth.
   2. *Second generation*: successional teeth.
E. Dental structures have double origin.
   1. From *surface ectoderm*: tooth enamel is produced from oral epithelium of *ectodermal origin*.
   2. From *cephalic ectomesenchyme*: the rest of the dental structures and insertion periodontium are formed from cephalic ectomesenchyme, i.e. from mesenchymal tissue originating from *neural crest* cells of *neuroectodermal origin*.
F. During the development and formation of the tooth, numerous events of *induction* occur between the various tissues that form the tooth. In fact, tooth formation begins

by induction processes between the ectodermal epithelium of the oral cavity and the underlying cephalic ectomesenchyma.

G. Morphogenesis is a continuous process, but in order to study it, dental development has been divided into several phases:
   1. *Crown formation*. Several phases can be distinguished in this case:
      a. *Bud stage*.
      b. *Cap stage*. An initial and a terminal phase can be distinguished.
      c. *Bell stage*. Here too, it is possible to distinguish between an initial and an advanced bell.
      d. *Appositional stage*. In many books, it is considered the continuation of the advanced bell phase.
   2. *Root formation and eruption*.

13.3 Origin of Dental Primordia

A. *Primary epithelial band* formation (Fig. 13.4).
   1. The first histological sign of dental development, prior to the formation of the dental epithelium, is the *condensation of ectomesenchyme* and the appearance of a capillary network beneath the epithelium that will later form the dental epithelium.
   2. From the fifth or sixth week of embryonic development, there is a thickening of the epithelium of the oral cavity over the condensation of mesenchyme:
      a. This thickening is called the *primary epithelial band*.
         (1) In the lower part of the oral cavity, the thickening is in the form of a continuous arch.

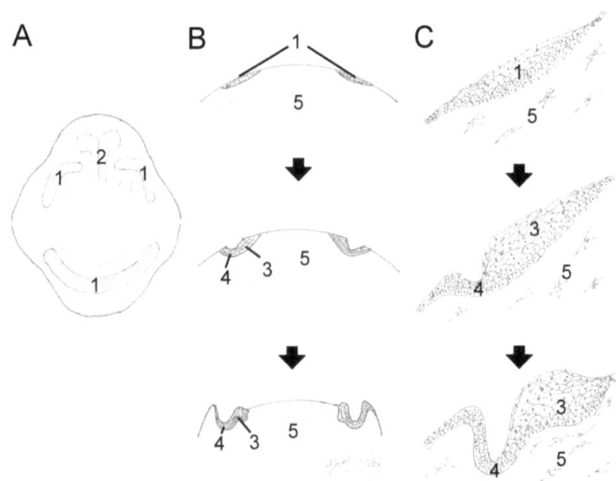

Fig. 13.4 Location of primary epithelial band, dental placodes and the dental lamina in the oral cavity (**A**), section of the oral wall showing the development of the dental lamina at low magnification (**B**) and at higher magnification (**C**). (1) dental placode; (2) nasobuccal fissure; (3) dental lamina; (4) vestibular lamina; (5) mesenchyme

(2) In the upper part of the oral cavity, the thickening of the epithelium appears in the form of an arch divided into two halves, one on each side of the central midline.

b. The formation of thickening at certain points of this epithelial band, called *dental placodes,* has been described. These placodes are an important point in tooth formation, as defects in placode formation lead to defects in dental development, such as teeth with aberrant morphology, small teeth, large teeth, and supernumerary teeth.

B. Formation of the *dental lamina* (Fig. 13.4).

1. In the seventh week of development, two proliferating processes occur in the primary epithelial band and invaginate into the underlying mesenchyme:
   a. The outward process is the *vestibular lamina*, which is divided into two layers separated by a space that will form the *vestibule* of the mouth, i.e. the space between the lips and cheeks and the teeth.
   b. The rest of the epithelial band forms the *dental lamina.*

C. Formation of the primordia of the deciduous teeth (Fig. 13.5).

1. Between the eighth and seventeenth week of gestation, five *outlines* or *primordia* of the future teeth are formed in each hemiarch (twenty outlines in total) from the dental lamina, also called the *general lamina*, on its free edge and in the labial direction.
2. These primordia give rise to the teeth of the deciduous dentition, which begin to develop into the bell stage.
3. The primordia are located on the labial side of the general plate and are connected to it by a short *lateral plate*.

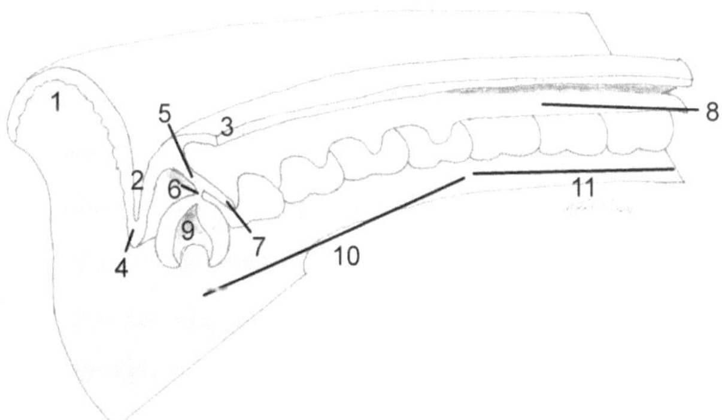

Fig. 13.5 Hemiarch of the developing lower jaw to show the general organisation of odontogenesis. This is a theoretical model, as it shows structures at different stages of development appearing at different times. (1) lip; (2) vestibule; (3) oral epithelium; (4) vestibular lamina; (5) dental lamina or general lamina; (6) lateral lamina; (7) successional lamina (will form the tooth germs of the successional teeth); (8) accessional lamina (accessional teeth); (9) tooth germ; (10) tooth germs of the deciduous teeth; (11) tooth germs of the accessional teeth

D. Formation of the primordia of permanent teeth (Fig. 13.5). The primordia of the permanent teeth are formed in early stages of development.
   1. *Successional teeth* are formed from the fifth month of intrauterine life until the tenth month of postnatal life.
      a. A prolongation of the general lamina is formed in the lingual (lower arch) or palatal (upper arch) direction with respect to the primordia of the deciduous teeth.
      b. This extension is called the *successional lamina.*
      c. From the successional lamina arise the primordia of the successional teeth.
   2. The *accessional teeth* are formed from three and a half to four months of foetal life.
      a. There is a backward elongation of the general lamina, which forms the so-called *accessional lamina.*
      b. First, the primordium of the first permanent molar is formed (three and a half to four months of development).
      c. The second molar primordium appears around the 6th month after birth.
      d. The third molar primordium may appear as early as 4 or 5 years of age.

E. As a general rule, the front teeth develop earlier than the back teeth.
   1. This applies to both deciduous and permanent teeth.
   2. This phenomenon is related to the fact that the jaws are growing, lengthening during development, and new teeth may appear in back places as the necessary space is created.

F. The following table summarises the dental development schedule:

	Deciduous	Permanent			
		Successional	Accessional		
			1st molar	2nd molar	3rd molar
Week 5–6	Epithelial band				
Week 7	Dental lamina				
Week 8	Primordia				
Week 11	Early cap				
Months					
3rd month	Early bell				
4th month	Appositional (mineralisation)	Permanent incisor outlines	Primordium		
Birth		Calcification	Calcification		
1st year	Final crown formation ----- Eruption and root formation begins			Primordium (6th month)	
Year 2					

(continued)

	Deciduous	Permanent			
		Successional	Accessional		
			1st molar	2nd molar	3rd molar
Year 3	End of eruption and root formation		Crown formation		
4th year					Primordium
6th year		Incisor eruption	Eruption		
10th year		End eruption of incisors ---- Eruption of premolar and canines	End of root formation	Eruption	Crown calcification
11th year				Eruption	
12th year		End of eruption of premolar and canines		Eruption	
16th year		End of root formation		End of root formation	
17th year and subsequent years					Eruption?

13.4 Crown Formation

A. During crown formation, a series of phases or stages can be distinguished to facilitate the study. It must be emphasised that the process is continuous, so there are transitional periods between the stages in which the tooth germ may show characteristics of two stages.

B. Microscopic observation of a histological preparation of a tooth germ can be complicated, as the tooth germ can be sectioned in different planes, thus showing the appearance of a different stage.

C. *Bud stage* (Fig. 13.6).
   1. It is so called because it is marked by the appearance, from the dental lamina, of solid *epithelial buds*. These buds will form the future enamel organs.
   2. Around the bud, the mesenchyme condenses.

D. *Cap stage*.
   1. As the shoot grows and changes its shape, an elongation occurs in the general lamina that joins the shoot to the lamina. This extension is called the *lateral lamina*.
   2. Histological sections sometimes show that the tooth germ is attached to the dental lamina by two epithelial extensions.
      a. These extensions border a region of mesenchyme called the *enamel niche*.

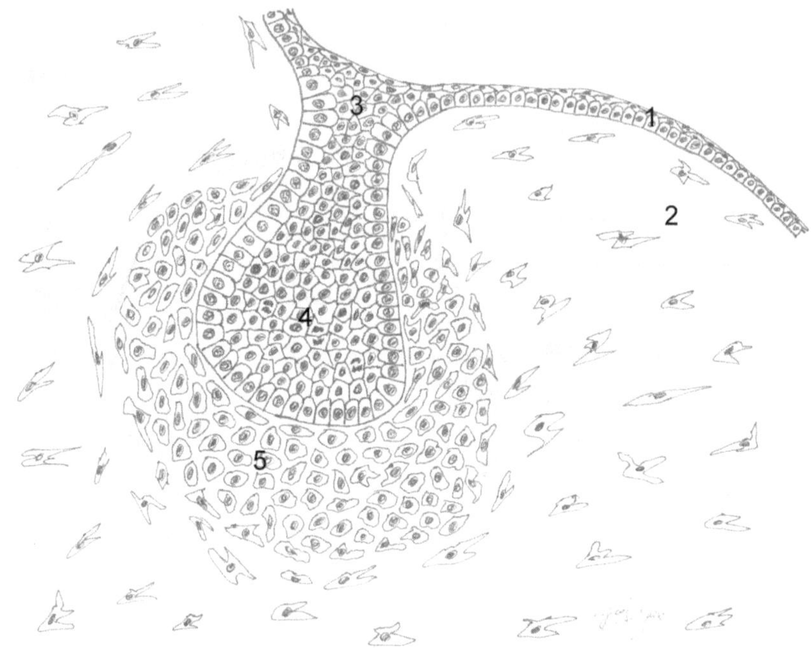

Fig. 13.6 Bud stage. (1) oral epithelium; (2) mesenchyme; (3) dental lamina; (4) bud; (5) condensed mesenchyme

b. In reality, this is a fictitious effect produced because the dental lamina is curved, and the cutting plane crosses the plate at two points.

3. This cap stage can be divided into an early and a late phase.
   a. *Early cap phase* (Fig. 13.7).
      (1) Formation of the *enamel organ*.
         (a) The bud grows, widens and an invagination occurs at its free end, and the bud takes the form of a *cap*. The bud becomes the *enamel organ*.
         (b) The epithelium of the enamel organ is differentiated. The layer of more tightly bound outer cells surrounds more loosely bound cells in the central part:
            1. The epithelial layer that lines the outside of the convex part of the cap is called the *outer dental epithelium*.
            2. The part that lines the concavity of the cap is called the *inner* or *preameloblastic dental epithelium*, as its cells will develop into the *ameloblasts* (enamel-producing cells).
            3. The central region is called the *stellate reticulum*:
               a. Between their cells, many spaces are progressively created and the cells acquire a stellate appearance and are joined together by the ends of their cytoplasmic extensions.

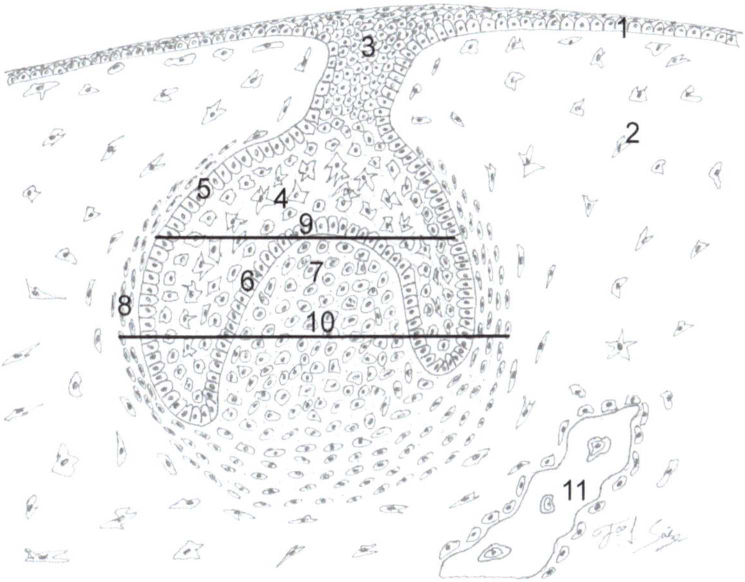

Fig. 13.7 Tooth germ at the early cap stage. (1) oral epithelium; (2) mesenchyme; (3) dental lamina; (4) stellate reticulum; (5) outer dental epithelium; (6) inner or preameloblastic dental epithelium; (7) dental papilla; (8) dental sac or dental follicle; (9) enamel organ; (10) tooth germ; (11) intramembranous ossification

b. The spaces between the cells fill with a viscous fluid rich in *glycosaminoglycans* such as *hyaluronic acid.*
c. These cells produce type I, II and III *collagen.*
d. In their cytoplasm, there are *tonofilaments* that are anchored to the *desmosomes* that link cells to each other. There are also *gap junctions.*
e. They produce chemical mediators that activate *osteoclasts* in their vicinity in order to remodel the bone surrounding the developing tooth.
f. They can produce *alkaline phosphatase*, an enzyme involved in tissue mineralisation.

(c) The entire enamel organ, which is an epithelium, is surrounded by a *basal lamina.*

(2) The enamel organ includes at its free end part of the condensed mesenchyme that surrounded it.
(a) This region of mesenchyme forms the *dental papilla.*
(b) The dental papilla begins to vascularise at this stage (the first *blood vessels* appear).

(3) The mesenchymal cells surrounding the cap and papilla form the *dental follicle* or *sac.*

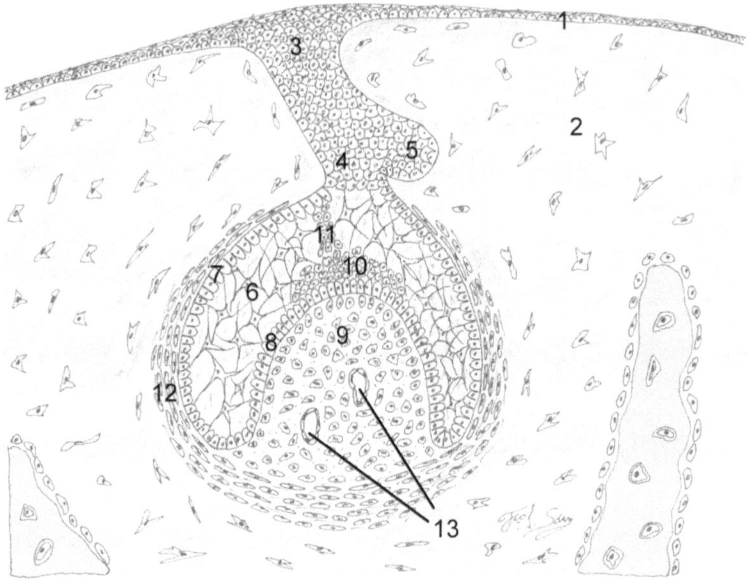

Fig. 13.8 Tooth germ at late cap stage. (1) oral epithelium; (2) mesenchyme; (3) dental lamina; (4) lateral lamina; (5) successional lamina; (6) stellate reticulum; (7) outer dental epithelium; (8) inner dental epithelium; (9) dental papilla; (10) primary enamel knot; (11) enamel cord; (12) dental follicle; (13) blood vessels

(4) The enamel organ, the dental sac and the dental papilla together constitute the *tooth germ.*

b. *Late* or *advanced cap stage* (Fig. 13.8). In this stage, the cap is larger and is distinguished by the appearance of two transitional structures, the cells of which die by *apoptosis.*

(1) *Primary enamel knot.*

(a) A cellular condensation of the stellate reticulum in contact with the inner dental epithelium. It is formed at the future site of the cusp or cusps of the tooth. Therefore, in incisors and canines, the primary knot is unique, but in premolars and molars, knots are produced on each cusp.

(b) It is made up of non-dividing cells and has been shown to express specific genes. Enamel knots are believed to be organisational centres for tooth shape and to produce molecules that regulate tooth development.

(2) *Enamel cord,* formed by a cord of cells that contact the enamel knot with the outer epithelium in its region opposite the inner dental epithelium. It is thought to be involved in the development of the shape of the enamel organ.

E. *Bell stage.*

1. The enamel organ grows and takes on the shape of a bell that surrounds the dental papilla. The edges of this bell are called *cervical loops.*

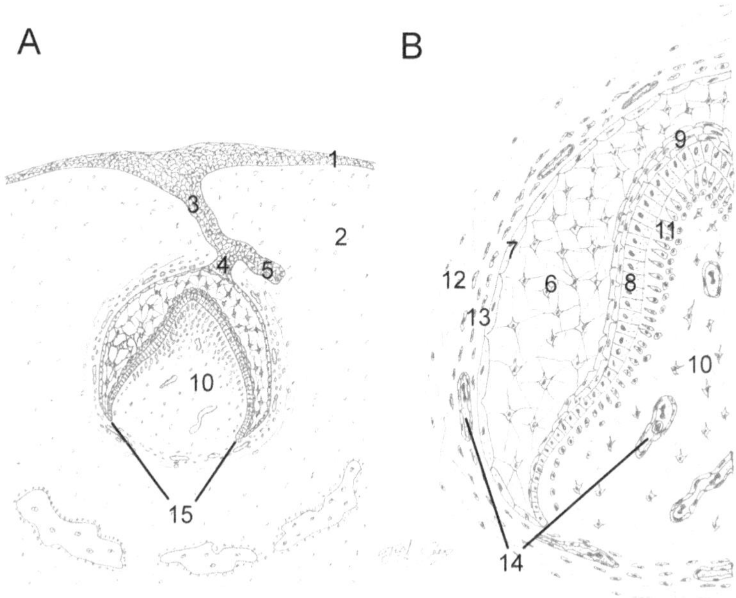

Fig. 13.9 Tooth germ in early bell stage (**A**), with an enlarged detail (**B**). (1) oral epithelium; (2) mesenchyme; (3) general lamina; (4) lateral lamina; (5) successional lamina; (6) stellate reticulum; (7) outer dental epithelium; (8) inner dental epithelium (preameloblasts); (9) stratum intermedium; (10) dental papilla; (11) preodontoblasts; (12) fibrous layer of dental sac; (13) cellulovascular layer of dental sac, (14) blood vessels; (15) cervical loops

2. The dental lamina gradually degrades, except where there are developing tooth germs or the primordia of permanent teeth (*succession lamina* and *accessional lamina*). As a result, the enamel organ is connected to the dental lamina and the oral epithelium by an elongated structure called the *dental pedicle*.
3. It is possible to distinguish between an early and a late bell.
   a. *Early bell stage* (Fig. 13.9). It is characterised by the following facts:
      (1) Two layers are observed in the dental sac, an inner layer with abundant cells and blood vessels (*cellulovascular layer*) and an outer *fibrous layer*.
      (2) The enamel knot and the enamel cord disappear.
      (3) The *stratum intermedium*, a new structure between the inner dental epithelium and the stellate reticulum, appears.
         (a) It consists of several layers, up to four or five, of flattened cells with elongated nuclei and no visible polarity.
         (b) They produce alkaline phosphatase, suggesting that they may be involved in enamel mineralisation.

(4) As the tooth germ grows, the stellate reticulum becomes thinner and the stratum intermedium becomes closer to the outer epithelium and thus to the blood vessels of the dental sac, from which it receives its nutrients.
(5) The cells of the inner dental epithelium progressively develop into *preameloblasts*. This differentiation starts at the cusps and progresses to the cervical loops.
(6) No clear differentiation is observed in the dental papilla, except for the cells of the papilla in contact with the inner dental epithelium, which gradually develop into *preodontoblasts*.

b. *Late bell stage* (Fig. 13.10), This phase is characterised by:
(1) The cells of the dental papilla closest to the inner epithelium become *odontoblasts* and begin to synthesise *dentine*. This occurs first in the cusp regions, and cell differentiation and dentine synthesis progress to the cervical loops.
(2) The first dentine produced, called *predentine,* is not mineralised. *Mineralisation* begins shortly thereafter in the regions of dentine closest to the internal dental epithelium and progresses to those closest to the odontoblasts.
(3) *Innervation* begins in the dental papilla.
(4) The cells of the inner epithelium begin to transform into *young ameloblasts*.
(a) The young ameloblasts are separated from the odontoblasts or dentine by their basal lamina, which marks the point of future *DEJ*.

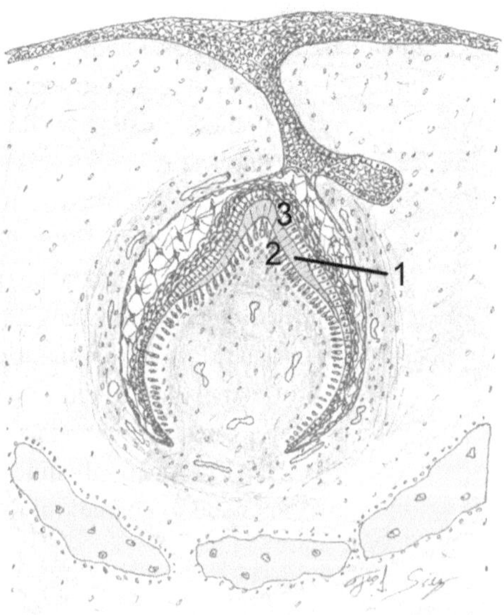

Fig. 13.10 Tooth germ in late bell stage. This stage differs from the previous one by the appearance of the first dentine. (1) dentine; (2) odontoblasts; (3) young ameloblasts.

(b) It is very important to note that young ameloblasts and preameloblasts can be nourished by the blood vessels of the papilla. However, as dentine apposition occurs between papilla and ameloblasts, this will become more difficult.

(5) The dental papilla develops into the *dental pulp*.

(6) During the development of the deciduous teeth, the *successional lamina* with the buds of the permanent teeth that will replace them are already visible at this stage.

(7) This phase is brief, as the next phase quickly starts. In fact, some authors consider both phases to be only one.

F. *Appositional stage* (Fig. 13.11). Some authors consider the late bell phase and this phase to be a single phase, and therefore it is sometimes included in the late bell phase. It is characterised by:

1. Young ameloblasts develop into *secretory ameloblasts* that produce *enamel*, which is deposited by apposition on dentine. This process starts at the cusps and progresses to the cervical loops.
2. The *dental lamina* (general lamina and lateral lamina) gradually disintegrates, although remnants called *Serres' pearls* remain.
3. The *stellate reticulum* thins and may disappear, allowing the stratum intermedium to contact the outer dental epithelium.

Fig. 13.11 Tooth germ in appositional stage (**A**), with an enlarged detail (**B**). (1) oral epithelium; (2) mesenchyme; (3) primordium of the successional tooth; (4) Serres' pearls; (5) dental papilla (dental pulp); (6) outer dental epithelium; (7) stellate reticulum; (8) stratum intermedium; (9) ameloblasts; (10) enamel; (11) dentine; (12) predentine; (13) odontoblasts

a. This process starts at the cusps and progresses to the cervical loops.
b. As a result of this thinning, the ameloblasts, which already find it difficult to obtain nourishment from the dental papilla, are now supplied with nutrients from the vessels of the dental sac.

4. The process continues until the entire crown is formed.
5. Eventually, the enamel organ atrophies and becomes the *reduced enamel epithelium*, formed by the fusion of inner dental epithelium, stratum intermedium and outer dental epithelium.

13.5 Root Formation

A. When crown formation is almost complete, root growth occurs. This is achieved by the proliferation of the cervical loops of the enamel organ (Fig. 13.12).
   1. The *cervical loops* of the enamel organ show the following organisation:
      a. They consist of an epithelium formed by two lamellae, one from the outer dental epithelium and the other from the inner dental epithelium.

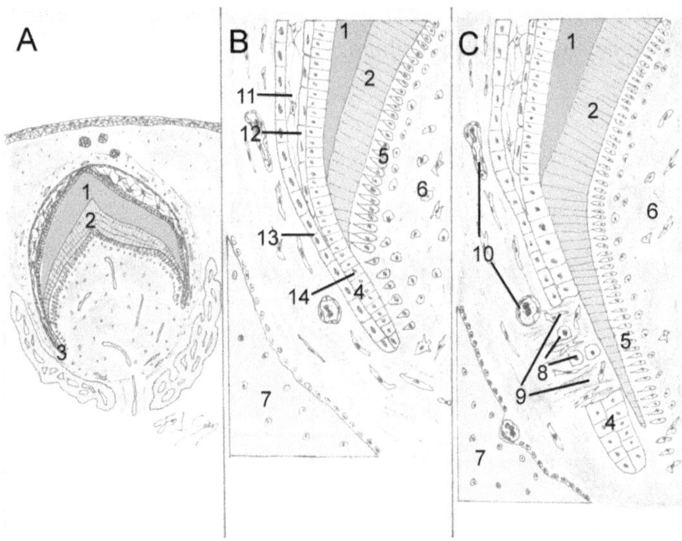

Fig. 13.12 Initial stages of root formation. From the already formed crown (**A**), it occurs the growth of the cervical loops of the enamel organ (**B**), and the synthesis of root dentine (**C**). (1) enamel; (2) dentine; (3) cervical loop; (4) Hertwig's epithelial root sheath; (5) odontoblasts; (6) dental papilla/pulp; (7) bone; (8) epithelial cell rests of Malassez; (9) mesenchyme of the dental sac invading the space left by the disintegration of Hertwig's sheath; (10) blood vessels; (11) stellate reticulum; (12) stratum intermedium; (13) outer dental epithelium; (14) inner dental epithelium

b. No stratum intermedium or stellate reticulum is observed between them.
c. This bilaminar structure is called *Hertwig's epithelial root sheath.*

2. The sheath proliferates and grows in an apical direction, enveloping the dental papilla.
   a. The dental papilla is surrounded by this sheath except in a small apical area where it communicates with the dental sac.
   b. In this apical area, the sheath leaves a small circular opening called the *epithelial diaphragm*, which indicates the site of the *apical foramen* of the future root.

B. In *single-rooted teeth*, the epithelial sheath in the diaphragm continues growing in an apical direction.
   1. As it grows, the sheath sends inductive signals to the papilla it surrounds, and the cells of the papilla closest to it become *odontoblasts* and begin to secrete *dentine* (the dentine of the future root).
   2. While Hertwig's epithelial sheath grows apically, in its cervical regions, those closest to the crown, disintegration of the sheath occurs. This occurs as dentine is deposited.
   3. In areas where Hertwig's epithelial sheath disintegrates, the root dentine is covered by:
      a. Remains of the sheath, which are called *epithelial cell rests of Malassez.*
      b. Tissue from the *dental sac*, which encompasses the epithelial cell rests of Malassez and comes into contact with the dentine. This tissue surrounding the root dentine and occupying the space of Hertwig's sheath will be transformed into the components of the *insertion periodontium.*
      c. The cells of the dental sac closest to the dentine become *cementoblasts*, which produce *cementum* that is deposited on the dentine.

C. In *multi-rooted teeth*, the process is essentially the same, except that Hertwig's root sheath divides into branches, one for each future root (Fig. 13.13).
   1. In a tooth with two roots:
      a. When the *epithelial diaphragm* and its apical foramen are produced, two opposing areas of the diaphragm emit extensions that are directed towards each other.
      b. When the two prolongations contact, they divide the apical foramen into two, each with its own epithelial diaphragm.
      c. In each of these foramina, the epithelial sheath progresses independently to form a root.
   2. In teeth with three roots, instead of two extensions, the diaphragm emits three extensions which, when joined together, form three epithelial diaphragms and three apical foramina.

D. Root growth occurs simultaneously with the *eruption* process of the tooth, which will be discussed in another chapter.

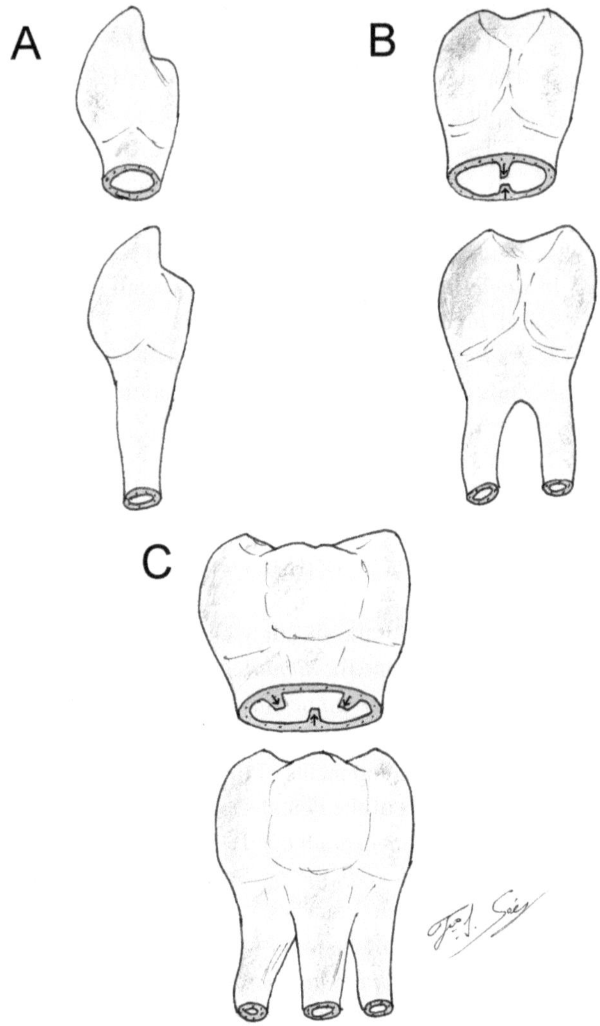

Fig. 13.13 Schematic representation of single-rooted (**A**), two-rooted (**B**) and three-rooted (**C**) tooth formation. The arrows indicate the direction of growth of the epithelial diaphragm extensions that divide the primary apical foramen and the diaphragm itself to produce the epithelial sheaths of each future root

13.6 Embryonic Origin of Dental Tissues and Periodontium

A. To summarise, this diagram shows the origin of each component of the tooth and the periodontium:

Surface ectoderm		Mesoderm	Neuroectoderm				
			↓				
↓		↓	Neural crests				
			↓				
Buccal epithelium		Mesenchyme	Ectomesenchyme				
↓	↓						
Dental lamina	Gingival epithelium	↓	↓		↓		
↓		Gingival chorion	Dental papilla		Dental follicle		
Enamel organ							
↓			↓	↓	↓		
Ameloblasts			Odontoblasts	Dental pulp	Insertion periodontium		
↓			↓		↓	↓	↓
Enamel			Dentin		Cement	Periodontal ligament	Alveolar bone

13.7 Induction Mechanisms During Odontogenesis

A. Dental development is regulated by multiple *induction* processes. Induction is one of the mechanisms of development that was studied in Chap. 2.

B. These processes are investigated with great interest as they can provide information for dentists in the field of tooth regeneration and repair.

C. Many *growth factors* and *transcription factors* have been shown to be involved in these induction processes, although the precise role of many of them remains to be elucidated. On the other hand, the fact that there are many molecules responsible for regulating the process ensures very fine control.

D. Many of these induction processes have been demonstrated experimentally. For the most part, these mechanisms are *epitheliomesenchymal interactions*, as structures derived from the *oral epithelium* and ectomesenchyme are involved in tooth development.

E. In the early stages of condensation of the ectomesenchyme and formation of the odontogenic epithelium (primary epithelial band), the *endoderm* of the anterior gastrointestinal tract is involved in an unclear way.

F. Induction processes between epithelium and mesenchyme take place through the exchange of information between the two tissues, which can occur in three ways:
   1. *Production of local chemical mediators*, i.e. substances that diffuse from one cell to another.
   2. *Contacts between molecules on cell surfaces*, i.e. molecules bound to cell plasma membranes. This implies that the cells must be in contact.
   3. *Extracellular matrix molecules*, which may be responsible for induction.

G. Experiments suggest that, at least in the early stages of tooth development, the induction processes between epithelium and mesenchyme are mainly driven by cell–cell contacts and the inductive action of the extracellular matrix, rather than by diffusible chemical mediators.
   1. In some experiments, mesenchyme and odontogenic epithelium were isolated at very early stages.

2. Once isolated, both tissues were cultured separately by a porous membrane that allows diffusion of substances, but not cell contact or contact with the extracellular matrix.
3. The result was that there was no differentiation of tooth germs.

H. During dental development, inductions occur in series, one after the other, as in a cascade of processes, so that when one of these processes occurs, the tissues formed induce a new process, creating a series of chained steps that give rise to the tooth.

I. Figure 13.14 shows, in a very simplified way, the main induction steps in the development of the tooth crown. In reality, some of the processes numbered in the figure may be the sum of several inductions that follow each other in rapid succession.

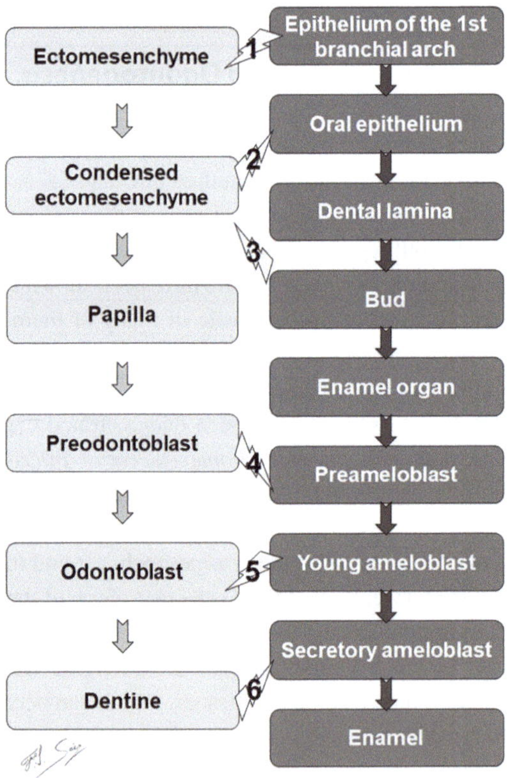

Fig. 13.14 Induction processes occurring during tooth development. The left column shows some of the stages through which the dental structures originating from the ectomesenchyme pass. The right column shows the structures originating from the epithelium. The arrows in each column indicate a differentiation process. The numbered rays between the columns indicate possible induction processes. For example, the epithelium of the first branchial arch induces (ray 1) the ectomesenchyme to condense. Once condensed, the mesenchyme induces the oral epithelium (ray 2) to produce the dental lamina. In this way, induction processes take place between the newly formed structures. The processes shown (numbered 1–6) are hypothetical, and there may be some more that have not been included here. Moreover, each induction process can be the sum of several simultaneous inductions

13.8 Development of the Dentition Pattern

A. The *pattern of dentition* is responsible for determining the type of tooth to be formed at each specific place in a dental arch, i.e. whether an incisor, a canine, etc., will be formed.

B. It is not clear how the pattern of dentition is regulated. Several hypotheses have been proposed:

1. *Odontogenic homeobox code hypothesis* (Fig. 13.15).
   a. According to this hypothesis, tooth pattern formation is controlled by the expression of certain *homeobox genes* in the odontogenic zone.

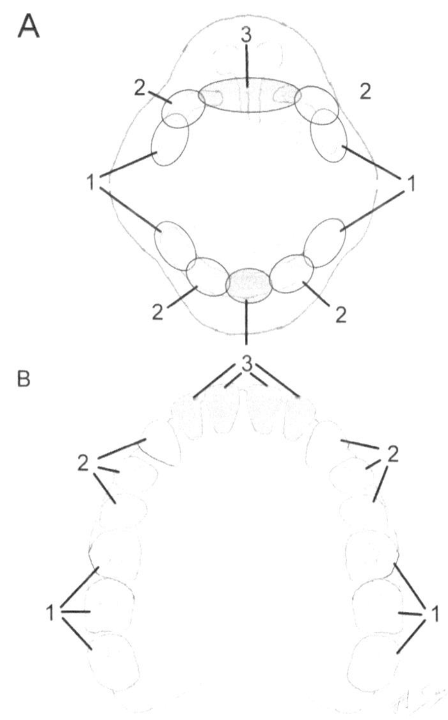

Fig. 13.15 Representation of the location of the odontogenic epithelium in the mouth of the embryo (**A**), as shown in Fig. 13.4. The shaded ovals indicate the regions where the ectomesenchymal cells express different homeobox genes, Barx1 (1), Msx2 (2) and Alx3 (3). In each of these regions, a tooth with a different shape is produced. Thus, in (**B**), the upper arch of the adult mouth has been represented, and each tooth has been shaded in the same tone as the ovals in (**A**). Molars are produced in the Barx1 expression zones (1), premolars and canines in the Msx2 expression zone (2) and incisors in the Alx3 expression zone (3). The expression of different homeobox genes is related to morphogen production. Thus, FGF-8 is more concentrated in zone 1, where Barx1 is expressed, and BMP-4 is present in region 3, where Alx3 is expressed. It should be noted that this scheme has been greatly simplified, as more homeobox genes seem to be involved

   b. Homeobox genes produce proteins that bind to nuclear DNA and regulate the expression of other genes.
   c. It has been shown that the ectomesenchyme in the region where the incisors are produced expresses certain homeobox genes, while the region where the molars are produced expresses different homeobox genes.
2. *Field model hypothesis* (Fig. 13.15). It is related to the previous one.
   a. According to this hypothesis, the homeobox genes would be responsible for the production of a different *morphogen* in each area.
   b. As explained in Chap. 2, a morphogen is a molecule that is produced at one site and diffuses into the tissue or adjacent tissues.
   c. Diffusion causes a *gradient of morphogen*, i.e. the concentration of this molecule will be higher near the point where it is produced than at distant points.
   d. The cells in each zone respond differently depending on the concentration of morphogen that reaches them.
   e. In the case of teeth, it has been shown that *FGF-8* and *FGF-9* are produced in the future molar regions, while *BMP-4* is produced in the incisor region.
3. *Clonal model hypothesis.*
   a. According to this hypothesis, all teeth are formed by the activity of a clone of *ectomesenchymal cells* and do not depend on morphogens present in the regions where teeth are produced.
   b. This clone may be determined to form teeth by the point of origin of the *ectomesenchymal cells* at the time of origin from the *neural crests.*
   c. This initial clone would settle and proliferate at a point in the future dental arch. At that point, the cells produce a type of tooth.
   d. As the jaw grows, the clone would proliferate and expand towards the back of the growing jaw. The moving cells, older than those that generated the clone, would have acquired the potential to create a different type of tooth. This would be repeated along the dental quadrant.
   e. In conclusion, the fact that a different tooth type is formed would not depend on local factors but on the age of the distally displacing cells.

Take-Home Message

Teeth consist of a crown protruding into the oral cavity and a root immersed in the gum. Both structures contact at the neck or cervical region. The tooth is formed by a mineralised tissue, dentine, which surrounds an internal cavity of connective tissue, the dental pulp, and is lined by other mineralised tissues, enamel in the crown and cementum in the root. The boundary between dentine and enamel is the dentinoenamel junction (DEJ), and the boundary between dentine and cementum is the cementum-dentine junction (CDJ). To describe dental tissues, we will use the terms proximal (when they are closer to the DEJ or CDJ) and distal (when they are further apart). Humans have two dentitions: the primary or deciduous dentition appears after birth, while the permanent dentition appears after the age of six. Permanent teeth can be either successional teeth (they replace primary teeth) or accessional teeth (they do not replace previous teeth). The crown is produced by induction mechanisms between the ectomesenchyma and the oral epithelium. A bud of the oral epithelium is produced and forms the enamel organ, which becomes bell-shaped. The mesenchyme enclosed in the bell forms the dental papilla, and the mesenchyme surrounding the bell forms the dental sac or follicle. The dental papilla will give rise to the pulp and dentine, and the enamel organ to the enamel. The epithelium of the enamel organ then grows apically and induces the formation of root dentine, while the dental sac produces cementum, periodontal ligament and alveolar bone. There are several hypotheses to explain why a different tooth is produced at each site: the odontogenic homeobox hypothesis, the field model hypothesis and the clonal model hypothesis.

Further Reading

Abramyan J, Geetha-Loganathan P, Sulcová M, Buchtová M (2021) Role of cell death in cellular processes during odontogenesis. Front Cell Dev Biol 9:671475

Bastos VC, Gomez RS, Gomes CC (2022) Revisiting the human dental follicle: from tooth development to its association with unerupted or impacted teeth and pathological changes. Dev Dyn 251:408–423

Bei M (2009) Molecular genetics of tooth development. Curr Opin Genet Dev 19:504–510

Berkovitz BKB, Holland GR, Moxham BJ (2018) Early tooth development. In: Oral anatomy, histology & embryology, 5th edn. Elsevier, pp 350–368

Chen J, Sun T, Lin B et al (2024) The essential role of proteoglycans and glycosaminoglycans in odontogenesis. J Dent Res 103:345–358

Chiego DJ (2023) Development of teeth. In: Essentials of oral histology and embryology, 6th edn. Elsevier, pp 59–74

Cobourne MT, Sharpe PT (2014) Tooth development. In: Hand AR, Frank ME (eds) Fundamentals of oral histology and physiology. Wiley Blackwell, pp 44–62

Fehrenbach MJ, Popowics T (2020a) Overview of dentitions. In: Illustrated dental embryology, histology, and anatomy, 5th edn. Elsevier, pp 197–210

Fehrenbach MJ, Popowics T (2020b) Tooth development and eruption. In: Illustrated dental embryology, histology, and anatomy, 5th edn. Elsevier, pp 52–76

Gómez de Ferraris ME, Campos Muñoz A (2009) Embriología dental (odontogénesis). In: Histología, Embriología e Ingeniería Tisular Bucodental, 3rd edn. Médica Panamericana, pp 113–136

Inubushi T, Nag P, Sasaki JI et al (2024) The significant role of glycosaminoglycans in tooth development. Glycobiology 34:cwae024

Jeong JK, Kim TH, Choi H, Cho ES (2024) Impaired breakdown of Herwig's epithelial root sheath disturbs tooth root development. Dev Dyn 253:423–434

Kierszenbaum AL, Tres LL (2019) Upper digestive segment. In: Histology and cell biology – an introduction to pathology, 5th edn. Elsevier, pp 511–542

Kovacs CS, Chaussain C, Osdoby P et al (2021) The role of biomineralization in disorders of skeletal development and tooth formation. Nat Rev Endocrinol 17:336–349

Lowe JS, Anderson PG (2015) Blood and lymphatic circulatory systems and heart. In: Stevens & Lowe's human histology, 4th edn, pp 143–165
McCollum M, Sharpe PT (2001) Evolution and development of teeth. J Anat 199:153–159
Nanci A (2013) Development of the tooth and its supporting tissues. In: Ten Cate's oral histology, 8th edn. Elsevier, pp 70–85
O'Dowd G, Bell S, Wright S (2023) Oral tissues. In: Wheater's functional histology. Elsevier, pp 241–252
Pawlina W (2024) Digestive system I: oral cavity and associated structures. In: Histology a text and atlas with correlated cell and molecular biology, 9th edn. Wolters Kluwer, pp 588–631
Rao PC, Jing JJ, Fan Y, Zhou CC (2023) Spatiotemporal cellular dynamics and molecular regulation of tooth root ontogeny. Int J Oral Sci 15:50
Smith TM, Tafforeau P (2008) New visions of dental tissue research: tooth development, chemistry, and structure. Evol Anthropol 17:213–226
Sui BD, Zheng CX, Zhao WM et al (2023) Mesenchymal condensation in tooth development and regeneration: a focus on translational aspects of organogenesis. Physiol Rev 103:1899–1964
Thesleff I, Humerinta K (1981) Tissue interactions in tooth development. Differentiation 18:75–88
Warshawsky H (1988) The teeth. In: Weiss L (ed) Cell and tissue biology – a textbook of histology, 6th edn. Urban & Schwarzenberg, pp 594–640
Yamanaka A (2022) Evolution and development of the mammalian multicuspid teeth. J Oral Biosci 64:165–175

Oral Mucosa

14

What You Will Learn in This Chapter
This chapter explains the histological organisation of the wall of the oral cavity, with special attention to the structure of the different types of epithelium. The cell types present in the epithelium and the processes of keratinisation and cornification are described. The three basic types of mucosa present in the oral cavity and where they are found are listed. Finally, the specialised mucosa of the tongue, with its lingual papillae and taste buds, is described in detail.

14.1 General Concepts

A. The *stomatognathic system* is made up of all the organs and tissues associated with the *jaws* to carry out the functions of the initial phase of the digestive process, such as:
   1. Ingestion.
   2. Salivation.
   3. Chewing.
   4. Tasting.
   5. Deglutition.
B. The stomatognathic system consists of:
   1. Both jaws.
   2. The articulation of the lower jaw with the base of the skull.
   3. The musculature associated with these bones.
   4. Teeth and their supporting structures.
   5. Salivary glands.
   6. The tissues of the mouth wall, including its blood vessels and nerves.
   7. Any structure found in the mouth (tongue, tonsils, etc.).
C. The stomatognathic system is important because it is involved in many functions: digestion, speech, facial expression, breathing, coughing, sneezing, yawning, kissing, etc.

F. J. Sáez, I. Badiola, *Essential Oral Histology*, Learning Materials in Biosciences,
https://doi.org/10.1007/978-3-032-03165-5_14

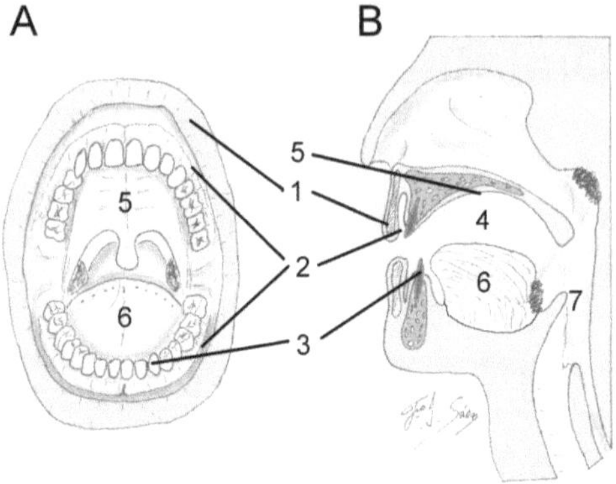

Fig. 14.1 View of the open mouth (**A**) and medial section of the head (**B**). (1) lips; (2) vestibule; (3) teeth; (4) oral cavity; (5) palate; (6) tongue; (7) pharynx

D. The *mouth* is where food enters the digestive tract (Fig. 14.1). It is a cavity with the following boundaries:
   1. The entrance to the mouth is limited by the *lips*.
   2. The passage to the oropharynx is marked by the *isthmus of the fauces (oropharyngeal isthmus)*.
   3. The *cheeks* laterally limit the oral cavity.
   4. At the top is the *palate*.
   5. At the bottom, the *floor of the mouth*, where the *tongue* is located.

E. Inside the mouth, the two *dental arches*, installed in the upper and lower jaws, delimit two spaces:
   1. The *buccal vestibule*, a space located externally to the teeth, between the teeth and the inner walls of the cheeks and lips.
   2. The mouth proper, also called the *oral cavity*. This latter cavity, when the jaws are closed, is almost entirely occupied by the tongue.

F. With the opening of the jaw that separates the teeth, the vestibule and the oral cavity are connected.

G. The organs of the digestive tract and others, such as the respiratory tract, are tubes that communicate with the external environment and are formed by a wall in which several layers can be distinguished:
   1. *Mucous membrane* or *tunica mucosa*. The innermost layer, which lines the lumen of these organs.
      a. There are usually associated mucous or serous glands, whose secretions moisten and lubricate the surface, which is therefore moist and more or less viscous.
      b. The tunica mucosa of these organs usually consists of:
         (1) An *epithelium*, of a different nature depending on the organ, which lines the lumen of the duct.
         (2) A *connective tissue* below the epithelium, towards the outside of the organ wall, called the *lamina propria*.[1]

[1] Sometimes called chorion (from the Greek χόριον chorion 'skin, leather, membrane').

2. *Submucosal layer* or *tunica submucosa*. A region of connective tissue different from the previous one and located externally to the tunica mucosa.
3. Other more external layers or tunics usually include a muscularis, a layer typically of smooth muscle, and other layers of connective tissue. This is variable in each organ.

H. The usual organisation of the walls of the digestive tract is partially present in the mouth but differs from that of other parts of the digestive tract depending on the area being studied. Its organisation can be summarised as follows:
   1. The *tunica mucosa*, which lines the inside wall of the mouth.
   2. Sometimes there is a *tunica submucosa*.
   3. In some cases, as in the hard palate, the submucosa is in direct contact with the *bone*. When there is no submucosa, the mucosa may rest directly on the bone (*mucoperiosteum*).
   4. In other areas, such as the cheeks, the submucosa contacts the *skeletal muscle* that forms the cheeks. Sometimes there is no submucosa, and the mucosa may rest directly on the muscle.

14.2 General Characteristics of the Oral Mucosa

A. The oral mucosa is histologically very similar to the *skin* but is more coloured due to several causes: degree of vascularisation, thickness of the epithelium, degree of keratinisation and amount of melanin.

B. The oral mucosa normally has a pinkish hue, but when inflammation occurs, it usually shows a red colour.

C. The mucosa of the oral cavity maintains its characteristic moisture thanks to the secretions of numerous glands called *salivary glands* (from the Latin ‘salīva’).

D. The oral mucosa consists of:
   1. An *epithelium*.
   2. A *lamina propria* of connective tissue.
   3. A *basal membrane is* situated between the epithelium and the lamina propria. Its structure is typical of basal membranes:
      a. *Basal lamina*, consisting of:
         (1) *Lamina lucida*.
         (2) *Lamina densa*.
      b. *Lamina reticularis*.

E. The embryonic origin of epithelium and lamina propria is different:
   1. The oral epithelium originates from the *surface ectoderm*.
   2. The lamina propria is of *mesenchymal* and *ectomesenchymal* origin, i.e. derived from *mesoderm* and *neural crest* cells.

F. The contact surface between lamina propria and epithelium is not flat and is often wavy in sections, which is a reflection of its three-dimensional organisation:

1. The epithelium has evaginations towards the lamina propria, called *epithelial ridges*.
2. Between these epithelial projections are regions of the lamina propria called *connective tissue papillae*.

14.3 Oral Mucosal Epithelium

A. The oral mucosa always has a *stratified squamous epithelium*, which may show a variable degree of *keratinisation*.
B. It is an epithelium very similar to that of the skin, with the difference that the latter is always keratinised. Consequently, much of what is known about the epithelium of the oral cavity is based on studies of the epithelium of the skin.
C. Three types of epithelia can be distinguished in the oral mucosa according to their degree of keratinisation (Fig. 14.2):
   1. *Keratinised (or orthokeratinised) stratified squamous epithelium.* Also called *cornified epithelium (horny epithelium).* Its cells are organised in four strata (each consisting of one or more layers of cells), which from the basal to the apical zone are:
      a. *Stratum germinativum (basal layer).*
      b. *Stratum spinosum (spinous layer* or *prickle cell layer).*
      c. *Stratum granulosum (granular layer).*
      d. *Stratum corneum (keratinised* or *cornified layer).* Its cells have the following characteristics:

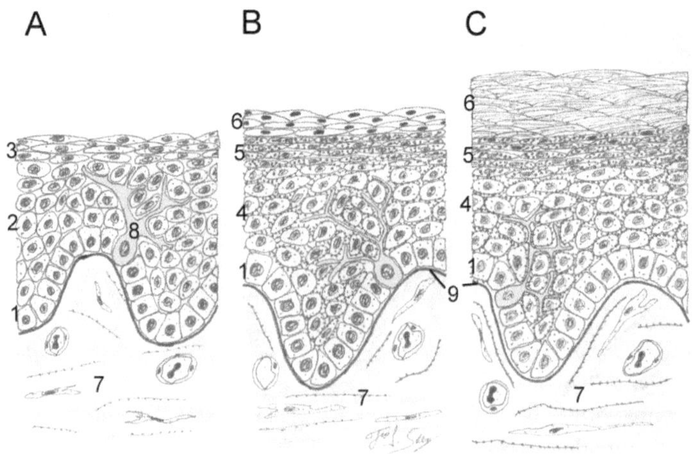

Fig. 14.2 Non-keratinised (**A**), parakeratinised (**B**) and keratinised or orthokeratinised (**C**) stratified squamous epithelium. (1) stratum germinativum; (2) intermediate layer; (3) superficial layer; (4) stratum spinosum; (5) stratum granulosum; (6) stratum corneum; (7) lamina propria; (8) melanocyte; (9) basal lamina

(1) They are dead cells without nuclei or other organelles.
(2) They are *cornified*, completely filled with *keratin* (from the Greek κερατίνη keratínē, 'horny, horn-like') and other associated substances, and bound together by a waterproofing lipoprotein-like material.
(3) They are very acidophilic.

2. *Parakeratinised stratified squamous epithelium.*
   a. It has the same strata as the keratinised epithelium.
   b. However, the stratum corneum, also called here *parakeratinised*, is characterised by cells with the following features:
      (1) They have a *pyknotic* nucleus (from the Greek πυκνός pyknós, 'dense'), i.e. a small, condensed and very dark nucleus, due to highly condensed chromatin.
      (2) They have a lower degree of keratinisation and cornification.
      (3) They are acidophilic (to a lesser extent than those of the keratinised epithelium).
3. *Non-keratinised stratified squamous epithelium.* It lacks a stratum granulosum and has only three strata:
   a. *Basal.*
   b. *Intermediate.* Equivalent to stratum spinosum.
   c. *Superficial.* It consists of flattened cells with very little acidophilic cytoplasm and little keratinisation, with an oval nucleus of normal appearance.

D. There are several cell types present in the oral epithelium that can be grouped into two categories:
   1. *Intrinsic cells.* These are the cells of the epithelial tissue.
      a. They are called *keratinocytes.*
      b. They are produced in the stratum germinativum, from *reserve stem cells*, and rise to the upper stratum corneum, where they die and are lost through desquamation.
      c. During this ascent, their characteristics and appearance change.
   2. *Permanent extrinsic cells.* These are cells originating from other areas of the body that settle in the epithelium. They are:
      a. *Langerhans cells.*
      b. *Merkel cells.*
      c. *Melanocytes.*

E. The organisation of the keratinised epithelium is described in detail below, as it is similar to that of the skin and is the best studied. The other types of epithelia have similar characteristics. However, some of the characteristics of the non-keratinised epithelium will be indicated.

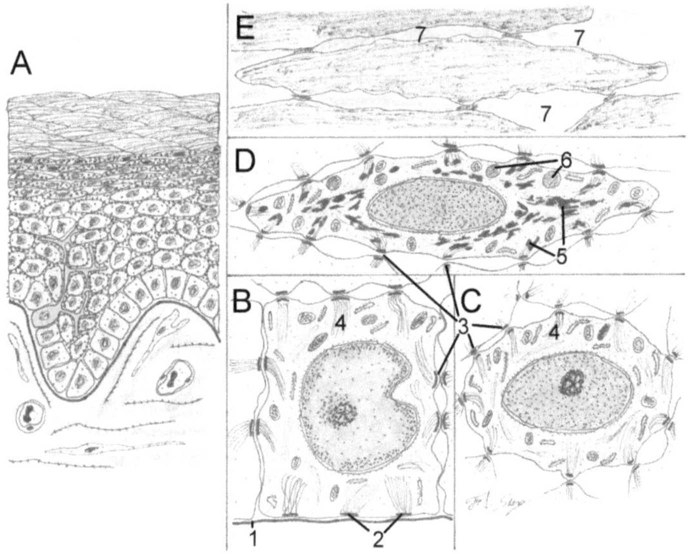

Fig. 14.3 Keratinised stratified squamous epithelium (**A**) and keratinocytes of the stratum germinativum (**B**), stratum spinosum (**C**), stratum granulosum (**D**) and stratum corneum (**E**). (1) basal lamina; (2) hemidesmosomes; (3) desmosomes; (4) tonofilaments; (5) keratohyalin granules; (6) keratinosomes (lamellar or Odland bodies); (7) hydrophobic envelope

14.4 Keratinocytes

A. Keratinocytes have different characteristics in each stratum, so we must study each one separately (Fig. 14.3).

B. *Keratinocytes of the stratum germinativum.* The basal layer consists of a single layer of cubic cells, varying in height, characterised by:
   1. The nucleus may be rounded or slightly oval.
   2. Due to the presence of many ribosomes and rough endoplasmic reticulum, the cytoplasm is highly basophilic.
   3. The proteins synthesised include those of the cell itself and those of the basal lamina, such as *type IV collagen.*
   4. Many *hemidesmosomes* are formed between the cells and the basal lamina.
   5. Between cells, there are *tight junctions*, *desmosomes*, and *gap junctions*.
   6. In the cytoplasm, there are many intermediate keratin filaments or *tonofilaments* formed by *cytokeratins 5 and 14*.
   7. They have high proliferation capacity. This is why this cell layer is called the *stratum germinativum.*

C. *Keratinocytes of the stratum spinosum.* In the stratum spinosum, keratinocytes are arranged in several layers; their shape is varied, cubic towards the basal zone (*parabasal cells*), more flattened in the more apical layers. They are characterised by:

1. In many histological preparations, filaments can be seen passing through the sparse intercellular space and connecting spiny evaginations on the surface of the cells, as if linking the cells together.
   a. It is because of this spiny appearance of the cells that it is called stratum spinosum. In reality, this appearance is an *artefact* due to the preparation of the tissue.
   b. What actually happens is that the cells are connected to each other by abundant *desmosomes*, to the cytosolic face of which numerous *tonofilaments* are anchored.
2. The cell cytoplasm is slightly basophilic.
3. The tonofilaments are *cytokeratin 1* and *10* in the lower layers and *cytokeratin 2* and *11* in the apical layers.
4. In the upper layers, the cells show 0.25 μm diameter vesicles in their cytoplasm. These granules contain lipids and have a lamellar appearance inside. They develop in the following layers and form the *keratinosomes*.
5. The stratum spinosum and the stratum germinativum together have been called *stratum Malpighi*.

D. *Keratinocytes of the stratum granulosum*. In the stratum granulosum, there are two or three layers of cells larger than those of the lower strata, which in histological cross-section appear spindle-shaped, but are in fact flattened, squamous cells. They are characterised by:
   1. Its nucleus is small and fairly dense.
   2. Very few organelles are observed in the cytoplasm.
   3. Granules 0.5–1.0 μm in diameter, called *keratohyalin granules,* are very abundant. These granules lack a membrane and are aggregates of proteins synthesised by surrounding ribosomes. They are highly basophilic and are intimately associated with the tonofilaments. They contain several proteins:
      a. *Profilaggrin*, a protein precursor of *filaggrin* (keratin *fil*ament *agg*regating pro*tein*). The latter is a cementing substance that binds the tonofilaments.
      b. *Loricrin.*
      c. *Trichohyalin.*
   4. *Tonofilaments* are very abundant and more phosphorylated than in the basal and spinous layers.
   5. *Keratinosomes, lamellar bodies* or *Odland bodies*. They are larger than in the stratum spinosum.
      a. They are usually oval vesicles 0.1–0.3 μm in diameter, with numerous lamellar structures inside.
      b. These vesicles fuse with the plasma membrane by exocytosis. The lipoprotein material they contain is incorporated into the cell membrane.
   6. Proteins such as *loricrin* and *involucrin* begin to be added to the inside of the plasma membrane, increasing the thickness of the membrane.
   7. They show intense synthetic activity, which is surprising because they themselves trigger their own death and within a few hours, they become dead cells.

E. *Keratinocytes of the stratum corneum.* In the stratum corneum, there are several layers of squamous cells, the *corneocytes*, without nuclei in keratinised epithelium or with pyknotic nuclei in parakeratinised epithelium. They are characterised by:
   1. Cytoplasms are highly acidophilic.
   2. Corneocytes are the end result of transformations of keratinocytes, which begin in the stratum spinosum and are accentuated in the stratum granulosum. In total, the cells undergo the following transformations:
      a. *Keratinisation.* The following events occur:
         (1) Keratohyalin granules produce *filaggrin* and trichohyalin. Both cause aggregation of cytokeratin intermediate filaments, leading to collapse and flattening of keratinocytes.
         (2) The cytoplasm loses its organelles and is filled with *tonofilaments* enveloped by material from the keratohyalin granules.
         (3) The density of filaments is variable, and there are *dark corneocytes* and *pale corneocytes*.
         (4) The cells are very dehydrated.
      b. *Aggregation of proteins to the inner side of the plasma membrane.* Certain proteins produced in the cytosol intertwine and associate with intermediate filament proteins just beneath the plasma membrane, causing a thickening on the inside of the membrane. This thickening involves:
         (1) *Loricrin.* It plays a major role (75%).
         (2) *Involucrin.* It makes up only 5% of the coat material, but its involvement is important as it also provides binding sites for the lipid constituents of the outer envelope of the membrane.
         (3) *Trichohyalin.* This protein binds intermediate filaments to each other and other proteins.
         (4) *Small proline-rich proteins* (*SPR proteins*).
      c. *Addition of lipids to the membrane.* This produces a much thicker plasma membrane than usual. The process is as follows:
         (1) Lamellar bodies are rich in *ceramides*, *free fatty acids* and *cholesterol.*
         (2) When these lamellar bodies fuse with the plasma membrane (*exocytosis*), these lipid materials are added to the plasma membrane.
         (3) The plasma membrane becomes a thick hydrophobic envelope.
   3. All these transformations turn the keratinocyte into a corneocyte, which is why this process is called *cornification.* Cornification includes:
      a. Keratinisation.
      b. Loss of organelles.
      c. The loss of the nucleus.

   d. The thickening of the plasma membrane by the incorporation of lipid and protein substances produced by the cells themselves. The complex structure that is produced has been called the *cell envelope (cornified cell envelope).*
4. In the keratinised (orthokeratinised) epithelium of the oral mucosa, the cornification processes are not as intense as in the epithelium of the skin.
5. The cells of the stratum corneum are bound together by modified desmosomes (*corneodesmosomes*), which gradually degrade to allow desquamation.
6. In the mouth, the stratum corneum is usually thick, up to 20 cells thick, more than is usual in most areas of the skin.

F. *Keratinocytes in the non-keratinised epithelium.* Keratinocytes in the non-keratinised epithelium show most of the characteristics described in the keratinised epithelium. However, some differences can be highlighted:
1. The granular and cornified layers are replaced by a *superficial layer.* The cells of this superficial layer are not as acidophilic as in the keratinised epithelium and are alive, with nuclei and organelles.
2. No or very few keratohyalin granules are observed, and they are not associated with the tonofilaments.
3. The tonofilaments are dispersed and do not cluster or condense as in keratinised epithelium, due to the absence of filaggrin.
4. *Lamellar bodies* are small and rounded. Their interior is not lamellar, but shows a dense mass. Consequently, the surface layer of this epithelium is much more permeable than that of the keratinised epithelium.
5. The cells lack loricrin but contain *involucrin.*
6. Cells of the stratum germinativum can produce *cytokeratin 19*, in addition to cytokeratin 5 and 14, which are also produced in the keratinised epithelium.
7. The cells of the intermediate and superficial layers possess *cytokeratins 4* and *13*, instead of the cytokeratins 1 and 10 or 2 and 11 of the keratinised epithelium.

G. *Differentiation of keratinocytes.* The keratinocytes produced in the stratum germinativum ascend and change their properties. As they ascend, they differentiate:
1. New keratinocytes are produced from *reserve stem cells.*
2. These stem cells are indistinguishable from other cells and are arranged in niches in the basal layer, at the points of contact with the apexes of the papillae of the lamina propria and perhaps at the apexes of the epithelial ridges.
3. The cells are arranged in *proliferative epithelial units* described by Allen and Potten in the epidermis (Fig. 14.4). Each unit is organised in the form of a column with a hexagonal outline and consists of three cell types:
   a. *Stem cells*, located in the niches of the stratum germinativum. They make up 10% of the cells of the stratum germinativum.
   b. *Transit amplifying cells (TACs).* These are cells derived from dividing stem cells.
      (1) They actively divide and move laterally through the stratum germinativum, around the stem cell niche.

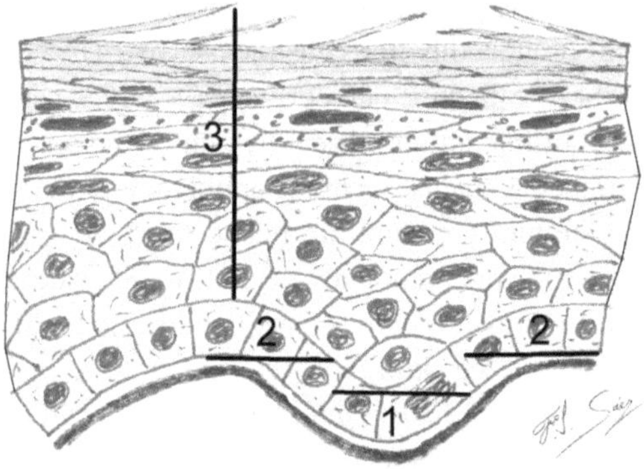

Fig. 14.4 Proliferative epithelial unit, according to Allen and Potten. Cells proliferate and ascend in columnar regions of the epithelium. In the stratum germinativum, in the centre of the column, are the niches with the stem cells (1). Surrounding them, in the same stratum germinativum, are transit amplifying cells (2), which divide actively. Above these two zones, in the remaining layers of the epithelium, are cells in differentiation and functional activity (3), which do not divide and ascend towards the surface

(2) Finally, they undergo critical changes in their hemidesmosomes, which cause them to separate from the basal lamina, stop dividing and ascend to the stratum spinosum.

c. *Cells in differentiation and functional activity.* These are cells that are no longer dividing and are progressively differentiating as they move upwards in the epithelium. They are therefore found in the stratum spinosum, stratum granulosum and stratum corneum.

4. There are an estimated 1400 proliferative units per square millimetre.

14.5 Extrinsic Cells

A. In addition to keratinocytes, other cells can be found in the epithelium of the oral mucosa that do not originate from the epithelium itself. Two types can be distinguished:
   1. *Permanent extrinsic cells.* These are cells that are always present in the epithelium.
   2. *Transient extrinsic cells.* These are cells of the immune system that can infiltrate the epithelium if inflammation occurs.

B. The main characteristics of permanent extrinsic cells are shown in Fig. 14.5 and described below.

C. *Melanocytes.*
   1. Under brightfield microscopy with the usual stains, melanocytes are rounded, shrunken cells with a halo around them.
   2. They are located in the *stratum Malpighi (stratum germinativum and stratum spinosum)*, and even in the *lamina propria.*

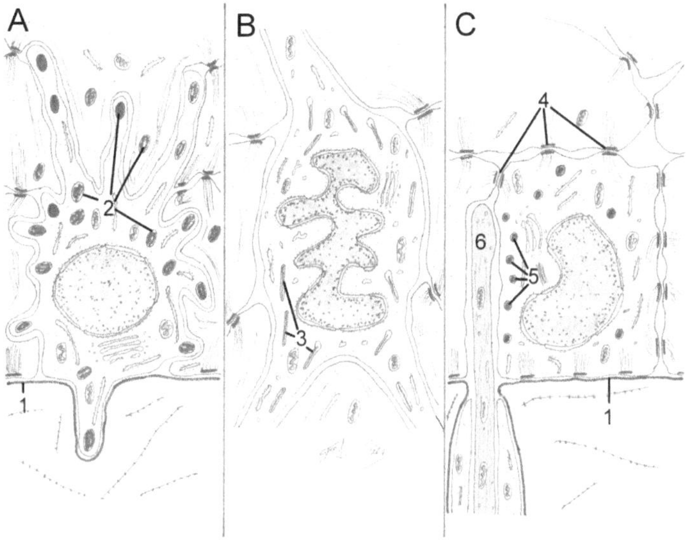

Fig. 14.5 Ultrastructure of the melanocyte (**A**), Langerhans cell (**B**) and Merkel cell (**C**). (1) basal lamina; (2) melanosomes; (3) Birbeck granules; (4) desmosomes; (5) dense vesicles; (6) free sensory ending

3. These cells have very complex cytoplasmic branching with dendritic appearance.
4. They are located between keratinocytes, to which they do not bind, as melanocytes lack desmosomes. Due to the prolongations of their cytoplasm, each melanocyte can contact 30–40 keratinocytes.
5. They produce vesicles derived from the Golgi apparatus called premelanosomes.
   a. In these vesicles, the amino acid tyrosine is oxidised and converted to 3,4-dihydroxyphenylanine (*DOPA*), from which various forms of *melanin* are produced:
      (1) *Eumelanin*, a brownish-black pigment.
      (2) *Pheomelanin*, of a redish-yellow colour.
   b. Melanin synthesis is stimulated by solar radiation.
6. As more melanin is produced, the premelanosomes darken and eventually develop into *melanosomes*, elongated structures of 0.2–0.6 μm. Sometimes clusters of melanosomes, called melanin *granules,* are seen under a bright-field microscope.
7. Melanosomes are transferred to keratinocytes by a process called *cytocrine secretion*. What actually happens is that the keratinocytes phagocytose the ends of the dendritic extensions of the melanocytes.
8. Melanin has a high capacity to absorb ultraviolet light, which it removes in the form of heat.
9. The areas of the oral mucosa where pigmentation is most noticeable are the gingiva, hard palate, tongue and cheeks.

10. The epithelium of the oral mucosa has twice the ratio of melanocytes to keratinocytes compared to the epithelium of the skin.
11. There is a relationship between skin pigmentation and oral mucosa pigmentation, as people with more pigmented skin have more pigmentation in the oral mucosa.
12. Much evidence suggests that melanocytes are produced from *neural crest cells.*
    a. *Melanocyte stem cells* are produced from these cells and migrate to the skin and other mucous membranes of the body between the 3rd and 6th month of gestation.
    b. These cells are long-lived and can divide, although they divide more slowly than keratinocytes.
    c. In the skin, melanocyte stem cells are located in hair follicles. The niche where these stem cells reside in the oral mucosa is unknown.

D. *Langerhans cells.*
1. Langerhans cells are cells present throughout the epithelium, but mainly in the upper layers of the *stratum spinosum.*
2. They are stellate or dendritic cells. Their prolongations are inserted between keratinocytes, but do not attach to them because they do not have desmosomes.
3. Under the light microscope, the nucleus is brightly stained, while the cytoplasm is clear.
4. Under electron microscopy, the nucleus is irregular, with many indentations. The cytoplasm lacks melanosomes and tonofilaments, and has tennis racket-shaped granules, the *Birbeck granules*, 150–500 nm long and 40 nm wide, with a dense central line in the middle. These granules are formed by invagination of the membrane and are then part of the *endosome* system.
5. As explained in Chap. 4, these cells are *interdigitating dendritic cells* originating from *yolk sac wall* precursors that during embryonic development settle into the epithelium, maintain themselves and self-renew.
6. Their mission is to act as *antigen-presenting cells.* Such antigens are internalised and processed in Birbeck granules.
7. Langerhans cells are associated with the *diffuse lymphatic tissue* of the *MALT (mucosa-associated lymphatic tissue) system*, which consists of lymphocytes that accumulate in the lamina propria and can invade the epithelium.

E. *Merkel cells.*
1. Merkel cells are found in the stratum germinativum and are similar to keratinocytes, with which they can bind via *desmosomes*, something that melanocytes cannot do.
2. Another difference from melanocytes is that they lack dendritic extensions.
3. The Merkel cell population is maintained and self-renewing.
4. The nucleus has an invagination that usually contains an inclusion formed by short parallel filaments.

5. Although perinuclear and peripheral *tonofilaments* are present in the cytoplasm, they are not as abundant as in the keratinocyte. They consist of *cytokeratins 8, 18 and 20*.
6. They have characteristics of *neurosecretory cells*. Under electron microscopy, their cytoplasm has dense vesicles, 80–180 nm in diameter, similar to the catecholamine secretion vesicles of some endocrine gland cells.
7. Associated with these cells are *myelinated axons* of sensory nerves:
   a. The nerve ending loses its myelin sheath and passes through the basal lamina of the epithelium. There, it widens to form a flattened end called the *receptor disc*, which is in close proximity to the Merkel cell.
   b. The assembly formed by the Merkel cell and the receptor disc is called *Merkel corpuscle*.
   c. Merkel corpuscles are *mechanoreceptors*, probably responsible for *pressure* perception.
8. Two explanations have been given for its origin, because they possess components typical of epithelia and nerve cells:
   a. *Neuroectodermal origin*, from precursors derived from neural crests:
      (1) Merkel cell precursors would infiltrate the epithelium during embryonic development.
      (2) In support of this theory is the presence of neurosecretory vesicles.
   b. *Ectodermal origin*, from the *surface ectoderm*.
      (1) According to this theory, Merkel cells are derived from cells of the epithelium itself and are modified epithelial cells.
      (2) One support for this theory is that they possess *tonofilaments* and not neurofilaments.
      (3) This theory is the most widely accepted today. However, they are still classified as extrinsic cells.

14.6 Lamina propria

A. The *lamina propria* is a loose, semi-dense or dense connective tissue, depending on the area, innervated and vascularised.
B. Two zones are usually distinguished:
   1. *Superficial* or *papillary layer*, located adjacent to the epithelium, between the epithelial ridges, forming the *connective tissue papillae*. It is usually a somewhat loose tissue.
   2. *Deep* or *reticular layer*,[2] of denser tissue.

[2] Despite the name "reticular", type III collagen usually accounts for only 8% of the total, while 90% is type I collagen. The name actually refers to the cross-linked organisation of the collagen fibres.

C. In all cases, *type I collagen* predominates. *Elastic fibres* and, in some regions, *oxytalan fibres* may be present.
D. On the upper lip and cheeks, three-quarters of adults have *sebaceous glands*, located close to the salivary glands in the submucosa. Their structure is similar to that of the sebaceous glands in the skin, but the function of their secretion is unclear.

14.7 Submucosa of the Oral Cavity

A. The *tunica submucosa* consists of *loose connective tissue*, but of variable density.
B. Sometimes, as when the mucosa attaches directly to the palatal bone, there is no submucosa, and the mucosa is called *mucoperiosteum.*
C. The submucosa is present in areas where movement is required and which are not exposed to mastication.
D. In some regions, it has *minor salivary glands.*

14.8 Vascularisation and Innervation of the Oral Mucosa

A. *Blood vessels of the oral mucosa.*
   1. The blood vessels supplying the oral mucosa are numerous and originate from arteries running parallel to the surface in the submucosa or, if there is no submucosa, in the deeper part of the reticular layer of the mucosa.
   2. The vessels branch through the lamina propria and produce a *capillary network* in the papillary layer of the lamina propria that extends through the connective tissue papillae.
   3. The capillary network of the lamina propria is much richer than that of the connective tissue of the skin, which influences the pink colour of the oral mucosa.
   4. Unlike the skin, there are no arteriovenous anastomoses in the oral mucosa.
B. *Innervation of the oral mucosa.*
   1. The oral cavity is highly innervated, including the mucosa, teeth and periodontium.
   2. Several types of nerve endings are distinguished:
   3. Autonomic innervation for salivary glands and blood vessels.
      a. *Parasympathetic fibres* from cranial nerves VII (facial) and IX (glossopharyngeal) synapse with neurons of the submandibular and otic ganglia, which innervate the oral cavity via the lingual and auriculotemporal nerves.
      b. *Sympathetic fibres* occur in the thoracic spinal cord, synapse with neurons in the superior cervical ganglion that send axons to the mouth.
   4. *Somatic innervation.* The oral cavity is innervated by the *V cranial nerve* or *trigeminal nerve* (from Greek tri- τρι- ‘three’ and Latin geminum ‘twin’, ‘paired’).

This nerve has three branches, ophthalmic, maxillary and mandibular. The oral cavity is innervated by the last two:

a. *Maxillary branch*. All its fibres are *sensory*. Innervates upper lip, upper cheek, upper jaw, upper teeth and palate.
b. *Mandibular branch*. It has two types of fibres:
   (1) *Sensory*. Innervate lower lip, lower cheek, lower jaw, lower teeth, anterior two-thirds of the tongue and floor of the mouth.
   (2) *Motor*, which innervates the muscles of mastication and other muscles related to the mouth.

5. The sensory nerve fibres innervating the oral cavity lose their myelin and form a network in the reticular layer that ends in a *subepithelial nerve plexus*.
6. Sensory nerve endings can be of three types:
   a. *Free nerve endings*. They are found in the lamina propria and epithelium.
      (1) The endings penetrating the epithelium are introduced between the keratinocytes to the middle or upper layers of the epithelium. Sometimes the keratinocytes engulf the nerve fibres and a *mesaxon* is formed.
      (2) They can respond to chemical, temperature or touch stimuli.
   b. *Nerve endings associated with Merkel cells*.
      (1) They are large myelin fibres that produce naked disc-shaped endings (*Merkel discs*) that contact Merkel cells.
      (2) Merkel cells respond to tactile stimuli and transmit the signal to nerve endings.
   c. *Encapsulated nerve endings*. They are abundant in the papillary layer of the lamina propria.
      (1) Meissner's corpuscles, Ruffini corpuscles, Krause's end bulbs and mucocutaneous terminal organs have been described.
      (2) *Meissner's corpuscles* pick up low-frequency tactile stimuli.
      (3) Other corpuscles may be specialised in sensing some kind of sensation, although it is not clear that there is strict specialisation.
7. A special case is the reception of *taste*, which is carried out by specialised structures called *taste buds*. They can be found at various points of the oral mucosa, but as they are predominant in the specialised mucosa of the dorsal surface of the tongue, we will describe them in the section devoted to it.

14.9 Oral Mucosa Types and Anatomical Distribution

A. The oral mucosa is organised differently in each region of the mouth. Three types of oral mucosa can be distinguished:
   1. *Lining mucosa*.
      a. Its function is to protect areas that are not subject to masticatory pressure and friction.

   b. Its epithelium is non-keratinised.
   c. It is found on the floor of the mouth, the lips, the cheeks, the ventral surface of the tongue and the soft palate.
2. *Masticatory mucosa.*
   a. Its function is to protect those regions directly subjected to friction and masticatory pressure.
   b. Its epithelium can be keratinised (hard palate) or parakeratinised (gingiva).
   c. The mucosa of the dorsum of the tongue has the characteristics of a masticatory mucosa, but due to the presence of other special structures, we consider it a specialised mucosa.
3. *Specialised mucosa.*
   a. These are the mucous membranes of regions that have specialised functions and have a peculiar organisation to perform that function.
   b. There are two very different specialised mucous membranes:
      (1) *Mucosa of the dorsal surface of the tongue*:
         (a) It specialises in *taste* perception.
         (b) It has structures called *lingual papillae*.
         (c) Its epithelium is parakeratinised.
      (2) *Mucosa of the root of the tongue*:
         (a) It covers the *lingual tonsil*, a structure of the *lymphatic* or *immune system*.
         (b) Its epithelium is not keratinised.

B. The *lips*, which form the entrance to the oral cavity, have three anatomical regions:
   1. The *skin of the lips* is on the outer surface of the lips and has skin characteristics such as the presence of *sebaceous glands* and *hair follicles*.
   2. The *vermilion zone* is also located on the outer side but lacks sebaceous glands and hair follicles. Its boundaries are well marked. Some authors consider this region of the lip to be a specialised mucosa.
   3. The *intermediate zone* is the junction zone between the vermilion zone and the labial mucosa. It has a parakeratinised epithelium without stratum granulosum. In children, it can be thick and form the suckling pad.
   4. The *labial mucosa* is the inner surface of the lips. It has the characteristics of a *lining mucosa*.

C. The anatomical distribution and histological organisation of the different types of mucosa in the wall of the oral cavity can be summarised in the following table:

Type of mucosa	Place	Epithelium	Lamina propria	Tissue connective papillae	Submucosa
Masticatory	Gingiva	Keratinised or parakeratinised	Dense, poor vascularisation		No. Mucosa attached to the periosteum (mucoperiosteum)
	Hard palate	Keratinised, sometimes parakeratinised, thickened	Dense, medium vascularisation	Long	Mucoperiosteum at midline, submucosa with adipocytes and salivary glands in some lateral areas
Lining	Alveolar mucosa	Non-keratinised, thin	Rich in elastic fibres, highly vascularised	Short and absent	Loose with thick elastic fibres attached to the periosteum. Salivary glands
	Floor of the mouth	Non-keratinised, very thin	Collagenous fibres and some elastic fibres, highly vascularised	Short and wide	Loose with adipocytes and salivary glands, attached to muscle
	Lips (inner side)	Non-keratinised, very thick	Dense with some elastic fibres, highly vascularised. Sometimes sebaceous glands.	Short and irregular	Dense with adipocytes and salivary glands. Attached to muscle
	Cheeks	Non-keratinised, very thick	Dense with some elastic fibres, highly vascularised and sometimes sebaceous glands	Short and irregular	Dense with adipocytes and salivary glands. Attached to muscle
	Ventral side of the tongue	Non-keratinised, thin	Thin layer with collagen fibres, some elastic fibres, salivary glands, poorly vascularised	Short and numerous	Thin and irregular, attached to muscle, with adipocytes. Sometimes absent, and mucosa attached to muscle connective tissue.
	Soft palate	Non-keratinised, thick, with taste buds	Fine, elastic fibres forming elastic lamina. High vascularisation	Short	Loose tissue with numerous salivary glands.

(continued)

Type of mucosa	Place	Epithelium	Lamina propria	Tissue connective papillae	Submucosa
Specialised	Dorsal side of the tongue	Keratinised (anterior zone) and non-keratinised (posterior zone), thick, forming lingual papillae, some with taste buds	Collagen fibres and some elastic fibres. Salivary glands in the posterior area, highly vascularised and abundant innervation.	Long (anterior region), short or absent (posterior region)	No, the mucosa is attached to the connective tissue of the muscle.
	Lingual root	Non-keratinised	Fused lamina propria and submucosa, invaded by lymphatic tissue	–	–
	Lips (intermediate zone)	Parakeratinised, fine (thick in breastfed infants)	Dense, not very vascularised, collagen and elastic fibres	Long and irregular	–
	Lips (vermilion zone)	Keratinised, fine	Collagen fibres and some elastic fibres, some sebaceous glands. High vascularisation	Long and narrow	No, mucosa directly attached to the muscle

D. Each of the regions listed in the table will not be described in detail. Only the most important features of the regions lined by specialised mucosa, i.e. the dorsal mucosa of the tongue (which will be explained below) and the mucosa of the lingual tonsil (explained in the next chapter), will be studied.

14.10 Specialised Mucosa of the Tongue

A. The *dorsal side of* the tongue is covered by a specialised mucosa.
   1. Functionally, it is a masticatory mucosa but is characterised by the presence of structures called *lingual papillae* (from the Latin papilla 'teat nipple').
   2. Lingual papillae are outgrowths that arise from the mucosal surface.

B. *Anatomy of the tongue*. Several areas can be distinguished on the tongue:
   1. The *terminal groove or sulcus terminalis* is a V-shaped line with the apex directed towards the oropharynx. It is situated towards the posterior third of the tongue and separates the anterior two-thirds of the tongue from the posterior third.

2. The *body*, also called the *palatal surface*, consists of the area of the tongue anterior to the *sulcus terminalis*. It originates from the *first pharyngeal arch*. The lingual papillae are located here.
3. The *base* or *root of the tongue* is the posterior third to the terminal groove. It originates from the *third pharyngeal arch*. It lacks lingual papillae. *Lymphatic follicles* are present in this area, but this type of specialised mucosa will be discussed in the next chapter.

C. There are four different types of papillae (Fig. 14.6):
   1. *Filiform papillae*.
      a. Arranged on the dorsal side of the whole body of the tongue, in rows parallel to the terminal groove.
      b. Each papilla is conical in shape and 2–3 mm long.
      c. The epithelium at the tips may be thick and highly keratinised, while the epithelium lining the spaces between papillae may be thinner and little or no keratinised.
      d. The lamina propria forms a *primary connective tissue papilla,* which may have minor *secondary papillae.*
      e. They lack taste buds.

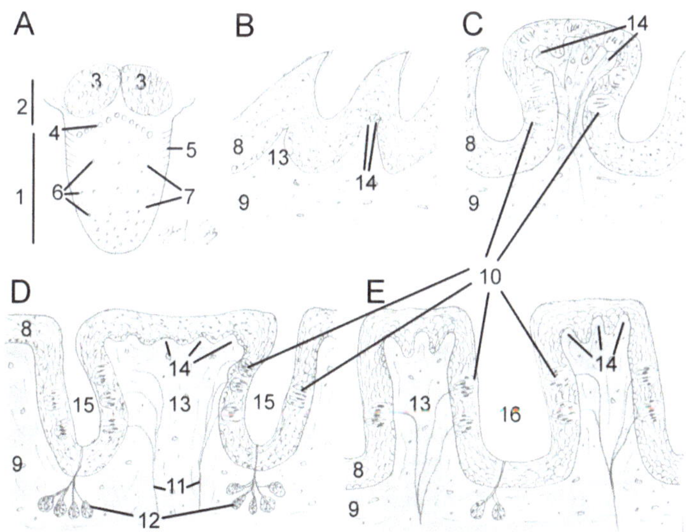

Fig. 14.6 View of the tongue (**A**), and histological sections of filiform (**B**), fungiform (**C**), circumvallate (**D**) and foliate (**E**) papillae. (1) body of the tongue; (2) base or root of the tongue; (3) lingual tonsils; (4) circumvallate papillae; (5) foliate papillae; (6) fungiform papillae; (7) rows of filiform papillae; (8) epithelium; (9) lamina propria; (10) taste buds; (11) sensory fibres; (12) salivary glands of von Ebner; (13) connective tissue papillae; (14) secondary connective tissue papillae; (15) circular groove; (16) interpapillary groove

f. Their abundance turns the dorsal surface of the tongue into a rough, abrasive surface, which, when in contact with the epithelium of the hard palate, has the function of a masticatory mucosa.

2. *Fungiform papillae.*
   a. Scattered and isolated, among the filiforms, and more abundant on the tip of the tongue.
   b. They have a short stem and a wider top, 150–400 μm in diameter.
   c. The surface of the epithelium is blunt, smooth, but inside the lamina propria forms a primary papilla and several secondary connective tissue papillae only on the upper flat part of the lingual papilla.
   d. The lamina propria is highly vascularised, which is why these papillae appear red.
   e. The epithelium has *taste buds*.
3. *Circumvallate papillae.*
   a. There are between 10 and 12 of them, and they are located in front of and adjacent to the sulcus terminalis.
   b. They are sunken in the mucosa and surrounded by a circular trench-like groove, so that the papilla is not higher than the rest of the lingual surface.
   c. The upper epithelium is smooth, keratinised and the epithelium of the lateral sides of the papilla is non-keratinised and has *taste buds*, which can also be found in the epithelium of the opposite wall of the circular groove surrounding the papilla.
   d. These papillae are usually related to serous salivary glands present in the lamina propria and between the lingual muscle, called *von Ebner glands*, whose excretory ducts open into the circular groove surrounding the papilla.
4. *Foliate papillae.*
   a. They are found in pairs on the dorsolateral side of the areas closest to the sulcus terminalis.
   b. They are abundant in some mammals, more difficult to detect in humans. They are visible in the newborn, but in the adult, they are rare and rudimentary.
   c. They are in the form of ridges separated by *interpapillary grooves*, giving them the appearance of folds, sheets or blades perpendicular to the edge of the tongue, separated by the grooves.
   d. Its dorsal side has a keratinised epithelium, but its lateral sides have a non-keratinised epithelium.
   e. They usually have one *primary connective tissue papilla* and three *secondary papillae*.
   f. Humans do not have many *taste buds*, which are very abundant in the foliated papillae of other animals, such as rabbits. Taste buds are found in the lateral walls.
   g. They are related to *serous glands*.

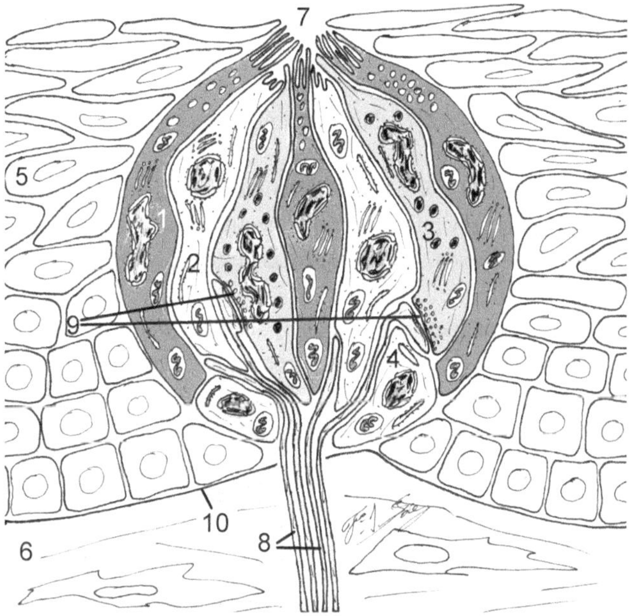

Fig. 14.7 Ultrastructure of the taste bud. (1) type I or dark cells; (2) type II, light or receptor cells; (3) type III, intermediate or presynaptic cells; (4) type IV or basal cell; (5) epithelium; (6) lamina propria; (7) taste pore; (8) sensory axons; (9) synapses; (10) basal lamina

D. *Taste buds*. The lingual papillae, except the filiform papillae, have taste buds (Fig. 14.7).
   1. Isolated taste buds can also be found in the pharynx, epiglottis, soft palate, larynx and oesophagus.
   2. They are barrel-shaped structures present within the mucosal epithelium.
      a. In histological sections, they are usually seen as oval corpuscles of about 70 μm, lighter than the rest of the epithelium.
      b. They extend from the basal lamina almost to the surface, where they are covered by a thin layer of epithelium that is perforated by an external *taste pore*.
   3. In the corpuscle, there are 50–100 cells, some basal rounded, most of them spindle-shaped, extending from the base to the apex of the corpuscle. Several cell types are distinguished[3] :
      a. *Type I* or *dark cells*. Approximately 60% of the cells.
         (1) Their cytoplasm is dark when viewed under the electron microscope.
         (2) They are fusiform, with an elongated and irregular nucleus, *elongated microvilli* at their apex, sometimes branched and apical *granules* in their cytoplasm.

[3]The student who consults different Histology books may notice differences and contradictions between the descriptions in some texts and others.

(3) They resemble glia cells in that they emit thin *laminar processes* that surround other cells, regulate ion levels in the extracellular medium and remove neurotransmitters released by other cells.
(4) It is not clear whether they transmit signals to the nerve endings or to the other cells of the bud.

b. *Type II, light* or *receptor cells*. 30% of the total.
(1) They appear very pale under the electron microscope.
(2) They are fusiform, with a rounded nucleus, short and thick *microvilli* at the apex.
(3) In their plasma membranes, they have *G-protein-associated receptors*[4] .
(4) Each cell has receptors for a particular taste, but not for others, and in a taste bud, there are cells that perceive different tastes.

c. *Type III, intermediate* or *presynaptic cells*. 7% of the bud cells.
(1) They have an intermediate density in the cytoplasm between type I and type II cells.
(2) They are fusiform, with an elongated nucleus with deep invaginations, a single thick *apical process*, and dense *perinuclear vesicles*.
(3) They show conventional *synapses*—with *synaptic vesicles*—at points of contact with sensory nerve endings.
(4) They produce some proteins that are typical of neurons.

d. *Type IV* or *basal cells*.
(1) Small, rounded cells at the base of the bud, with an irregular spheroidal nucleus and *intermediate filaments*.
(2) They are thought to be the *progenitor cells* that proliferate to produce new type I, II and III bud cells.
(3) Some authors believe that these cells do not divide, but originate from basal cells of the epithelium close to the bud.

e. *Perigemmal cells (type V cells)*, described by some authors.
(1) They are located at the periphery and separate the bud cells from the rest of the cells of the epithelium.
(2) It is thought that they may be stem cells of taste bud cells.

4. The taste bud cells are short-lived and constantly renewed.
   a. Type II cells live for about 8 days.
   b. The longest-living cells are type III cells, with a half-life of 22 days.
5. Renewal occurs through the proliferation of type IV basal cells of the taste bud itself, basal cells of the epithelium close to the bud, or perigemmal cells.

---

[4] G protein-associated receptors are a large family of proteins found in the plasma membranes of cells. They are proteins with 7 transmembrane domains. They are signal transducers, as they are stimulated by the arrival of a molecule and transmit the signal into the cell. The dentistry student will study these receptors in cell biology or biochemistry courses.

6. Taste stimuli are produced by the *adsorption*[5] of molecules to receptors on the surface of taste cell membranes. This causes membrane polarisation, which results in the release of substances that stimulate the nerve endings.
7. The cells of the taste bud can produce neurotransmitters, autocrine mediators and paracrine mediators, indicating that the cells of the bud regulate each other. The following neurotransmitters or regulatory substances have been found to be produced in the taste bud:
   a. *Type I cells*. Produce *GABA*. It is unclear whether they can stimulate nerve endings or perform paracrine control of type II cells.
   b. *Type II cells*. They produce:
      (1) *ATP*. It is thought to stimulate nerve endings and control the activity of type III cells and type II cells themselves.
      (2) *Acetylcholine*. It appears to perform autocrine control.
      (3) *Adenosine*. Its function is autocrine regulation.
   c. *Type III cells*. They produce:
      (1) *Serotonin*. They may stimulate nerve endings and perform paracrine control of type II cells.
      (2) *GABA*. Like serotonin, it can stimulate sensory fibres and control type II cells.
      (3) *Noradrenaline*.
8. The taste-sensing cells are type II and type III cells. Sensory nerve endings surround the type II and III cells.
   a. Type III cells have *synapses* with nerve endings. Stimulus transmission therefore appears to occur synaptically.
   b. It is not clear how type II cells transmit the stimulus. There are two hypotheses:
      (1) *Indirect transmission*. The impulse caused in the type II cell is transmitted to the type III cell, which transmits it to the nerve ending via synapses.
      (2) *Direct transmission*. The impulse is transmitted directly from the type II cell to the nerve ending, by means as yet unknown.
9. Taste cells can distinguish five types of *tastes*:
   a. *Saltiness*, which is mainly due to Na^+ ions.
   b. *Sourness*, produced by H^+ ions.
   c. *Sweetness*, produced by various molecules, mainly *carbohydrates*, but also other types of macromolecules, such as *saccharin*.
   d. *Bitterness*, produced by many substances.
   e. *Umami* or *savoriness*, a word that comes from the combination of the Japanese words うまい umai, 'delicious' and 味 mi, 'flavour'.
      (1) This taste is produced by:
         (a) *Glutamate*. It is an amino acid. It can be found:

[5] Adsorption means that it binds to the surface, and should not be confused with absorption, which means that it is incorporated into the interior.

1. Naturally occurring in breast milk, meats and vegetables.
2. In many substances used as a condiment or sauce. For example, the *garum* (from Latin garum, and this from Ancient Greek γάρονο, 'mackerel'') was a condiment produced from fish scraps that was rich in glutamate and was widely used by the upper classes of ancient Roman society.

(b) Some *ribonucleotides*, for example:

1. *Inosinic acid (inosine monophosphate* or *IMP).* IMP is mainly found in meat.
2. *Guanylic acid (guanosine monophosphate* or *GMP).* GMP is found in vegetables.

(2) Monosodium glutamate, IMP (or its salts) and GMP are currently used by the food industry to enhance the flavour of processed products.

E. Specialisation in taste capture:
   1. Some type II cells have sweet taste receptors, some have bitter taste receptors and some have umami taste receptors. Type II cells specialised for each of these tastes are mixed in a corpuscle.
   2. Type III cells are the sour taste receptor cells.
   3. The identity of salt-tasting cells remains a subject of current research.
F. Although we only perceive five different tastes, the combination with *smell* perception allows us to distinguish a wide range of flavours.

Take-Home Message

The wall of the mouth consists, from the oral cavity outwards, of a tunica mucosa, a tunica submucosa, sometimes absent, and skeletal muscle in some areas and bone in others. The epithelium of the mouth is stratified squamous, and can be keratinised, parakeratinised and non-keratinised. In the keratinised and parakeratinised epithelium, the basal, spinous, granular and cornified layers can be distinguished. In the non-keratinised, basal, intermediate and superficial strata are observed. Keratinocytes (cells of the epithelium) form from stem cells in the basal layer and ascend as they differentiate to die and be eliminated in the stratum corneum or stratum superficial. During their differentiation, the cells fill with keratin, lose nucleus and organelles and their membrane thickens by addition of proteins and lipids. In the epithelium, there are other cell types: melanocytes (pigmented cells), Merkel cells (sensory) and Langerhans cells (interdigitating dendritic cells). Three types of mucosa can be found in the mouth: lining, masticatory and specialised. On the dorsal surface of the tongue, there is a specialised mucosa that has lingual papillae, in some of which the epithelium has taste buds, which are taste receptors.

Further Reading

Adams D (1976) Keratinization of oral epithelium. Ann R Coll Surg Engl 58:351–358

Alonso L, Fuchs E (2003) Stem cells of the skin epithelium. PNAS 100(suppl_1):11830–11835

Arvidson K, Grafstrom RC, Pemer A (1988) Scanning electron-microscopy of oral-mucosa invivo and invitro – a review. Scanning Microsc 2:385–396

Barret AW, Cruchley AT, Williams DM (1996) Oral mucosal Langerhans' cells. Crit Rev Oral Biol Med 7:36–58

Barret AW, Scully C (1994) Human oral mucosal melanocytes – a review. J Oral Pathol Med 23:97–103

Boulais N, Misery L (2007) Merkel cells. J Am Acad Dermatol 57:147–165

Brand A, Havav AH, Clausen BE (2023) Langerhans cells in the skin and oral mucosa: brothers in arms? Eur J Immunol 53:2149499

Chandvarkar V, Mishra MN, Sangeetha R, Premalatha BR (2020) The current understanding on Langerhans' cells and its role in oral lesions. Contemp Clin Dentist 11:211–216

Chiego DJ (2023) Oral Mucosa. In: Essentials of oral histology and embryology, 6th edn. Elsevier, pp 168–183

Cui YZ, Man XY (2023) Biology of melanocytes in mammals. Front Cell Dev Biol 11:1309557

Dale BA, Salonen J, Jones AH (1990) New approaches and concepts in the study of differentiation of oral epithelia. Crit Rev Oral Biol Med 1:167–190

Eisenberg E, Natarajan E, Formaker BK (2014) Oral Mucosa and Mucosal sensation. In: Hand AR, Frank ME (eds) Fundamentals of oral histology and physiology. Wiley Blackwell, pp 165–190

Farbman AI (1988) The oral cavity. In: Weiss L (ed) Cell and tissue biology – a textbook of histology, 6th edn. Urban & Schwarzenberg, pp 573–594

Fehrenbach MJ, Popowics T (2020) Oral Mucosa. In: Illustrated dental embryology, histology, and anatomy, 5th edn. Elsevier, pp 105–123

Feller L, Masilana A, Khammissa RAG et al (2014) Melanin: the biophysiology of oral melanocytes and physiological oral pigmentation. Head Face Med 10:8

Frank ME (2014) Chemoreception and perception. In: Hand AR, Frank ME (eds) Fundamentals of oral histology and physiology. Wiley Blackwell, pp 191–220

Gartner LP (2017) Digestive system: oral cavity. In: Textbook of histology, 4th edn, Elsevier, pp 419–434

Gómez de Ferraris ME, Campos Muñoz A (2009) Cavidad bucal. In: Histología, Embriología e Ingeniería Tisular Bucodental, 3rd edn. Médica Panamericana, pp 137–176

Karring T (1973) Mitotic-activity in oral epithelium. J Periodontal Res 8:11–47

Kierszenbaum AL, Tres LL (2019) Upper digestive segment. In: Histology and cell biology – an introduction to pathology, 5th edn. Elsevier, pp 511–542

Krstić RV (1991) Digestive apparatus. In: Human microscopic anatomy. An atlas for students of medicine and biology. Springer Verlag, pp 161–255

Larjava H, Koivisto L, Häkkinen L, Heino J (2011) Epithelial integrins with special reference to oral epithelia. J Dent Res 90:1367–1376

Lucarz A, Brand G (2007) Current considerations about Merkel cells. Eur J Cell Biol 86:243–251

Moll I, Roessler M, Brandner JM et al (2005) Human Merkel cells – aspects of cell biology, distribution and functions. Eur J Cell Biol 84:259–271

Nanci A (2013) Oral Mucosa. In: Ten Cate's oral histology, 8th edn. Elsevier, pp 278–310

O'Dowd G, Bell S, Wright S (2023) Oral tissues. In: Wheater's functional histology. Elsevier, pp 241–252

Sidhu GS, Chandra P, Cassai ND (2005) Merkel cells, normal and neoplastic: an update. Ultrastuct Pathol 29:287–294

Watanabe L-S (2004) Ultrastructures of mechanoreceptors in the oral mucosa. Anat Sci Int 79:55–61

Salivary Glands and Tonsils

15

What You Will Learn in This Chapter

In this chapter, you will learn that the major salivary glands are made up of the parenchyma (the epithelium) and the stroma (the other tissues). You will study the organisation of these tissues and how saliva is produced in the salivary gland. You will see that saliva has many important functions. Finally, you will learn that there are several types of tonsils in the mouth and pharynx, and what is their microscopic structure.

15.1 General Characteristics of Salivary Glands

A. Saliva.

1. Numerous *extraepithelial compound exocrine glands* called *salivary glands* flow into the oral cavity.
2. Some of these glands are *serous,* some are *mucous* and some are *mixed.*
3. The secretion of all these glands constitutes *saliva*, a viscous, colourless liquid consisting mainly of water in which proteins, antibodies, small organic molecules and electrolytes are dissolved.
4. Saliva is produced in the *adenomeres* of the salivary glands, but is modified in the first *excretory ducts* leaving the adenomeres (*intercalated ducts* and *striated ducts*, which are *intralobular ducts*).
    a. As a result, the final product that reaches the mouth is not the same as the one that leaves the adenomere.
    b. For this reason, the ensemble formed by the adenomere, intercalated duct and striated duct is called *salivon*, and is considered to be the functional saliva-producing unit (Fig. 15.1).

F. J. Sáez, I. Badiola, *Essential Oral Histology*, Learning Materials in Biosciences,
https://doi.org/10.1007/978-3-032-03165-5_15

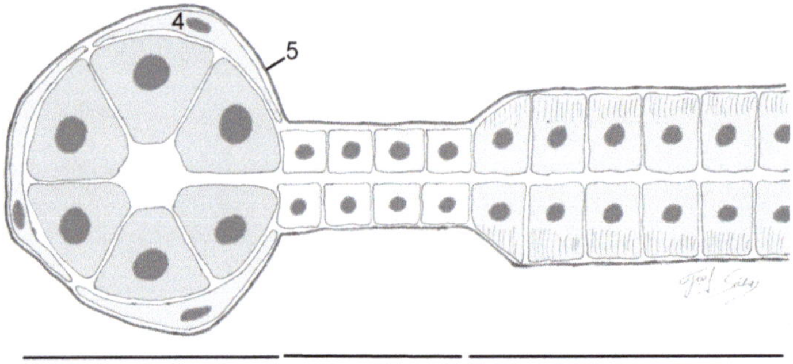

Fig. 15.1 Simplified diagram illustrating the organisation of the salivon. (1) adenomere; (2) intercalated duct; (3) striated duct; (4) myoepithelial cells; (5) basal lamina

B. Types of salivary glands.
1. *Minor* or *accessory salivary glands*:
   a. They are small salivary glands that constantly secrete saliva and are located in the *mucosa* or *submucosa* of the oral cavity wall, without a connective tissue capsule around them.
   b. It is thought that there may be between 600 and 1000 of these glands.
   c. Most are of mucous or mixed secretion.
   d. Its secretion may be increased by nervous stimuli.
   e. They can be found all over the wall of the mouth, except on the gingiva and the anterior part of the hard palate.
   f. They are usually named according to their location, and are as follows:
      (1) Lingual or von Ebner's glands, with serous secretion.
      (2) Buccal or jugal glands, mixed secretion (seromucous).
      (3) Labial glands, mixed.
      (4) Molar glands, seromucous secretion.
      (5) Palatine glands, mucous.
      (6) Pharyngeal glands, mucous.
2. *Major salivary glands*.
   a. There are three types of large glands that only secrete saliva when they are stimulated.
   b. Each type of major gland occurs as paired organs that are surrounded by a connective tissue *capsule*.
   c. These glands are (Fig. 15.2):
      (1) *Parotid glands*.
         (a) Of serous secretion.
         (b) They are the largest.
         (c) Located behind the jaw, they extend to the middle ear.
         (d) Each one of them discharges saliva through a duct that ends in front of the second maxillary molar.
         (e) They are innervated by cranial nerve IX.

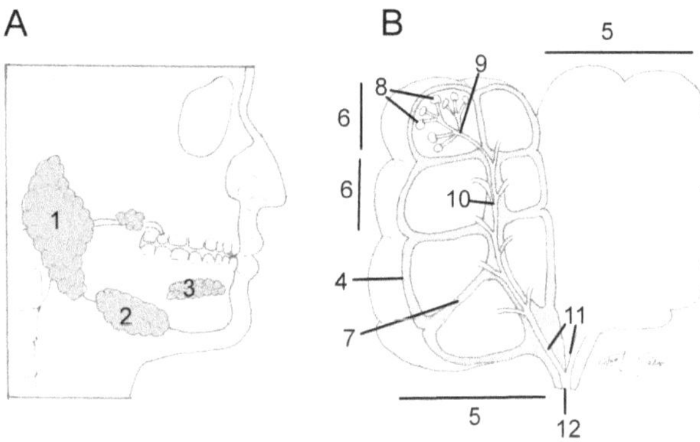

Fig. 15.2 Location of salivary glands in the head relative to the mouth (**A**) and basic anatomical organisation of a major salivary gland (**B**). (1) parotid gland; (2) submandibular gland; (3) sublingual gland; (4) capsule; (5) lobe; (6) lobule; (7) interlobular septa; (8) adenomeres; (9) intralobular duct; (10) interlobular duct; (11) lobar duct; (12) main duct

(2) *Submandibular glands.*
  (a) Mixed secretion.
  (b) They are located at the back of the floor of the mouth.
  (c) Each one pours its saliva into a duct that runs under the tongue.
  (d) They are innervated by cranial nerve VII.
(3) *Sublingual glands.*
  (a) Mixed secretion.
  (b) They are the smallest.
  (c) They are located in the anterior part of the floor of the mouth, just under the oral mucosa.
  (d) They have many excretory ducts that end independently into the oral cavity under the tongue.
  (e) They are innervated by cranial nerve VII.

d. The parotid and submandibular glands are located outside the wall of the mouth, between the muscles of the face and communicate with the oral cavity by long excretory ducts. The sublingual glands, on the other hand, are located just below the mucosa and have short ducts.
e. The parotid and submandibular glands produce saliva exclusively in response to nervous stimuli. The sublingual glands produce some saliva constitutively, but their secretion is greatly increased by nervous stimuli.

C. Embryonic development.
1. Salivary glands begin to form between the fourth and twelfth week of embryonic development.
   a. The parotid glands begin to develop between the fourth and sixth week of embryonic life.
   b. The submandibular glands by the 6th week.

   c. Sublingual and minor glands between the 8th and 12th weeks.
2. They form from the oral epithelium. Because the epithelium of the oral cavity undergoes major displacements during this period, it is not possible to determine exactly whether they originate from the ectoderm or the endoderm:
   a. The parotid glands are thought to originate from the ectoderm and the submandibular and sublingual glands from the endoderm.
   b. However, as already indicated in Chap. 2, some authors claim that all salivary glands are ectodermal.
3. The process of gland formation is regulated by *epithelial-mesenchymal interactions*, similar to that described in tooth development.
4. They originate as *buds* from the oral epithelium that grow into the underlying *mesenchyme (ectomesenchyme)*. The bud grows and branches. Subsequently, in the centre of the epithelial structure, the *lumen* of the gland opens (Fig. 3.13A–D).
5. Epithelial cells reach maturity in the last two months of gestation.
6. Subsequently, the glands grow and the volume of adenomeres increases in relation to the volume of ducts and stroma.

D. Basic anatomical and histological organisation of the salivary glands.
1. In glands, the epithelial glandular tissue is called *parenchyma*, while the other tissues (connective tissue, blood vessels and nerves) surrounding the parenchyma form the *stroma*.
2. Organisation of the stroma:
   a. In the major salivary glands, a *capsule* of dense connective tissue surrounds the entire gland.
   b. The capsule can divide the gland into several more or less independent portions, which we call *lobes* (from the Greek λοβός lobós 'sack').
   c. Dense connective tissue *septa* (singular *septum*, from Latin saeptum ‘a fence, enclosure, partition’) emerge from the capsule inwards. These partitions separate small portions of each lobe, which we call *lobules* (from the Greek λοβός lobós ‘sack’ and the Latin diminutive suffix -ŭlus).
   d. Within each lobule, a *loose connective tissue* encloses the glandular epithelia. This tissue is called *intralobular stroma*.
3. Organisation of the parenchyma:
   a. Numerous adenomeres are found in the lobules.
   b. Epithelial excretory ducts arise from the adenomeres, which initially have a narrow lumen and run through the loose stroma of the lobule.
      (1) These ducts may differ in appearance and organisation, but as they run through the lobule they are called *intralobular ducts* (from the Latin prefix, intra-, ‘within’).
      (2) In the salivary glands, there are two types of intralobular ducts, the *intercalated duct* and the *striated duct*.

c. The striated ducts converge into larger ducts that leave the lobule and run through the septa that separate them, thus forming the *interlobular ducts* (from Latin inter-, 'between').
d. The interlobular ducts converge into larger ducts. Normally, secretion from one lobe may converge into a single duct called a *lobar duct.*
e. In the parotid and submandibular ducts, the lobar ducts converge into a single *main duct.*
f. Excretory ducts have simple epithelia with low cells in their initial segments. At greater distances from the adenomere, the cells are taller and may eventually form stratified epithelia, sometimes with some *goblet cells*.
   (1) The intralobular ducts have *simple cuboidal epithelia.*
   (2) The interlobular ducts have *high cuboidal, columnar* and even *pseudostratified columnar epithelia.*
   (3) The lobar duct has *cubic* or *columnar stratified* epithelium.
   (4) Finally, in the main duct, especially near its opening in the oral cavity, we can find *non-keratinised stratified squamous epithelium* similar to that of the oral mucosa.

15.2 Histological Organisation of the Salivons of the Major Salivary Glands

A. Salivary glands can be mucous, serous or mixed:
   1. *Serous glands* are made up of serous acini (singular acinus).
   2. *Mucous glands* have mucous tubules, but because they often look like acini in histological section, they are also called mucous acini or mucous tubuloacini.
   3. *Mixed glands* include serous acini, mucous tubuloacini and mixed tubuloacini (mucous tubules lined by a cap of serous cells called *serous demilune* or *crescent of Giannuzzi*).
B. Serous acini.
   1. Serous acini are spherical structures consisting of 8–12 cuboidal cells, which are actually pyramidal in shape, surrounding a small central lumen.
   2. Cell structure (Fig. 15.3A):
      a. Before releasing their secretion, the cells are large and filled with secretory granules. After releasing their secretion, the cells decrease in size and do not show so many granules.
      b. The spherical nucleus is usually located in the basal half of the cell.
      c. Organisation of the cytoplasm:
         (1) There is abundant basal rough endoplasmic reticulum surrounding the nucleus.
         (2) The Golgi apparatus is usually supra- or paranuclear.

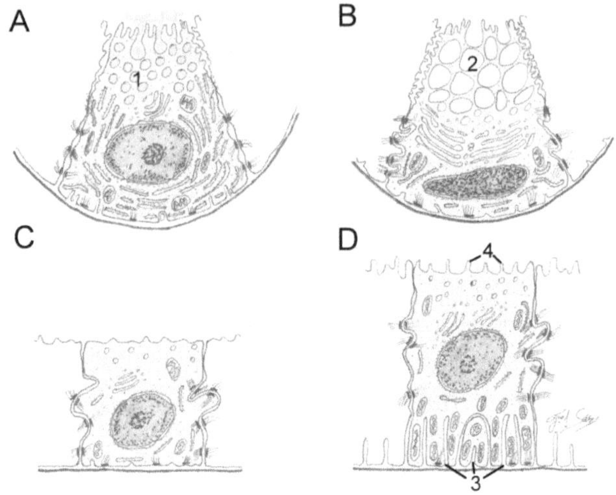

Fig. 15.3 Ultrastructure of serous (**A**) and mucous (**B**) glandular cells and of intercalated (**C**) and striated (**D**) duct cells. (1) zymogen granule; (2) mucinogen granule; (3) basal labyrinth; (4) microvilli

(3) Mitochondria are scattered between the endoplasmic reticulum and the apical supranuclear zone.

(4) Some lysosomes, tonofilaments and microtubules may be present in the cytoplasm.

(5) The apical cytoplasm is usually filled with large, 1 μm diameter secretory vesicles, the *secretory granules*, which are called *zymogen granules* (from the Greek ζύμη zýmē, 'ferment' and -γενής -genḗs, 'producing') because of their serous nature.

(6) The basal region of the cell is basophilic, while the apical region is acidophilic due to zymogen granules.

d. The apical surface of the cell has small *microvilli*, which also line small *canaliculi* between the apical regions of neighbouring cells. These intercellular canaliculi are like digitating invaginations that start from the lumen of the acinus and go about 5 μm between the cells of the epithelium.

e. The basal surface of the cell may have folds that sometimes form a small *basal labyrinth*.

3. Cells are bound to neighbouring cells by junctional complexes:

a. There are *tight junctions,* located apically, except when there are intercellular canaliculi, in which case they are located at the bottom of the canaliculi.

(1) These junctions isolate the lumen of the acinus from the tissue environment surrounding the acinus:

(a) They are selectively permeable to small molecular weight substances (water and certain ions).

(b) They can become partially permeable for large molecules by the action of neurotransmitters released by nerve endings.

(2) These junctions divide the cellular interior into two *domains*, the *apical* domain responsible for secretion, and the *basolateral* domain responsible for coordination with neighbouring acinar cells and the stroma.

b. There are also *adherens junctions* surrounding the cells, under the tight junctions, and *desmosomes*.

4. Release of the granules occurs by *eccrine secretion*, i.e. exocytosis, which releases the contents of the granules into the lumen of the adenome.
5. Some of the substances produced by these cells are *α-amylase*, *lysozyme*, *lactoferrin*, *peroxidase*, various *proteases* and *growth factors* (NGF and EGF). *Lipase* is produced in infants.

C. Mucous tubuloacini.

1. Mucous tubules have a wider lumen than serous acini and are made up of cubic cells.
2. The cells are filled with *mucinogen granules* (from the Latin muccus, 'mucus'), which are so named because their secretion is called *mucin*.
3. The material in these granules is often destroyed during tissue processing, so that in the usual histological preparations, the cells have clear appearance, look empty or are filled with large void spaces surrounded by a loose network of poorly stained cytoplasm.
4. The mucous cell has mucinogen granules in the apical two-thirds and the nucleus and organelles appear as if they are compressed in the basal portion (Fig. 15.3B):
   a. The nucleus is in a basal position and usually has a flattened, compressed appearance.
   b. Under the electron microscope, some cisternae of rough endoplasmic reticulum, the supranuclear Golgi apparatus and some mitochondria can be seen around this nucleus.
5. When the cell has finished expelling its secretion, the nucleus and organelles expand and the volume they occupy increases, to the point that the cells can be mistaken for serous cells. Under normal physiological conditions, mucous cells do not empty all their granules, so they are rarely seen in this state.
6. The cells have *junctional complexes* similar to those of serous cells.
7. The secretion is released by exocytosis (*eccrine secretion*) and consists of *mucins*, a family of highly glycosylated proteins, which form a viscous substance called *mucus*.

D. Mixed tubuloacini.

1. A mixed tubuloacinus consists of a mucous tubuloacinus surrounded by one or more serous cell caps, close to the blind end of the tubule.
2. These caps are called *serous demilunes* or *crescents of Giannuzzi.*
3. Freeze preparations show that mucous and serous cells are intermingled in the tubular wall. Therefore, the image seen in the usual preparations may be an *artefact* because the preparation techniques swell the mucous cells, which displace the serous cells to the outside.

4. The structure of mucous and serous cells does not differ from that of these cells in serous acini or mucous tubules.

E. Myoepithelial cells.
   1. All salivary gland adenomeres are surrounded by a basal lamina.
   2. Between it and the glandular cells, there are the *myoepithelial cells* or *basket cells*.
   3. These cells can also be found in the intercalated and striated ducts.
   4. They are thin cells, so they are difficult to see and usually only the nucleus and rarely a thin acidophilic cytoplasm are visible.
   5. They are stellate in shape, and their extensions, which embrace the epithelium of the adenomere or duct, are packed with *actin* and *myosin filaments*.
   6. They have contractile capacity. Their contraction facilitates the release of secretion.
   7. There are *caveolae* in the plasma membrane, which are thought to be involved in the mechanisms that trigger contraction.
   8. They are linked to each other and the epithelial cells by *desmosomes*, and to the basal lamina by *hemidesmosomes*.
   9. Although they resemble smooth muscle cells, they are of epithelial origin.

F. Intercalated or intercalary ducts.
   1. The first ducts to receive adenomere secretion are the *intercalated ducts*. They are small, with little lumen and a low cubic epithelium, making them difficult to distinguish.
   2. There are *myoepithelial cells* and a *basal lamina*.
   3. The intercalated ducts leading from the adenomeres converge into slightly larger intercalated ducts, before flowing into the striated ducts.
   4. In serous glands, these ducts are long and in mucous glands they are shorter and difficult to observe.
   5. The electron microscope shows (Fig. 15.3C):
      a. A central nucleus and few organelles, namely a sparse, preferably basal, rough endoplasmic reticulum, a small, supranuclear Golgi apparatus, small granules and some mitochondria.
      b. The cells are joined by apical *junctional complexes*, scattered *desmosomes* and some *gap junctions*.
      c. Cells are attached to the basal lamina by *hemidesmosomes*.
      d. Cells show small *interdigitations* with neighbouring cells.
   6. The granules of these cells are released by exocytosis into the duct lumen. They contain substances that are incorporated into saliva, such as *lysozyme* and *lactoferrin*, and perhaps other unidentified compounds.

G. Striated ducts
   1. *Striated ducts* are larger in diameter, with a larger lumen than intercalated ducts.
   2. Its epithelium is high cuboidal or columnar, with a highly acidophilic cytoplasm. The following cellular structures are observed (Fig. 15.3D):
      a. A rounded, centrally positioned nucleus.
      b. Basal striations are perpendicular to the surface. This is a *basal labyrinth*.

(1) It is formed by abundant and deep *invaginations* of the plasma membrane.
(2) Invaginations delimit regions of the cytoplasm in which *mitochondria* are found arranged vertically, parallel to the invaginations and perpendicular to the basal lamina.
(3) The basal labyrinth is typical of epithelia in which *active transport* of electrolytes takes place.

c. The Golgi apparatus and endoplasmic reticulum are very poorly developed.
d. There is some smooth endoplasmic reticulum in the apical cytoplasm, accompanied by a few *secretory vesicles*. These vesicles contain *kallikrein* and probably other proteins, which are added to saliva.
e. *Endocytosis vesicles* remove certain substances from saliva.
f. Some lysosomes and peroxisomes, and perinuclear *glycogen* deposits are also observed.
g. In the apical zone, the cell surface has small *microvilli*.
h. Cells are joined together by *junctional complexes*, but lack gap junctions.
i. In the basolateral area, there are *interdigitations* with neighbouring cells.

3. In the striated ducts and some interlobular ducts, basal cells are scattered among the columnar cells of the epithelium. These may be stem cells that regenerate the epithelium.

H. Stroma

1. The stroma of the major salivary glands consists of:
   a. The connective tissue *capsule* that surrounds the lobes and forms the boundary of the gland, separating it from neighbouring tissues and organs.
   b. Connective tissue *septa*, which derive from the capsule and run inwards, dividing each lobe into several lobules. Blood vessels, nerves and interlobular ducts that carry saliva into the oral cavity are located inside the septa.
   c. *Intralobular stroma*, which is the loose connective tissue found within each lobule, surrounds the adenomeres and the intercalated and striated ducts.
2. The intralobular stroma is a loose connective tissue containing:
   a. The *blood vessels* of the microcirculation and the finer *nerve endings* that innervate the adenomeres, ducts and blood vessels.
   b. Few *lymphatic vessels* are distinguishable.
   c. There are fibroblasts and macrophages, and sometimes adipocytes isolated or in small groups.
   d. Some lymphocytes may be present.
   e. Abundant *plasma cells* producing *immunoglobulin A (IgA)* are seen (Fig. 15.4):
      (1) IgA is released in the form of dimers.
      (2) The dimers bind to a *receptor protein* on the surface of the plasma membrane of the basal surface of epithelial cells (both adenomeric and ductal). This receptor protein binds to IgA through one of its regions, called the *secretory component*.

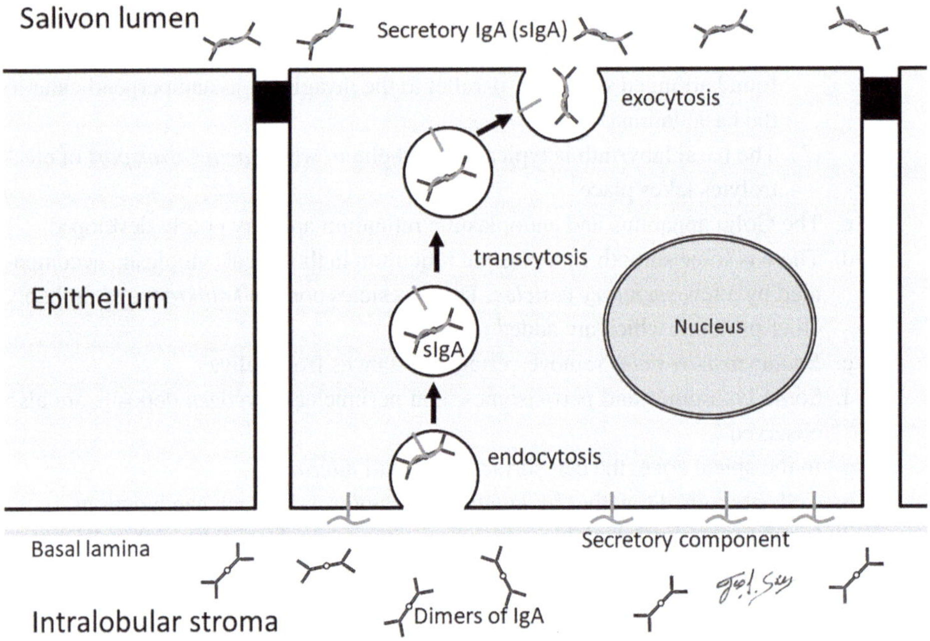

Fig. 15.4 Simplified diagram of the process of salivary secretion of immunoglobulin A produced by plasma cells. Cellular and molecular structures are not drawn to scale

(3) Once IgA is bound to the receptor, the complex is introduced into the epithelial cells by *endocytosis*, producing intracellular vesicles with the IgA-receptor complex on the inner side of their membrane.

(4) These vesicles travel through the cytoplasm, are directed to the apical side of the cell by *transcytosis* and are released at the apical pole by *exocytosis*.

(5) At that point, the receptor-IgA complex breaks down:

   (a) One part is anchored to the membrane, and the other fragment, which is the secretory component bound to IgA, is released.

   (b) The released molecule, consisting of IgA and the secretory component, is called *secretory immunoglobulin A (sIgA)* and is soluble, so it mixes with the components of saliva.

(6) Small amounts of IgG and IgM are also released.

3. Innervation.
   a. *Nerve fibres* penetrate into the glands accompanying the blood vessels and eventually form a plexus of *amyelinated fibres* attached to the arterioles, adenomeres and ducts, in which the fibres are protected by their Schwann cells.
   b. These fibres regulate secretion and are derived from the sympathetic and parasympathetic nerve systems.
   c. Axon*s* contact epithelial cells in two ways:

    (1) *Hypolemmal relationship.* The axon leaves the nerve, is released from its Schwann cell, crosses the basal lamina of the epithelium and forms a *varicosity* that contacts the basolateral membrane of the epithelial cells, from which it remains at a distance of 10–20 nm.
    (2) *Epilemmal relationship.* The axon does not leave the nerve and is protected by the Schwann cell, except where it forms a *varicosity* approaching the epithelial cell, from which it is separated by 100–200 nm. In this case, between the varicosity and the epithelial cell are the *basal laminae* of the Schwann cell and the epithelium.
  d. In both cases, *synaptic vesicles* are observed in the varicosities, but no further differentiations typical of synapses are seen in either the axon or the epithelial cell.
  e. The stimulatory effect is on glandular cells and blood vessels, which affects the concentration and composition of saliva:
    (1) Stimulation of the *parasympathetic* through release of *acetylcholine* and *vasoactive intestinal peptide* produces saliva rich in water and salts, but poor in organic compounds.
    (2) Stimulation of the *sympathetic nervous system* by *noradrenaline* release leads to the production of a small amount of dense saliva with a high concentration of organic molecules.
    (3) The parasympathetic system also innervates myoepithelial cells.
4. Vascularisation:
  a. The glands are richly vascularised, which is necessary since 99% of saliva is water.
  b. Different arterioles supply blood to the capillaries surrounding the adenomeres and ducts.
  c. The capillaries and postcapillary venules are *fenestrated.*
  d. Sometimes there are *arteriovenous anastomoses*.

15.3 Specific Characteristics of the Major Salivary Glands

A. *Parotid glands*:
  1. They consist exclusively of *serous acini.*
  2. The intercalated ducts are usually long.
  3. The striated ducts are numerous.
  4. *Adipocytes* are usually present.
  5. *Lymphatic follicles* are sometimes seen, which are not seen in the rest of the major salivary glands.

B. *Sublingual glands*:
   1. *Mucous acini* predominate, but there are *mixed acini* and some *serous acini*, although what are seen as serous acini could be cross-sections of crescents of Giannuzzi of mixed tubuloacini.
   2. The intercalated ducts are short and sparse.
   3. The striated ducts are also sparse, and sometimes the basal striation is not well distinguishable.

C. *Submandibular glands*:
   1. They show a predominance of *serous acini*, although there are some *mucous acini* and some *mixed acini.*
   2. The intercalated and striated ducts are less numerous than in the parotid glands.

15.4 Saliva Secretion

A. The saliva that reaches the mouth differs from that produced in the adenomere:
   1. The saliva produced in the adenomere is called *primary saliva.* Primary saliva is *iso* or *hypertonic* with respect to plasma.
   2. Primary saliva is modified in the salivon ducts and becomes *secondary saliva.* Secondary saliva is *hypotonic.*

B. Composition of saliva. Saliva includes numerous substances:
   1. Water, which makes up 99% of saliva.
   2. Proteins and peptides:
      a. *α-amylase*. Degrades starch.
      b. *Salivary lipase* or *lingual lipase*, which degrades milk lipids.
      c. *Proline-rich proteins*. They accumulate on the surface of the tooth, forming the *acquired pellicle*. They facilitate mineralisation of the tooth and resistance to caries.
      d. *Mucins*. Highly glycosylated proteins. They protect, facilitate digestion of the bolus and have a defensive function.
      e. *Histatins*. *Histidine-rich* peptides. There are several histatins in saliva that are located in the tooth-acquired pellicle, where they have bactericidal and fungicidal action.
      f. *Cystatins*. A family of *cysteine-containing* phosphoproteins. They are inhibitors of proteases produced by bacteria.
      g. *Peroxidase*. It uses *hydrogen peroxide* to produce substances that are toxic to bacteria.
      h. *Lysozyme*, which breaks down bacterial walls.
      i. *Lactoferrin*. Protein that binds *iron* and inhibits bacterial metabolism.
      j. *Defensins*. Small peptides (15–20 amino acids) that permeabilise bacterial membranes.
      k. *Secretory-IgA (sIgA)*. Antibodies.
      l. *EGF*. Epidermal growth factor promotes wound healing of the oral mucosa.

m. *NGF*. Involved in wound healing. It is produced by the submandibular gland.
n. *Insulin*.
o. *cAMP-binding proteins*.
p. *Serum albumin*.

3. Small organic molecules:
   a. Glucose.
   b. Amino acids.
   c. Urea.
   d. Uric acid.
   e. Lipids.
4. Electrolytes:
   a. Na^+.
   b. K^+.
   c. Cl^-.
   d. HCO_3^-.
   e. Ca^{2+}.
   f. Mg^{2+}.
   g. HPO_4^{2-} (monohydrogen phosphate).
   h. Thiocyanate ($N{\equiv}C\text{-}S^-$, SCN^-).
   i. F^-.

C. Many of these substances are produced in the adenomeres, but some, such as sIgA, are produced in the stroma and pass through adenomeres and ducts.

15.5 Secretion and Modification of Saliva

A. Most of the saliva secretion in the major glands is produced by external stimuli that provoke a reflex action.
   1. The external stimuli that trigger the reflex act on:
      a. Olfactory and taste *chemoreceptors*.
      b. *Mechanoreceptors* of mastication, including *periodontal ligament* mechanoreceptors.
   2. External stimuli are picked up by sensory endings and transmitted to the central nervous system, where a nerve impulse is generated in the *sympathetic* and *parasympathetic systems* and directed to the salivary glands.

B. Primary saliva has a different composition than the secondary saliva that enters the mouth. The concentration of some substances, such as electrolytes, is modified in the ducts. In fact, many electrolytes are secreted or reabsorbed in this part of the saliva.
   1. Discharge into the adenomere (Fig. 15.5):
      a. Proteins are produced in the endoplasmic reticulum, transported to the Golgi apparatus, where they are packaged into vesicles and released by *exocytosis*. Exocytosis is stimulated by the release of *noradrenaline* from sympathetic endings.

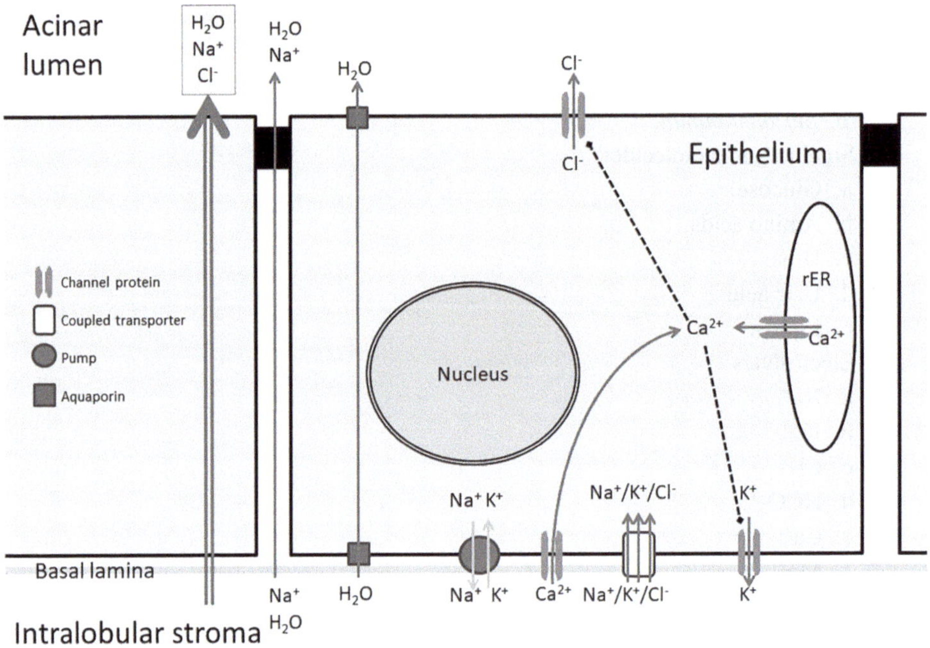

Fig. 15.5 Simplified diagram of water and electrolyte transport in salivary gland acinar cells. Nerve stimuli cause an increase in the concentration of calcium ions in the cytosol, which stimulates the outflow of Cl^- into the lumen and potassium into the stroma. The outflow of Cl^- drives the paracellular passage of Na^+ and transcellular (via aquaporins) and paracellular water transport. The net result is the passage of water, chloride and sodium into the saliva

b. The release of water and electrolytes is controlled primarily by the parasympathetic system and to a lesser extent by the sympathetic system.
   (1) The release of *acetylcholine* by the parasympathetic leads to a release of Ca^{2+} into the cytoplasm of the epithelial or duct cell (stored in the endoplasmic reticulum). This calcium release can also be stimulated by noradrenaline from the sympathetic system.
   (2) The increased Ca^{2+} concentration stimulates *exocytosis* of secretory vesicles, but its main effect is to open *Cl^- channels* at the apical pole and *K^+ channels* at the basolateral pole.
   (3) The flow of Cl^- into the lumen pulls Na^+ by *paracellular transport*[1] .
   (4) The increase of NaCl in the adenomere lumen entrains water flux by *osmosis*. Water passes through epithelial cells by *aquaporins* in their membranes, and perhaps also by paracellular transport.

[1] Paracellular transport is transport through the space between cells. It is controlled by the tight junctions between epithelial cells.

c. To maintain the necessary ionic balance in the adenomere cell, there is a $Na^+/K^+/Cl^-$ coupled transporter on the basolateral side of the cell, which brings these ions into the cell from the stromal tissues.
d. For all this ion transport to take place, it is necessary to maintain a high intracytosolic concentration of K^+ and a low intracytosolic concentration of Na^+. This is achieved by a *Na^+/K^+ pump* located on the basolateral side of the cell, which brings K^+ in and Na^+ out. All ion transport thus requires energy input, which is consumed as ATP in the Na^+/K^+ pump.

2. In the intercalated duct, the composition of the saliva is partially modified:
   a. Some *lysozyme* and *lactoferrin* stored in its secretory vesicles are secreted.
   b. In addition, in serous and mixed glands, HCO_3^- is secreted and Cl^- is reabsorbed.
3. The main modifications of saliva occur in the striated duct:
   a. *Kallikrein* is secreted.
      (1) It is a protein that modifies *cystatins* and *proline-rich proteins* produced in the adenomere.
      (2) Unlike other proteins secreted in saliva, it does not accumulate in secretory vesicles that are released in response to nerve stimuli, but is transported in small vesicles from the Golgi apparatus to the cell surface and released as it is produced, without nerve stimuli.
   b. The cells of the striated duct perform *endocytosis* to take up some molecules that are removed from the saliva.
   c. The main changes in saliva that occur in the striated duct are changes in the *electrolyte* composition (Fig. 15.6):
      (1) These cells have the following components responsible for the process:
         (a) The enzyme *carbonic anhydrase*, which produces HCO_3^- and H^+ from carbon dioxide and water.
         (b) The following transmembrane transporter proteins are found in the basolateral membrane of these cells:
            1. *Na^+/K^+ ATPase (sodium/potassium pump)*, which expels Na^+ and introduces K^+.
            2. *Cl^- channel*, which carries these ions out of the cell.
         (c) Other transport proteins are found in the apical membrane:
            1. *Na^+ channel*, which removes these ions from the saliva and introduces them into the cytosol.
            2. *Cl^- channel*, which brings chloride into the cell.
            3. *Bicarbonate channel*, which brings this ion into the lumen of the duct.
            4. *K^+ channel*, which allows this ion to exit.
            5. *HCO_3^-/Cl^- coupled transporter*, which expels HCO_3^- and introduces Cl^-.
            6. *Na^+/H^+ coupled-transporter*, which expels H^+ into the lumen and introduces Na^+ into the cytosol.

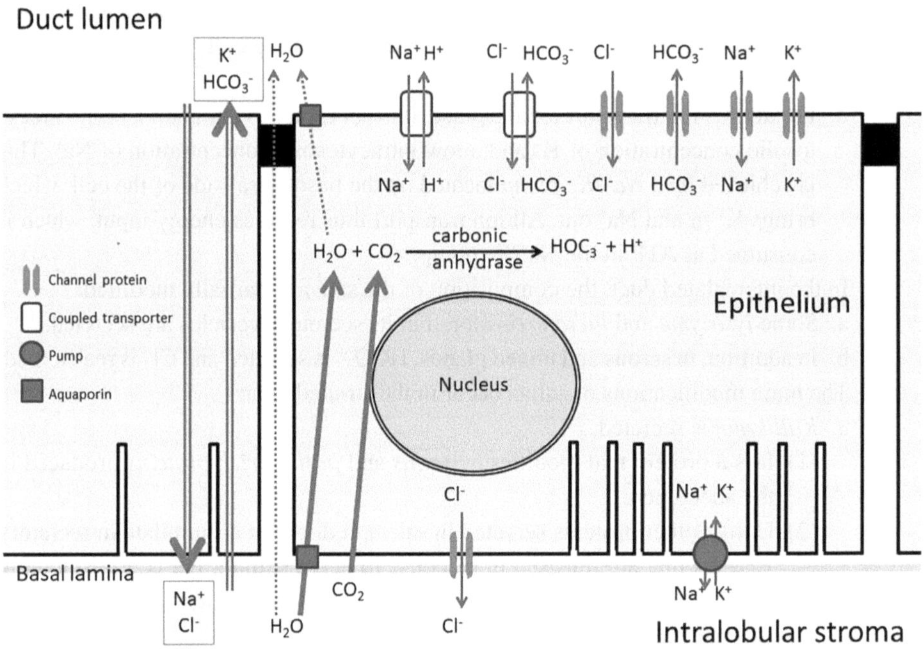

Fig. 15.6 Simplified diagram of water and electrolyte transport in the cells of the striated duct. Carbonic anhydrase produces bicarbonate, which is released into the duct lumen. K^+ ions are also expelled. At the same time, Cl^- and Na^+ are absorbed. Water transport may occur, but no net flow is usually noted

(2) The energy input that drives the transport that takes place in these cells is a result of:
   (a) Intracellular production of HCO_3^- and H^+.
      1. The increase in the intracellular concentration of HCO_3^- causes it to escape from the cell into the lumen, which acidifies the cytosol.
      2. Acidification of the cytosol allows the uptake of sodium from the lumen by the Na^+/H^+ coupled transporter.
   (b) Basolateral sodium/potassium pump activity:
      1. As Na^+ is expelled into the interstitium, it decreases its concentration in the cytosol and drags down the capture of the ion present in saliva.
      2. The output of K^+ restores electrical neutrality.

(3) There are many more transporters in both membranes involved in the process, which is much more complex than summarised here.

(4) The end result of the activity of these proteins is:
   (a) A *net reabsorption of* Na^+ *and* Cl^-, resulting in hypotonic saliva.
   (b) A *net secretion of* K^+ *and* HCO_3^-.
   (c) *Paracellular transport of water* into the duct lumen may occur, which may also move through aquaporins in cell membranes, although no net flow of water in either direction is usually observed.

C. The final composition of saliva varies greatly depending on the flow.
   1. If saliva flow is high:
      a. Na^+ and Cl^- concentration increases and K^+ concentration decreases.
      b. In addition, high saliva flow increases HCO_3^- levels in saliva.
   2. If saliva flow is low, the reverse effect occurs.

15.6 Functions of Saliva

A. Saliva is a complex fluid that performs multiple functions:
   1. *Food processing*. It is carried out through several actions:
      a. Moisten the food to form the *bolus*.
      b. Digesting certain substances, thanks to salivary enzymes such as *amylase* or *ptyalin*, which degrade starch, and *salivary lipase*.
      c. Facilitate the activity of the taste buds by dissolving the substances that produce *flavours*.
   2. *Protection*, which is carried out through various mechanisms:
      a. *Lubrication*, mainly thanks to *mucins*. This facilitates tongue movements, enables phonation and protects the oral mucosa.
      b. *Mechanical cleaning* or *rinsing*, produced by the flow of liquid through the mouth.
      c. Protection against *pathogenic microorganisms*. This is due to many factors:
         (1) *Mucins* facilitate the agglutination of bacteria, which prevents them from adhering to mucous membranes.
         (2) *Secretory IgA*, which binds to bacteria so that they are phagocytosed by leukocytes and macrophages.
         (3) *Lysozymes*, which break down bacterial walls.
         (4) *Defensins*, which permeabilise bacterial membranes.
         (5) *Sialoperoxidases*, which produce bactericidal substances.
         (6) *Lactoferrin*, which captures iron, preventing certain bacteria that need iron from using it.
         (7) *Histatins*, which are histidine-rich peptides that have antifungal capacity.
   3. *Regulation* of various processes:
      a. *Regulation of oral pH*, which is usually close to neutral pH.
         (1) An acidic pH facilitates bacterial action that leads to dental caries.
         (2) Saliva controls the pH by means of a *bicarbonate/carbonic acid buffer*.
         (3) *Histidine-rich peptides* and *phosphates* in saliva also play a role.
      b. Intervention in the *integrity of the tooth*. In addition to pH regulation, which prevents caries, saliva is involved in the maintenance of tooth structure by several mechanisms:
         (1) *Maturation of the enamel*, the hardness of which may increase after tooth eruption.
         (2) *Remineralisation*, by two mechanisms:

(a) Forming *fluorapatite* crystals in the enamel, more resistant to caries.
(b) Repairing enamel and dentine stains caused by demineralisation with Ca^{2+} and PO_4^{3-} ions.

(3) *Formation of the acquired pellicle*, an acellular organic layer covering the teeth and consisting of glycoproteins of salivary origin. This pellicle is protective and lubricating.

c. Involvement in *excretion* and *water balance*. The contribution of saliva to these processes is minor. For example:
(1) Alcohol is excreted through saliva, although it may be reabsorbed in the stomach.
(2) Lack of fluids can be compensated by a decrease in salivary secretion.

B. As saliva is easy to collect, it can be used for medical purposes. Substances that are incorporated into saliva during production in the salivary gland or that pass into saliva in the oral cavity can be detected:
1. *Diagnosis*: Saliva is currently used in pharmaceutical (steroid) and narcotics screening tests.
2. *Forensic techniques*:
   a. The *blood group* of 80% of people can be determined, as the oligosaccharides of the mucins can have the same structure as the oligosaccharides responsible for the blood groups.
   b. Cells shed from the oral mucosa in saliva are used as a source of *DNA*.

15.7 Tonsils

A. The *fauces* or *isthmus of the fauces* is the opening that connects the mouth to the pharynx. In this area, the mucosa may have:
1. *Diffuse lymphatic tissue*. It is a connective tissue infiltrated by many lymphocytes.
2. *Tonsils*. These are lymphatic organs formed by very evident and well-defined regions of lymphatic tissue, where the epithelium can invaginate to form *crypts*.

B. There are several types of tonsils that are arranged in the *oropharyngeal* region, as well as others that are located in the *rhinopharynx*, all of which constitute the *Waldeyer tonsillar ring* (Fig. 15.7), consisting of:
1. *Palatine tonsils*, two prominent oval-shaped bulges in the region between the glossopalatine and pharyngopalatine arches. They are usually called simply tonsils.
2. *Lingual tonsil*, located at the root of the tongue.
3. *Tubal tonsils*, in the lateral pharyngeal wall, at the entrance of the auditory tube.
4. *Pharyngeal tonsil* or *adenoid*, located in the roof of the nasopharynx.

C. The first two are in the oropharynx, the other two in the rhinopharynx.

D. Around the tonsils, connective tissue condenses to form a *capsule*. The presence of this capsule allows the tonsils to be surgically removed.

E. They are all characterised by spherical accumulations of lymphocytes called *lymphatic follicles* or *lymphatic nodules* in the lamina propria of the mucosa.

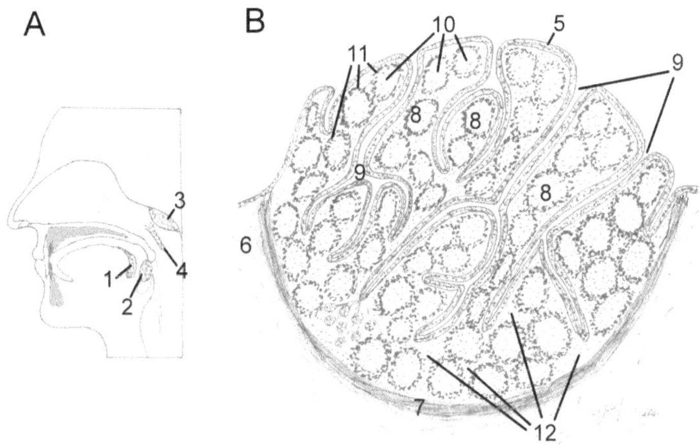

Fig. 15.7 Location of the tonsils in the oral cavity and pharynx (**A**) and histological organisation of the palatine tonsil (**B**). (1) lingual tonsil; (2) palatine tonsil; (3) pharyngeal tonsil or adenoid; (4) tubal tonsil; (5) lining epithelium; (6) lamina propria; (7) capsule; (8) lymphatic follicles or nodules; (9) crypts; (10) germinal centre; (11) mantle layer or corona; (12) parafollicular lymphatic tissue

1. The lymphatic follicle is a structure specific to lymphatic tissue that can be found in other lymphatic organs, such as the spleen or lymph nodes.
2. In the lymphatic follicle, contact between *B lymphocytes* and their antigen occurs. An accumulation of lymphocytes and other cooperating cells is observed, and a process of maturation and differentiation of the B lymphocytes takes place.
3. This process gives rise to new B lymphocytes capable of making more effective antibodies against the antigen.

F. Lymphatic follicles completely invade the connective tissue and displace and sometimes infiltrate the epithelium, making it difficult to distinguish a boundary between the epithelium and the lamina propria.

G. Lymphatic follicles can be:
1. *Primary follicles*: made up only of small lymphocytes, which are naïve lymphocytes that have not come into contact with an antigen.
2. *Secondary follicles*: these are follicles where contact with the antigen has occurred and lymphocyte maturation takes place. They are made up of:
   a. *Germinal centre*, in the central region, which is lightly stained in histological sections.
      (1) It occurs when a B lymphocyte that has recognised its antigen proliferates, and large lymphocytes (*immunoblasts* or *lymphoblasts*) accumulate.
      (2) These lymphocytes have dispersed euchromatin in their nuclei rather than heterochromatin as in small lymphocytes.
      (3) There are also *follicular dendritic cells (FDCs)* and *macrophages*.
   b. *The mantle layer* or *corona*, which is an outer ring or cap surrounding the germinal centre. It is made up of small B lymphocytes, which are *naïve lymphocytes* and *memory lymphocytes* that have been excluded from the germinal centre because they are not reacting with an antigen.

3. *Parafollicular lymphatic tissue*, consisting of *T lymphocytes*, is located around the lymphatic follicles.

H. There are *efferent lymphatic vessels* in the tonsils, which drain lymph to the lymph nodes.

I. All tonsils are similar but differ histologically by two structures that may vary:
   1. *Tonsillar crypts*. These are invaginations of the mucosal epithelium. They may be deep and branched, or simpler, and may even be absent.
   2. *Lining epithelium*, which depends on the area where it is located.
      a. In the oropharynx, it is usually *stratified squamous*.
      b. In the rhinopharynx, it is *pseudostratified columnar ciliated epithelium*, which is the typical epithelium of the upper airways (*respiratory epithelium*).

J. The main characteristics of the tonsils are summarised in the following table:

	Epithelium	Crypts	Capsule	Other features
Palatine tonsils	Non-keratinised stratified squamous	Deep and branched, abundant	Thick	Mucous glands outside the capsule
Lingual tonsil	Non-keratinised stratified squamous	Flattened and poorly branched, sparse	Poorly delimited	Mucous lingual glands flow into crypts
Pharyngeal and tubal tonsils	Pseudostratified columnar ciliated	No crypts, sometimes shallow folds	Attached to the periosteum	Mixed glands in the tonsil

Take-Home Message

Saliva is produced by salivary glands, which may be small glands scattered in mucosa and submucosa (minor glands) or may be encapsulated organs (major glands). In major glands, there are connective tissue septa separating several lobules, which together with the innervated and vascularised intralobular loose connective tissue, form the stroma. The parenchyma is formed by adenomeres and excretory ducts. Each adenomere with the first excretory ducts (intercalated duct and striated duct) forms the salivon, which produces saliva. Adenomeres can be serous acini, mucous tubuloacini and mixed tubuloacini. The parotid glands are exclusively serous, the sublingual and submandibular glands have serous, mucous and mixed adenomers. Primary saliva, iso or hypertonic, is produced in the adenomeres and is modified in the salivon ducts to form secondary saliva, which is hypotonic. Saliva contains enzymes for food processing, as well as lubricating, antibacterial, antifungal and pH-regulating substances. It also contains substances involved in the integrity of the tooth. The tonsils have mucosa infiltrated with lymphocytes that form lymphatic follicles, where B cells specialise to produce better antibodies against their antigen. The epithelium of the tonsils invaginates to form crypts.

Further Reading

Berkovitz BKB, Holland GR, Moxham BJ (2018) Salivary glands. In: Oral anatomy, histology & embryology, 5th edn. Elsevier, pp 303–321
Chibly AM, Aure MH, Patel VN, Hoffman MP (2022) Salivary gland function, development, and regeneration. Physiol Rev 102:1495–1552
Chiego DJ (2023) Salivary glands and Tonsils. In: Essentials of oral histology and embryology, 6th edn. Elsevier, pp 184–194
Emmelin N (1987) Neve interactions in salivary-glands. J Dent Res 66:509–517
Fehrenbach MJ, Popowics T (2020) Head and neck structures. In: Illustrated dental embryology, histology, and anatomy, 5th edn. Elsevier, pp 135–148
Gartner LP (2017) Digestive system: oral cavity. In: Textbook of histology, 4th edn, Elsevier, pp 419–434
Gómez de Ferraris ME, Campos Muñoz A (2009) Glándulas salivales. In: Histología, Embriología e Ingeniería Tisular Bucodental, 3rd edn. Médica Panamericana, pp 177–208
Hand AR (2014) Salivary glands, salivary secretion, and saliva. In: Hand AR, Frank ME (eds) Fundamentals of oral histology and physiology. Wiley Blackwell, pp 223–140
Jamieson JD (1988) The Exocrine Pancreas and salivary glands. In: Weiss L (ed) Cell and tissue biology – a textbook of histology, 6th edn. Urban & Schwarzenberg, pp 715–736
Kierszenbaum AL, Tres LL (2019a) Digestive glands. In: Histology and cell biology – an introduction to pathology, 5th edn. Elsevier, pp 575–606
Kierszenbaum AL, Tres LL (2019b) Upper digestive segment. In: Histology and cell biology – an introduction to pathology, 5th edn. Elsevier, pp 511–542
Krstić RV (1991) Digestive apparatus. In: Human microscopic anatomy. An atlas for students of medicine and biology. Springer Verlag, pp 161–255
Nanci A (2013) Salivary glands. In: Ten Cate's oral histology, 8th edn. Elsevier, pp 253–277
O'Dowd G, Bell S, Wright S (2023) Oral tissues. In: Wheater's functional histology. Elsevier, pp 241–252
Pawlina W (2024) Digestive system I: oral cavity and associated structures. In: Histology a text and atlas with correlated cell and molecular biology, 9th edn. Wolters Kluwer, pp 588–631
Redman RS (1994) Myoepithelium of salivary-glands. Microsc Res Tech 27:25–45
Redman RS (2025) Salivary gland carbonic anhydrases. Histol Histopathol 40:1675–1692
Shang YF, Shen YY, Zhang MC et al (2023) Progress in salivary glands: endocrine glands with immune functions. Front Endocrinol 14:1061235
Swid MA, Li LP, Drahnak EM et al (2023) Updated salivary gland immunohistochemistry. Arch Pathol Lab Med 147:1383–1389
Tandler B, Gresik EW, Nagato T, Phillips CJ (2001) Secretion by striated ducts of mammalian major salivary glands: Review from an ultrastructural, functional, and evolutionary perspective. Anat Rec 264:121 145
Tandler B, Nagato T, Toyoshima K, Phillips CJ (1998) Comparative ultrastructure of intercalated ducts in major salivary glands: A review. Anat Rec 252:64–91
Tandler B, Phillips CJ (2000) Organic secretion by striated ducts. Eur J Morphol 38:233–236
Tucker AS (2007) Salivary gland development. Semin Cell Dev Biol 18:237–244

Temporomandibular Joint Complex (TMJC) 16

What You Will Learn in This Chapter
The temporomandibular joint articulates the mandible with the temporal bone at the base of the skull. It is a diarthrosis with some special features. In this chapter, you will see its embryonic origin and its histological organisation, which differs from that typical of other diarthroses.

16.1 General Characteristics

A. The *temporomandibular joint (TMJ)* is the joint between the *head* or *condyle* of the *mandible* and the *temporal bone*, at the base of the *skull*.
B. The jaw articulates with the skull at two joints, right and left. Both are characterised by working in coordination with each other:
   1. If the TMJ moves on one side, the TMJ on the other side is also displaced in some way.
   2. It is impossible for the TMJ on one side to work independently of the TMJ on the other side.
C. In this joint, hinge movements (pivoting) are combined with sliding movements (gliding).
   1. Rotational or hinge movements in the sagittal plane make the TMJ a *ginglimoid joint*.
   2. The TMJ is also an *arthrodial joint*, as it performs translational or sliding movements.
   3. The combination of both types of movement allows the TMJ to be considered functionally as a *ginglimoarthrodial joint*.
D. The functional properties of the TMJ allow for various jaw movements:
   1. *Ascending* and *descending* of the mandible, leading to closing and opening of the mandible by rotational movements of the condyle.
   2. Forward displacement movements of up to 1.5 cm:

F. J. Sáez, I. Badiola, *Essential Oral Histology*, Learning Materials in Biosciences,
https://doi.org/10.1007/978-3-032-03165-5_16

   a. *Protrusion* is the forward displacement movement with tooth contact.
   b. *Propulsion* is displacement without tooth contact.
3. Backward displacement movements.
   a. *Retrusion* is the backwards displacement with tooth contact.
   b. *Retropulsion* is the backwards displacement without tooth contact.
4. *Centrifugal laterality* and *centripetal laterality*, which are lateral movements. When the mandible moves to one side, the condyle on that side moves centrifugally while the one on the other side moves towards the midline of the body (centripetally).
   a. During lateral movements, the TMJ on one side moves differently from the TMJ on the other side.
   b. The condyle that moves more is the one contralateral to the movement, i.e. in a displacement of the mandible to the right, the left condyle moves more than the right.

16.2 Anatomical Description of the Temporomandibular Joint Complex (TMJC)

A. The TMJ is associated with muscles and ligaments to form a complex structure that allows movement in various directions of the mandibular bone. This structure is called the *temporomandibular joint complex (TMJC)*. The description of this complex structure is the subject of anatomy, so it is briefly described here.

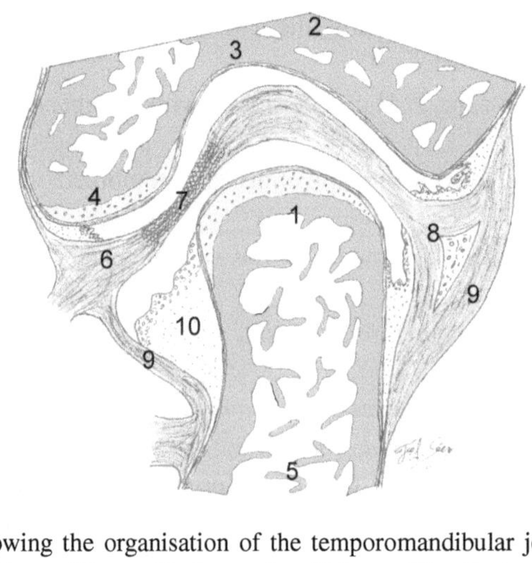

Fig. 16.1 Section showing the organisation of the temporomandibular joint. (1) condyle of the mandible; (2) temporal bone; (3) mandibular or glenoid fossa; (4) articular eminence; (5) mandibular bone; (6) anterior band of the articular disc; (7) intermediate zone of the articular disc; (8) posterior bilaminar articular disc; (9) capsular ligament; (10) synovial membrane

B. The TMJ is a *bicondylar*[1] *diarthrosis (synovial joint)*, consisting of (Fig. 16.1):
   1. *Mandibular condyle* or *condylar process*, which is the enlarged end of the mandibular bone that articulates with the temporal bone. In young people, the head of the condyle is more rounded than in adults.
   2. *Temporal bone*, in which there are two distinct surfaces that articulate with the mandibular condyle:
      a. *Mandibular fossa (articular* or *glenoid fossa)*, a concavity of the temporal bone.
      b. *Articular eminence*, a convex region situated anteriorly to the mandibular fossa.
   3. The *articular* or *synovial cavity*, located between the two bones and divided into two parts, upper and lower, by the articular disc.
   4. *Articular disc*, located between the mandibular condyle and the temporal bone.
      a. The articular disc divides the articular cavity into two separate cavities.
      b. Some authors mistakenly call it '*meniscus*', but it is not a meniscus, as menisci incompletely divide the joint cavities into two communicating cavities.
      c. Three regions are distinguished in this articular disc:
         (1) *Anterior band*, unilamellar.
         (2) Fine *intermediate zone.*
         (3) Thicker *posterior disc*, which eventually divides into two lamellae.
   5. *Capsular ligament*, which forms the *joint capsule* that constitutes the walls of the synovial cavity. As in other diarthroses, a *synovial membrane* is found on its inner side.

C. In addition, other ligaments and muscles are associated with this joint to form the TMJC. There are several muscles that can be involved in TMJ movements, but we are only listing the four most important ones:
   1. *Lateral pterygoid muscle*. Its superior head ends at the capsular ligament of the anterior region of the TMJ, while its inferior head contacts the periosteum of the mandible in the region of attachment to the anterior capsular ligament. It protrudes the mandible and pulls the articular disc forward.
   2. *Medial pterygoid muscle*. It runs from the pterygoid bone to the lower part of the ramus of the mandible. It brings forward and raises the mandible.
   3. *Temporalis muscle*. Raises and retracts the jaw.
   4. *Masseter muscle*. Raises the jaw.

D. The ligaments involved in the TMJC are grouped into two classes:
   1. *Major ligaments*:
      a. *Capsular ligament*. We have already mentioned it as a component of the TMJ.
         (1) This ligament restricts the movements of the joint and limits the distance the two bones can be separated without damage.
         (2) It consists of (Fig. 16.2):

[1] It is considered bicondylar because the two articulating bones have convex surfaces.

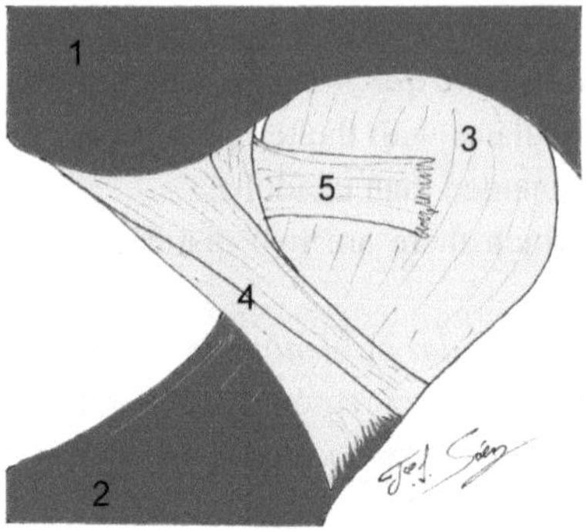

Fig. 16.2 Capsular ligament. The capsular ligament consists of the joint capsule (3) and the lateral or temporomandibular ligament, which has two parts: the external oblique portion (4) and the internal horizontal portion (5). (1) temporal bone; (2) mandible

(a) The *joint capsule*, which we have already mentioned, lines the articular cavity. Its outer part is fibrous, and its inner part is the synovial membrane.

(b) *Lateral* or *temporomandibular ligament*. It is a lateral thickening of the joint capsule with two parts:

1. An *oblique portion*, running from the articular eminence diagonally backwards and downwards to the outside of the condyle.
2. A *horizontal portion*, which originates from the articular eminence and inserts on the lateral part of the condyle.

b. *Collateral ligaments.*

c. *Temporodiscal ligament.*

2. *Accessory ligaments*:

a. *Pterygomandibular ligament.*

b. *Sphenomandibular ligament.*

c. *Stylomandibular ligament.*

16.3 TMJ Development

A. To understand TMJ, it is necessary to understand its development.

B. It begins in the eighth week of gestation (Fig. 16.3). Two condensations of mesenchyme form the *temporal blastema* and the *condylar blastema* (from the classical Greek βλάστημα blástēma, 'germ, shoot'), separated at some distance from each other.

C. The two blastemata eventually face each other by the 12th week. The separation between the blastemata decreases due to the growth of the condylar blastema.

D. An indentation then appears over the condylar blastema. This cleft will become the *lower articular cavity* of the TMJ.

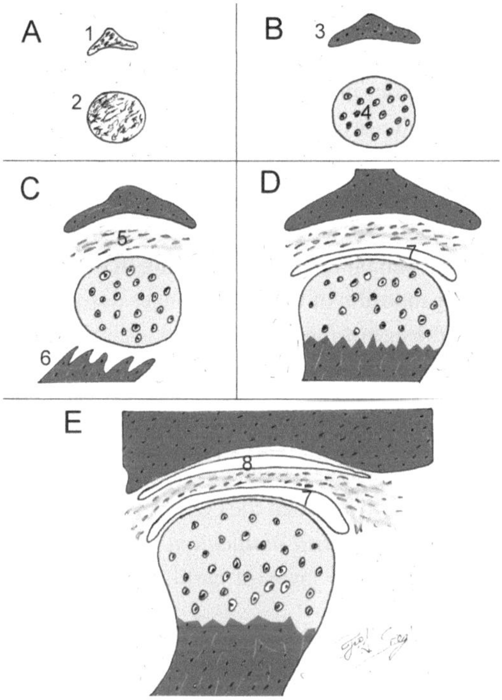

Fig. 16.3 Development of the temporomandibular joint. In the eighth week (**A**), two condensations of mesenchyme form the temporal blastema (1) and the condylar blastema (2). From the tenth week (**B**), intramembranous ossification begins in the temporal blastema (3) while the condylar blastema develops into cartilage (4). Shortly thereafter (**C**), a condensation of mesenchyme is observed between the two blastemae, which will form the articular disc (5). At the same time, the mandible (6), which is forming by intramembranous ossification, approaches the condylar blastema. In the twelfth week (**D**), a cleft appears between the future articular disc and the condyle, which will become the lower articular cavity (7). One week later (**E**), another cleft occurs between the temporal bone and the articular disc, which will give rise to the upper articular cavity (8)

E. Ossification begins earlier in the temporal blastema than in the condylar blastema:
   1. The temporal blastema undergoes *intramembranous ossification*.
   2. The condylar blastema becomes cartilage (*condylar cartilage*) and also contributes to the *joint capsule*.
F. At the same time, a second cleft is produced close to the ossifying temporal blastema, which will become the *upper articular cavity* of the TMJ.
G. Between the two clefts, the mesenchyme becomes the *articular disc*.
H. As a result, the outline of the mandibular condyle is formed by hyaline cartilage covered by a thin layer of fibrous mesenchyme. Subsequently, the condyle will undergo *endochondral ossification*, although a layer of cartilage will remain on the articular surface.

16.4 Histological Structure of the TMJ

A. There are very different descriptions of the histological organisation of the TMJ components. This may be due to several factors:
   1. There is a great deal of variation in different species. As many studies are done on animal models, the structure is not always identical.
   2. There is a large variation between different individuals. The histological organisation is different from person to person, resulting in different descriptions.
   3. The histological organisation of the TMJ changes during development and is therefore dependent on the age of the individual.

B. The histological features of the various components of the TMJ are described below.

C. *Mandibular condyle*. The histological organisation of the mandibular condyle changes during development and postnatal growth.
   1. Organisation of the mandibular condyle in the *foetus* and *neonate* (Fig. 16.4A):
      a. It has already been indicated that the condyle is formed from a *hyaline cartilage* mould in which *endochondral ossification* takes place.
      b. As the condyle ossifies, a layer of cartilage and other tissues remains on its surface.
      c. In the foetal and neonatal stages of development, the articular surface of the condyle is formed, from the articular cavity to the bone, by the following structures.
         (1) Superficial *articular zone*, a *fibrous connective tissue* covering with the following characteristics:
            (a) It is a continuous layer with the deep layer of the periosteum of the mandibular bone, while the superficial layer of the periosteum of the mandibular bone helps to form the joint capsule.
            (b) It has few blood capillaries. Vascularisation decreases with age.
            (c) It is made up of:

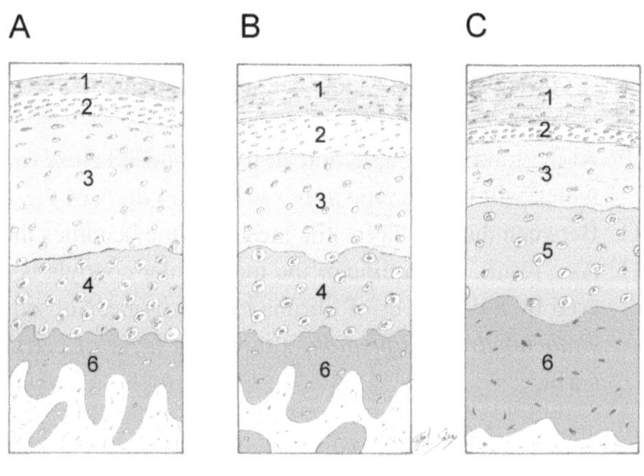

Fig. 16.4 Histological structure of the mandibular condyle in foetus and neonate (**A**), young individual (**B**) and adult individual (**C**). (1) superficial articular zone; (2) proliferative zone; (3) cartilage zone; (4) zone of hypertrophy, calcification and erosion; (5) calcified cartilage zone; (6) bone

1. *Type I collagen fibres*, oriented in an anteroposterior direction and parallel to each other.
2. *Fibroblasts.* Cellular components decrease with age.
3. *Elastic fibres*.

(2) *Proliferative zone*, a layer of undifferentiated or immature *mesenchymal cells* that gradually transform into chondroblasts and produce cartilage.

(3) *Cartilage zone*, which is *hyaline cartilage*.

(4) *Zone of hypertrophy, calcification and erosion*. As new cartilage is produced in the proliferative zone, the chondrocytes in the deeper zones undergo modifications:

(a) *Hypertrophy*.

(b) *Matrix mineralisation*.

(c) *Cartilage degeneration:* Cells become *necrotic*, there is vascular invasion, and *chondroclasts* are observed.

d. The erosion zone is the zone where *ossification* begins.

(1) *Osteoblasts* from the bone marrow of the mandibular bone are lodged in the spaces produced by cartilage degeneration, and growing *bone spicules* can be seen that are not parallel to each other, as occurs during the growth of long bones.

(2) This indicates that *multidirectional growth* occurs in the condyle.

(3) Bone in formation is usually spongy bone.

e. Some authors have compared the ossification process of the mandibular condyle with the ossification process of the epiphyseal plate of the long bones. The main characteristics of both are summarised in the following table:

	Cartilage	Coating	Chondroblasts	Growth	Cartilage resorption	Trabeculae of bone
Condylar cartilage	Hyaline	Fibrous connective	Disorderly	Multidirectional	By chondroclasts	Random
Epiphyseal cartilage	Hyaline	Uncoated	In columns	Unidirectional	Invasion of osteogenic capillaries	Verticals

2. Organisation of the mandibular condyle during *postnatal growth* (Fig. 16.4B): In the first two decades of postnatal life, the described layers are maintained. Some changes occur during this period:
   a. The *articular zone* and the *proliferative zone* are thickening.
   b. The *cartilage zone* (chondroblasts and chondrocytes) may be relatively thinner as hypertrophy appears early.
3. Organisation of the mandibular condyle in the *adult* (Fig. 16.4C).
   a. The following changes are observed in the adult:

(1) Proliferative activity ceases, although a layer of reserve cells with this capacity remains.

(2) Cartilage is converted into *fibrocartilage*.
(3) The area of hypertrophy is intensely calcified, so that the degree of calcification is even greater than in bone.

b. Therefore, in the adult condyle, we can find the following zones from the articular cavity towards the mandibular bone:
   (1) *Articular zone*. This is *fibrous connective tissue* forming a relatively thick layer.
      (a) It is rich in *collagen fibres* and may contain some *elastic fibres*.
      (b) *Fibroblasts* are scattered throughout the tissue.
      (c) Some *chondrocytes* can be seen, so it could be considered a form of fibrous cartilage.
      (d) Contains *superficial zone proteins (SZP)*:
         1. They facilitate the dissipation of frictional loads.
         2. They are synthesised by cells in the articular zone and can be released into the synovial fluid.
   (2) *Proliferative zone*. This is a thin zone rich in cells that retain the ability to proliferate.
      (a) If the occasion arises, these cells divide to renove the joint.
      (b) This may occur, for example, if there are functional changes or in certain orthodontic treatments.
   (3) *Fibrocartilaginous zone*. This layer is described as a fibrous layer of variable appearance.
      (a) In some texts, it is not strictly considered cartilage.
      (b) It is a very fibrous area with rounded cells that look like *chondrocytes*.
      (c) It has *type I collagen* and *type II collagen*, so it can be considered a *fibrous cartilage* layer.

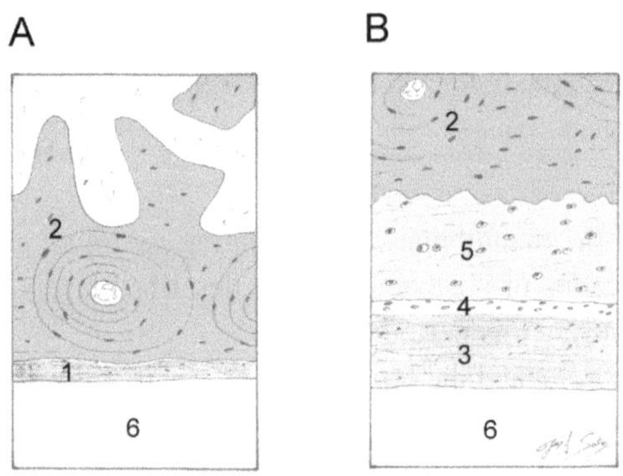

Fig. 16.5 Articular surfaces of the temporal bone: mandibular fossa (**A**) and articular eminence (**B**). (1) connective tissue similar to a periosteum; (2) bone; (3) articular zone of fibrous connective tissue; (4) proliferative zone; (5) fibrocartilaginous zone; (6) articular cavity with synovial fluid

(4) *Calcified cartilage zone*. Located in contact with the bone that, in this case, is usually a layer of *compact bone*. As we have said, its degree of mineralisation is very high.

D. *Mandibular fossa*. In this area, the temporal bone is covered exclusively by a thin layer of *connective tissue* similar to a periosteum (Fig. 16.5A).

E. *Articular eminence*.

1. In the newborn, there is no articular eminence, which develops later.
2. In the *child*, this structure develops in a similar way to the mandibular condyle, so it has the same layers as the growing condyle, from the articular cavity to the bone:
   a. Superficial fibrous *articular zone*.
   b. *Proliferative zone*. This zone actively produces chondrocytes.
   c. *Hyaline cartilage zone*.
   d. *Zone of hypertrophy, calcification and erosion*. In this zone, the chondrocytes die, the matrix calcifies and in the deep regions is replaced by bone matrix.
3. The process of formation and growth of the articular eminence is more rapid than that of the condyle, so that the proliferative capacity ceases earlier, and in the juvenile and adult the temporal bone of the articular eminence is covered from the articular cavity to the temporal bone by the following areas (Fig. 16.5B):
   a. Superficial *articular zone*, which is a *fibrous connective tissue* cover. As in the condyle, it produces SZP.
   b. *Proliferative zone*, which maintains little regenerative capacity.
   c. *Fibrocartilaginous zone*. It consists of *fibrous cartilage*, which is derived from the original hyaline cartilage.

F. *Articular disc*.

1. The articular disc, which conforms to the shape of the articular surfaces of the temporal bone and mandibular condyle, consists mainly of *dense fibrous tissue*. It contains *fibroblasts, type I collagen fibres* and *glycosaminoglycans*.
2. During foetal and postnatal development, the disc is highly vascularised. In the adult, the disc is innervated and vascularised in its peripheral regions.
3. Both the upper surface of the articular disc and the lower surface are concave, due to the fact that the anterior and posterior ends of the disc are thicker (up to 3 mm) than the central area (1 mm), which makes it possible to distinguish several zones in the disc:
   a. *Anterior zone*, thick, slightly anterior to the condyle. Its anterior part fuses with the wall of the joint capsule. It is characterised by:
      (1) Randomly disoriented *type I collagen fibres*.
      (2) Chondrocyte-like cells. Some authors call them *fibrochondrocytes*.
      (3) Isolated *adipocytes*.
      (4) Blood vessels and nerves.

b. Thin, *central zone* separating the anterior side of the condyle from the posterior side of the articular eminence. It is characterised by:
   (1) Have *type I collagen fibres* that are more densely arranged and form bundles, which are oriented in an anterior-posterior and medio-lateral direction. This orientation becomes more evident with age. They may adopt a wavy arrangement.
   (2) Being avascular and not innervated.
c. *Posterior zone*, occupying the space between the condyle and the roof of the glenoid fossa.
   (1) Like the anterior zone, it is characterised by:
      (a) Randomly arranged collagen fibres.
      (b) Chondrocyte-like cells.
      (c) Isolated adipocytes.
      (d) Blood vessels and nerves.
   (2) In its part closest to the capsule, it is divided into two lamellae. This is why it can be distinguished:
      (a) *Superior retrodiscal lamina*: this is relatively loose fibrous tissue with many *elastic fibres*, which only occurs in this region of the articular disc. This superior lamina contacts the wall of the joint capsule and the bones of the skull at the squamo-tympanic fissure.
      (b) *Inferior retrodiscal lamina*: consists exclusively of connective tissue without elastic fibres, and this lamina fuses with the periosteum of the neck of the condyle.
      (c) *Retrodiscal* or *retroarticular tissue*: located between the superior and inferior laminae of the bilaminar disc.
         1. It is a highly vascularised and innervated *very loose connective tissue.*
         2. If it is trapped between the condyle and the temporal bone, it causes a painful sensation.

4. During movement of the mandible, the disc moves passively to accommodate relative changes in the position of the condyle relative to the temporal bone (Fig. 16.6).
5. The disc moves with the condyle as the lower half of the joint capsule is closely attached to the condyle.

G. *Joint capsule.*
   1. The joint capsule encases the entire joint and creates an *articular cavity* divided into two by the articular disc and filled with *synovial fluid.*
   2. The joint sealing is achieved because the joint capsule is attached:
      a. At the top with the *periosteum of the temporal bone.*
      b. At the bottom with the *periosteum of the condyle.*
   3. The joint capsule has two layers:
      a. The *outer layer* is fibrous and is a modification of the fibrous outer layer of the bilaminar periosteum of the neck of the mandibular condyle.

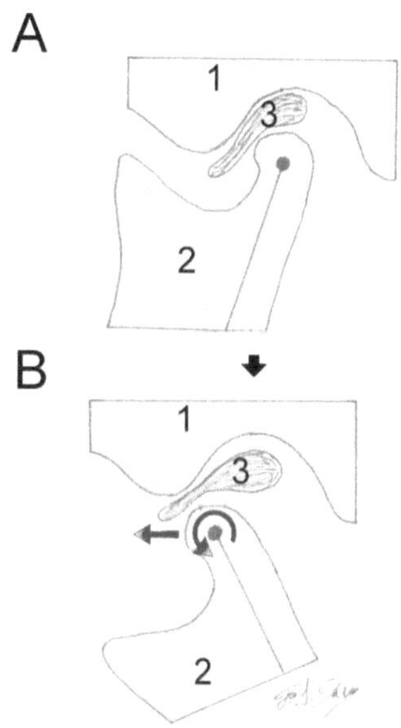

Fig. 16.6 Jaw movement: closed jaw (**A**) and open jaw (**B**). The long arrows indicate the direction of movement of the mandible. The articular disc deforms to adapt to the movements of the mandibular condyle. (1) temporal bone; (2) mandible; (3) articular disc

b. The *inner layer* is a *synovial membrane*. This synovial membrane is continuous with the osteogenic inner layer of the bilaminar periosteum that lines the neck of the condyle.
   (1) The synovial membrane covers the entire inner surface of the capsule and forms folds or *villi* that protrude into the synovial cavity. These villi increase with age.
   (2) The synovial membrane also lines the posterior bilaminar part of the articular disc, but not the rest of the disc.
   (3) It does not cover the mandibular condyle.

4. The histological organisation of the joint capsule is similar to that of other joint capsules of other synovial joints described in Chap. 7.
   a. Regarding the *subsynovial layer*, two of the three types of synovial membranes we have already studied have been described: *areolar* and *fibrous*. This depends on the joint area:
      (1) In the posterior region of the upper part of the joint capsule, it is areolar.
      (2) In the anterior region of the upper part, it is fibrous.
      (3) The presence of adipose synovial membrane has not been described, although with age and certain pathological conditions, there may be an increase in subintimal adipocytes.
   b. The *intimate layer* is characterised by:
      (1) It has both type A and type B synoviocytes.
         (a) *Type A synoviocytes*:
            1. Under the light microscope, they look like rounded cells located near the articular cavity.

2. Electron microscopy reveals *filopodia-like* surface folds, vesicles of various sizes and lysosomes. The nucleus has several small patches of heterochromatin and nucleoli are rare.
3 *Electron-lucent* and *electron-dense* type A synoviocytes have been described in the TMJ. The former have more vesicles and lysosomes and desmosomes with which they bind to other type A synoviocytes.
4. There is evidence that these cells are derived from *monocytes* and behave like *macrophages*, thus belonging to the *mononuclear phagocytic system*. They express *MHC type II*.

(b) *Type B synoviocytes*:
1. They are cells with well-developed rough endoplasmic reticulum and dense *secretory granules*. The nucleus is large and indented, with several nucleoli and little heterochromatin. *Caveolae* are observed in their plasma membrane. Sometimes these cells have large vesicles and can be confused with type A synoviocytes.
2. They produce *type I and II collagen*, *fibronectin* and glycosaminoglycans, such as *hyaluronan*, which appears to be a key component of synovial fluid.
3. These synoviocytes possess *oestrogen receptors*, which may explain why females have more TMJ disorders than males.

(2) The proportion of type A and type B synoviocytes in the synovial membrane is variable, depending on age, but in general, type B synoviocytes are more abundant.

(3) Origin of synoviocytes:
(a) Type B synoviocytes appear during the formation of the articular cavity and capsular ligament.
(b) Type A synoviocytes appear after birth.

(4) Type A and type B synoviocytes form an epithelial-like layer without basal lamina.
(a) To prevent the extracellular matrix of the synovial membrane from freely exiting into the articular cavity, type B synoviocytes possess cytoplasmic processes that extend superficially along the inner surface of the synovial membrane and form a lattice.
(b) Type A synoviocytes are located between the gaps left by the extensions of type B synoviocytes (see Fig. 7.10).

16.5 Vascularisation and Innervation of the TMJC

A. The vascularisation of the TMJC comes mainly from the *superficial temporal, anterior tympanic, deep auricular and ascending pharyngeal arteries*. All these arteries originate from the *external carotid artery*.

B. The nerve endings of the TMJC come from the *auriculotemporal nerve*, the *masseteric nerve* and the *deep temporal nerve*, all of which are branches of the *fifth cranial nerve*.
C. The TMJ shows free and encapsulated nerve endings, which, from most to least abundant, would be free endings, Ruffini corpuscles, Golgi tendon organs and Vater-Pacini corpuscles.
D. These terminations are arranged as follows:
   1. *Free endings*: scattered throughout the TMJ. They are even found on the synovial membrane and the edges of the articular disc. They are *pain* receptors.
   2. *Ruffini corpuscles*. Located in superficial layers of the joint capsule. They are mechanoreceptors that sense *posture*.
   3. *Golgi tendon organs*. Present in ligaments. They are mechanoreceptors that detect whether the joint is at the extremes of its capacity for movement.
   4. *Vater-Pacini corpuscles*. They are located in deep layers of the capsule. They are mechanoreceptors that detect *movement*.

Take-Home Message

The temporomandibular joint complex (TMJC) consists of the temporomandibular joint (TMJ) and associated ligaments and muscles. The TMJ articulates the mandible with the temporal bone at the base of the skull. It is a ginglimoarthrodial joint that enables various types of movement. It is an atypical diarthrosis, as the joint cavity is divided into two by an articular disc. The mandibular condyle is covered by several layers of tissue, one of which is cartilage, which with age becomes fibrocartilage. The condyle articulates with two areas of the temporal bone: the mandibular fossa (covered by a thin layer of connective tissue similar to periosteum) and the articular eminence (covered by several layers of tissue, including cartilage, which becomes fibrous with age). The articular disc shows an avascular and non-innervated central zone, which separates the condyle from the posterior part of the articular eminence, an anterior zone, vascularised and innervated, and a posterior zone, vascularised and innervated, divided into two laminae and a loose retrodiscal connective tissue between them. The joint capsule is similar to that of other diarthroses. During mandibular movements, the condyle rotates and moves from the mandibular fossa to the articular eminence, pressing on the central, non-innervated area of the articular disc. Abnormal displacements of the articular disc can cause the condyle to press on other regions of the articular disc, resulting in pain.

Further Reading

Barlattani A, Martelli M, Gargari M, Ottria L (2019) Articular disc of temporomandibular joint: an anatomical and histological study. Functional considerations. J Biol Regul Homeost Agents 33:199–208

Bender ME, Lipin RB, Goudy SL (2018) Development of the pediatric temporomandibular joint. Oral Maxillofac Surg Clin North Am 30:1

Berkovitz BKB, Holland GR, Moxham BJ (2018) Temporomandibular joint. In: Oral anatomy, histology & embryology, 5th edn. Elsevier, pp 293–302

Berkovitz BKB, Pacy J (2002) Ultrastructure of the human intra-articular disc of the temporomandibular joint. Eur J Orthod 24:151–158

Chiego DJ (2023) Temporomandibular joiing. In: Essentials of oral histology and embryology, 6th edn. Elsevier, pp 158–167

Constanza F, Bryan Q, Evelyn BE, Ramón F (2021) Anatomical and histological components of the bilaminar zone of the temporomandibular joint. A narrative review. Int J Morphol 39:477–483

Dijkgraaf LC, DEBont LGM, Boering G, Liem RSB (1996) Structure of the normal synovial membrane of the temporomandibular joint: a review of the literature. J Oral Maxillofac Surg 54:332–338

Fehrenbach MJ, Popowics T (2020) Temporomandibular joint. In: Illustrated dental embryology, histology, and anatomy, 5th edn. Elsevier, pp 275–284

Gómez de Ferraris ME, Campos Muñoz A (2009) Complejo articular temporomandibular (CATM). In: Histología, Embriología e Ingeniería Tisular Bucodental, 3rd edn. Médica Panamericana, pp 209–230

Griffin CJ, Hawthorn R, Harris R (1975) Anatomy and histology of the human temporomandibular joint. Monogr Oral Sci 4:1–26

Hinton RJ, Serrano M, So S (2009) Differential gene expression in the perichondrium and cartilage of the neonatal mouse temporomandibular joint. Orthod Craniofac Res 12:168–177

Kiga N (2012) Histochemistry for studying structure and function of the articular disc of the human temporomandibular joint. Eur J Histochem 56:56–57

Kuroda S, Tanimoto K, Izawa T et al (2009) Biomechanical and biochemical characteristics of the mandibular condylar cartilage. Osteoarthr Cartil 17:1408–1415

Mizoguchi I, Toriya N, Nakao Y (2013) Growth of the mandible and biological characteristics of the mandibular condylar cartilage. Japan Dent Sci Rev 49:139–150

Nanci A (2013) Temporomandibular joint. In: Ten Cate's oral histology, 8th edn. Elsevier, pp 311–327

Nozawa-Inoue K, Amizuka N, Ikeda N et al (2003) Synovial membrane in the temporomandibular joint – Its morpohology, function and development. Arch Histol Cytol 66:289–306

Piette E (1993) Anatomy of the human temporomandibular joint. An updated comprehensive review. Acta Stomatol Belg 90:103–127

Porto F (2014) Temporomandibular joint. In: Hand AR, Frank ME (eds) Fundamentals of oral histology and physiology. Wiley Blackwell, pp 149–161

Tsutusmi-Arai C, Tran A, Arai Y et al (2025) Mandibular condylar cartilage in development and diseases: a PTHrP-centric view. Orthod Craneofac Res. (in press). https://doi.org/10.1111/ocr.12936

Xu ML, Zhang XY, He Y (2022) An updated view on temporomandibular joint degeneration: insights from the cell subsets of mandibular condylar cartilage. Stem Cells Dev 31:445–459

Dental Pulp

17

What You Will Learn in This Chapter

In this chapter, you will learn how the connective tissue of the dental pulp is organised into various zones, which cells form it, and how the nerves and blood vessels that run through it are organised. You will see that odontoblasts, which form dentine, are found at the boundary of dentine and dental pulp, and when active are tall cells with many biosynthetic organelles. You will learn that there are multipotent stem cells that can produce many cell types and are being used for tissue regeneration studies. You will see that there are two types of sensory nerve fibres in the dental pulp: one is responsible for dentine sensitivity and the other for pulp sensitivity. Finally, you will study the functions performed by the pulp and learn that it can undergo some alterations, including the formation of pulp stones.

17.1 General Characteristics

A. The *dental pulp* (from the Latin pulpa, 'meat') is the soft, non-mineralised tissue that occupies the inner cavity of the teeth (*pulp cavity*), both at the crown and root. In colloquial terms, it is often referred to as '*nerve*', which is scientifically an incorrect term.

B. The dental pulp contains *odontoblasts*, the dentine-producing cells, so *dentine* and dental pulp are closely related and are considered the *dentine-pulp complex*. However, it is important to emphasise their differences:
   1. The dentine is mineralised, the pulp is a soft tissue.
   2. The cells that form dentine are of ectomesenchymal origin (from neural crests). In contrast, cells of different origins are mixed in the pulp.
   3. The dentine does not have vascularisation, the pulp does.

C. The dental pulp is formed from the *dental papilla*, a tissue of ectomesenchymal origin that is part of the tooth germ and originates from the cells of the *neural crests*. It is important to recall the development of the tooth discussed in Chap. 13.

F. J. Sáez, I. Badiola, *Essential Oral Histology*, Learning Materials in Biosciences,
https://doi.org/10.1007/978-3-032-03165-5_17

D. Shape of the pulp cavity:
   1. In teeth with a single root, the pulp cavity occupies the crown and root, with no visible boundary between the two regions. In the cusp area, there may be one (canines) or three (incisors) *pulp horns*, which are extensions of the chamber towards the cusp.
   2. In multi-rooted teeth (Fig. 17.1), the pulp cavity is divided into two regions:
      a. A coronary portion or *pulp chamber*, in which the following can be distinguished:
         (1) A roof from which *pulp horns* emerge towards the cusps.
         (2) A floor, in which there are openings that communicate with the root portions.
      b. *Root canals*, one for each root.
E. In the apical region, the pulp cavity of the root ends in an orifice called the *apical foramen*.
   1. At the foramen, the dental pulp communicates with the tissues of the periodontium, with no distinct boundary, so this region is called *Black's undifferentiated space* or *periapical space*.
   2. *Reserve (mesenchymal) stem cells* are located in the periapex.
   3. The blood vessels and nerves that supply and innervate the dental pulp pass through the apical foramen.

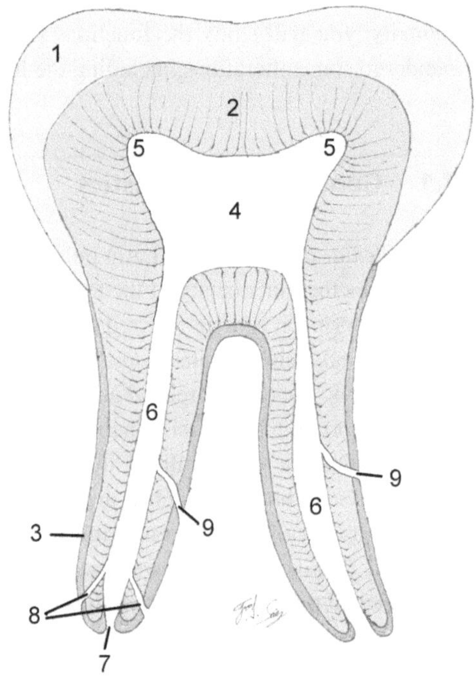

Fig. 17.1 Section of a tooth showing the distribution of tissues and the organisation of the pulp cavity. (1) enamel; (2) dentine; (3) cementum; (4) coronary portion of the pulp cavity or pulp chamber; (5) pulp horns; (6) root canals; (7) apical foramen; (8) accessory foramina; (9) lateral canals

4. Sometimes the apical part of the pulp chamber does not have a single orifice, but several. In these cases, the largest is the apical foramen, and the others are called *accessory foramina*.
5. Sometimes the pulp cavity of the root communicates with the periodontium through lateral canals, but this does not occur in all teeth.

F. Internal dentine deposition can occur throughout the life of the tooth, so that the pulp cavity decreases in volume with age and, consequently, the volume of dental pulp also decreases.

17.2 Tissue Organisation of the Dental Pulp

A. The dental pulp is a richly vascularised and innervated *loose connective tissue* that shows a zonal organisation, which is easier to see in the pulp of the crown than in the pulp of the root. The following zones can be distinguished (Fig. 17.2):

1. *Supraodontoblast zone*. This zone does not really exist and is therefore not shown in Fig. 17.2. However, some authors include it because when preparing the tissues for microscopic examination, there is a separation between the predentine and the odontoblasts. In this area are found:

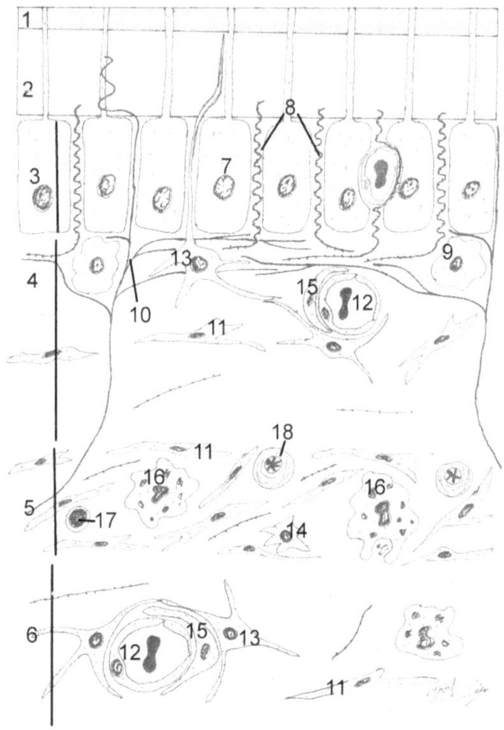

Fig. 17.2 Histological and topographical organisation of the dental pulp. (1) dentine; (2) predentine; (3) odontoblastic zone; (4) oligocellular or basal zone of Weil; (5) cell-rich zone; (6) central zone; (7) odontoblasts; (8) von Korff's fibres; (9) subodontoblastic Höhl cell; (10) nerve fibres (plexus of Raschkow); (11) fibroblasts; (12) blood vessels; (13) interdigitating dendritic cells; (14) dental pulp stem cells (DSPC); (15) pericytes; (16) macrophage; (17) lymphocyte; (18) plasma cell

a. *Marginal nerve plexus* or *predentinal plexus (plexus of Bradlaw)*. Not a true plexus, but a congregation of axons.
b. *Dendritic cells*.

2. *Odontoblastic zone*. This is a layer of cells that looks like an epithelium, although in reality it is not, as it has no basal lamina. It can be observed:
   a. *Odontoblasts*: cells of epithelial appearance, but of ectomesenchymal origin. The characteristics of odontoblasts will be explained below. They are usually arranged in a cylindrical epithelium, as if forming a palisade.
      (1) In the coronary zone, they usually have the appearance of *pseudostratified columnar epithelium.*
      (2) In the root zone, it is more reminiscent of a *simple columnar epithelium.*
   b. *Reticular fibres* (type III collagen).
   c. Blood capillaries.
   d. *Nerve fibre endings* that can penetrate dentine.
3. *Oligocellular* or *basal zone of Weil*. Also called the *subodontoblastic zone*. Located below the odontoblasts, it is a zone with very few cells. As there are few or no cells, few nuclei are observed, but there are many cytoplasmic extensions of cells located in the next layer, so it could be said to be a 'nuclear-poor' area.
   a. The oligocellular zone is well visible in the coronary region but is rarely distinguishable in the root region.
   b. In this region, we can distinguish:
      (1) Cells in contact with the odontoblasts, which are called *subodontoblastic Höhl cells*, are located at the border between this zone and the odontoblastic zone. Each of these Höhl cells originates from the last mitotic division of the cell that gave rise to an odontoblast.
      (2) A nerve plexus (*plexus of Raschkow*), more developed in the coronary than in the root portion of the pulp.
      (3) A capillary network (*subodontoblastic capillary plexus*).
      (4) Type III collagen fibres (*reticular fibres*) that enter the odontoblastic area (*von Korff's fibres*).
      (5) *Fibroblasts*, which contact odontoblasts and Höhl cells via gap junctions.
      (6) *Interdigitating dendritic cells* of the immune system (antigen-presenting cells), usually in perivascular position, whose prolongations enter between the odontoblasts and apparently enter the dentinal tubules.
4. *Cell-rich zone*. This is clearly seen in the crown, but is more difficult to see in the pulp of the root. Many cells are observed in this zone:
   a. The most numerous are *fibroblasts*.
   b. There are also *undifferentiated mesenchymal cells* (of ectomesenchymal origin).
   c. *Macrophages*.
   d. *Lymphocytes*.

5. *Central zone*. This is *loose connective tissue* with fewer cells than the cell-rich zone. It presents:
    a. *Fibroblasts*.
    b. *Macrophages*.
    c. Perivascular *interdigitating dendritic cells*.
    d. *Pericytes*. As mentioned above, pericytes are cells that surround small blood vessels (capillaries and postcapillary venules) and appear to be another form of mesenchymal reserve cell.

B. Some authors classify the odontoblastic, oligocellular and cell-rich zones in the so-called *odontogenic* or *peripheral zone*.

C. A *parietal nerve plexus* is sometimes identified between the cell-rich zone and the central zone.

17.3 Dental Pulp Cells

A. *Odontoblasts*. The odontoblast arises from the *ectomesenchymal cells* of the dental papilla, and their structure and function vary depending on when they are found during odontogenesis and in the adult tooth. These differences will be discussed when we look at dentinogenesis, in the chapter on dentine.
    1. Layout (Fig. 17.3):

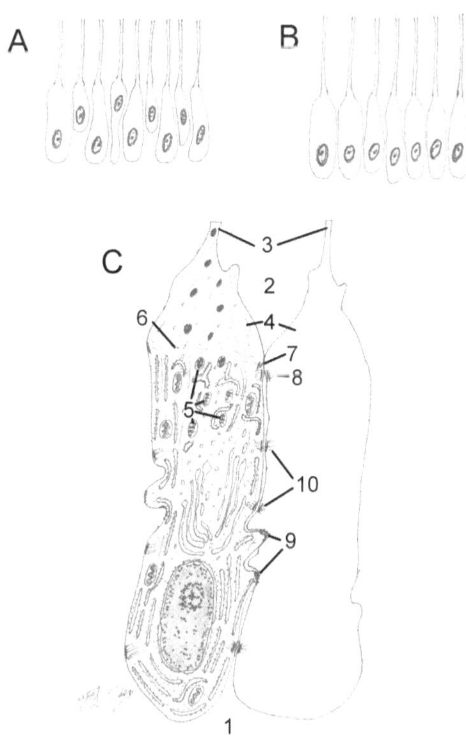

Fig. 17.3 Odontoblast organisation in the crown (**A**), in the root (**B**) and ultrastructure of the odontoblast (**C**). (1) distal pole; (2) proximal pole; (3) odontoblast process; (4) proximal cone; (5) abacus bodies; (6) terminal web; (7) tight junction; (8) banded desmosome; (9) gap junction; (10) desmosomes

a. As mentioned above, they are located at the periphery of the pulp, in contact with the dentine, forming the odontoblastic zone.
b. They form a layer of epithelial-like tissue that lacks a basal lamina.
   (1) This epithelial layer has a *pseudostratified columnar* appearance in the crown. The pseudostratified appearance may be an artefact produced by the plane in which the sections are made, as the odontoblasts tend to pack together because they are pushed into the pulp cavity as they synthesise dentine.
   (2) In the root, they appear as a *simple cuboidal* or *columnar epithelium.*
   (3) In the periapex, this epithelial layer is more *flattened* and less active.

2. Taking as a reference plane the *dentinoenamel junction (DEJ)* in the crown or the *cementum-dentine junction (CDJ)* in the root, the odontoblast has a *proximal pole*, in contact with the dentine, or more specifically with the predentine, and a *distal* or *basal pole*, in contact with the pulp.
3. *Resting odontoblasts (aged odontoblasts)* are small, tightly packed cells.
   a. Its nucleus is in a proximal position.
   b. In the distal cytoplasm are the organelles, which are few in number.
   c. The proximal, supranuclear cytoplasm has no organelles but lipid inclusions and cytoskeleton.
   d. Secretory vesicles are scarce or absent.
4. *Active odontoblasts* have a more complex structure (Fig. 17.3).
   a. They are tall, columnar cells, about 40–50 μm high and 5–10 μm wide, taller in the crown than in the cervical or root regions. If the odontoblasts are not actively secreting and are resting, they may have a more shrunken appearance.
   b. The proximal end of the cell is cone-shaped (*proximal cone*) and from its apex extends a cellular extension, the *odontoblast process*, which penetrates into tubular cavities in the dentine called *dentinal tubules.*
      (1) The length of the odontoblast processes is discussed. The dentinal tubules, which we will study in due course, can pass through the entire thickness of the dentine, but it is not clear whether the odontoblast process reaches the end of the tubule or only extends over part of its length. It is assumed that they reach at least one third of the dentine thickness, but there is also evidence that they may occupy the entire length of the dentine tubule.
      (2) In the odontoblast process, no large organelles (endoplasmic reticulum, mitochondria or Golgi apparatus) are observed, only secretory vesicles, cytoskeleton and pinocytosis vesicles.
   c. The nucleus is large, basal, with peripheral heterochromatin and several nucleoli.
   d. In the cytoplasm are observed:
      (1) Lots of *rough endoplasmic reticulum* and *ribosomes*. These organelles are spread throughout the cell, except in the cone on the proximal side. For this reason, the cytoplasm is very *basophilic.*
      (2) The *Golgi apparatus*, which is supranuclear.

(3) *Secretory granules*. All secretory granules are transported to the proximal cone and the odontoblast process. They are of various types:
    (a) Elongated secretory granules with a bead-like arrangement of material, which is why they are called *abacus bodies*. These granules transport and process collagen and other substances of the dentine matrix.
    (b) Rounded secretory granules known to contain *bone sialoprotein* and *osteocalcin*.
(4) Many mitochondria are scattered throughout the cytoplasm.
(5) Cytoskeleton:
    (a) *Microtubules*.
    (b) Abundant *vimentin filaments*, especially in the proximal cone and the odontoblast process, where they are arranged parallel to the axis of this cell extension.
    (c) Between the cell cone and the rest of the cytoplasm, there is a *terminal web*, formed by *actin microfilaments*, which contacts junctional complexes (*adherens junctions*) located in the membrane at that level.

e. Odontoblasts are joined together by *junctional complexes* on their proximal side.
    (1) These complexes are made up of:
        (a) *Tight junction*.
        (b) *Adherens junction*, under the tight junction.
    (2) These junctional complexes do not appear to form true belts surrounding the cells (zonulae), as is the case in true epithelia, so it is unclear whether they seal the intercellular space or only serve to give consistency to the cell assembly.
    (3) If the intercellular space is not well sealed, there may be a passage of substances from the dentine to the pulp and vice versa. In fact, some substances (e.g. serum proteins) have been shown to pass through the space between odontoblasts, but others do not, so there is some uncertainty on this issue.
f. The remaining lateral membranes of the cell are *interdigitated* and linked together by *desmosomes*.
g. There are *gap junctions*, both between odontoblasts and with other cells of the basal surface (Höhl cells and fibroblasts).

5. The known functions of the odontoblast are:
    a. Produce and mineralise *dentine* through secretions produced in the proximal cone and through the odontoblast process. One of its main products is collagen, which is one of the major components of dentine.
    b. Form a selective barrier that hinders the passage of toxins into the pulp and allows the outflow of fluid into the dentinal tubules. In addition, the odontoblast is thought to control the composition of this fluid.
    c. Respond to bacterial damage or toxins by producing *inflammatory mediators*, such as *interleukin-8* that attracts and activates neutrophils, or others that activate dendritic cells. They also produce *β-defensins* that kill bacteria.

   d. According to some theories, which will be discussed in the next chapter, the odontoblast may act as a *sensory receptor*. Although no synapses between odontoblasts and axon terminals have been described, the odontoblast may be able to affect the sensitivity of nearby axons.
6. The plasma membranes of odontoblasts express high levels of *transforming growth factor beta (TGF-β) receptors*. TGF-β has been shown to stimulate odontoblast activity during development and, in the adult tooth, during dentine repair.
7. The mature odontoblast is a cell incapable of dividing, so it is believed that it can live as long as the tooth.
   a. Approximately half of the odontoblasts die within four years of tooth eruption.
   b. However, if the odontoblasts in one area are damaged, the pulp can produce new odontoblasts.
   c. During odontogenesis, odontoblasts mature in response to *inducing stimuli* from the enamel organ (crown) or from Hertwig's epithelial sheath (root).
   d. In the adult tooth, the formation of new odontoblasts from reserve stem cells, when neither the enamel organ nor Hertwig's epithelial sheath is present, must be in response to different but unknown signals.

B. *Fibroblasts*. Connective tissue cells can be found in the pulp. The most numerous cells are fibroblasts. In some texts, they are called *pulpoblasts*.
   1. In the *young pulp*, *active fibroblasts* are usually abundant, with a lot of cytoplasm and well-developed organelles for protein synthesis (rough endoplasmic reticulum and Golgi apparatus).
   2. In the *mature pulp*, mostly *quiescent fibroblasts* with little cytoplasm and few biosynthetic organelles are observed.
   3. Fibroblasts can contact each other via *desmosomes*.
   4. These cells are replaced from time to time:
      a. They form in the central area of the pulp, migrate to the periphery and die there by apoptosis.
      b. They have been estimated to have a half-life of 24±11 days.

C. *Undifferentiated ectomesenchymal cells and stem cells*.
   1. There are undifferentiated ectomesenchymal cells in the pulp which, with appropriate stimuli, replace fibroblasts and odontoblasts.
      a. They are mainly found in the cell-rich area and in the central area, usually close to blood vessels.
         (1) They are usually large polyhedral cells with a large, clear central nucleus.
         (2) The cytoplasm is abundant and shows peripheral extensions.
      b. The number of these cells decreases with age.
   2. Some researchers have been able to isolate cells from dental pulp that, under certain conditions and with the addition of appropriate factors, multiply and can differentiate into odontoblasts, chondrocytes, osteoblasts, adipocytes, endothelial cells and neurons, i.e. they are *multipotent cells*.
      a. These cells have been called *dental pulp stem cells (DPSCs)*.

b. They are a potential source of tissue regeneration and are therefore being investigated in the field of *tissue engineering*.
c. It is unclear whether all undifferentiated ectomesenchymal cells in the dental pulp can act as stem cells, or only some of them.

D. *Defence cells*.
1. *Dendritic cells* can be found in the dental pulp.
   a. In the pulp of the preeruptive tooth, they are found in and around the odontoblastic zone.
   b. In erupted teeth, they are located under the odontoblastic layer.
   c. They are usually associated with blood vessels and nerve endings. They can infiltrate between odontoblasts, and their extensions penetrate the dentinal tubules.
   d. They are more abundant in teeth with *caries*.
   e. As explained above, its function is to present antigens to **T** *lymphocytes* to activate their immune response.
   f. These cells can enter the *lymph* and travel to the *lymph nodes* to stimulate the immune response there. In cases of caries, dendritic cells have been observed to reach the lymph nodes after 16 hours.
2. *Macrophages* are usually found in the central area.
   a. These cells are responsible for removing dead cell debris to allow cell renewal in the pulp.
   b. In case of inflammation, they act as *antigen-presenting cells* and phagocytose foreign agents.
3. *Lymphocytes* are found in small numbers in healthy dental pulp.
   a. They are mostly *T lymphocytes*, present in healthy and inflamed pulp.
   b. *B lymphocytes* are normally very rare in healthy pulp and in the early stages of inflammation but appear in more advanced stages.
4. *Plasma cells* are also found in the pulp.
5. *Mast cells* can sometimes be observed, usually in a perivascular position. Mast cells are difficult to observe, but they are responsible for the inflammatory processes in the dental pulp by releasing histamine, which causes oedema.
6. *Neutrophils* may be present, especially when caries occurs.
7. Some *eosinophils* may be found.

17.4 Extracellular Matrix of Dental Pulp

A. *Collagen* is found in the extracellular matrix of the pulp.
1. It is found in smaller proportion than in other connective tissues. In fact, collagen accounts for only 3–5% of the dry weight of the pulp.
2. The most abundant forms are *type I* and *III collagens*, in similar amounts (type I 55–60%—type III 40%). This proportion is maintained throughout life, but the total amount of collagen increases with age.

3. The reticular fibres:
   a. They are very fine fibres that are randomly distributed.
   b. In the subodontoblastic zone, they become very thick, take on a helical or cork-screw shape and are inserted between the odontoblasts, forming the *von Korff's plexus*, in which *type VI collagen* seems to be found.
4. There are also other collagenous varieties in smaller quantities (e.g. *collagen type V* and *type VI*).
5. In infancy, the collagen fibres are simple and scattered throughout the pulp.
6. In the adult pulp, the increased amount of collagen leads to the formation of fibre bundles.
7. Fibres are more abundant in the apical zone of the pulp, which is important in endodontic treatment.

B. *Elastic fibres* are very rare in the pulp, but are found in the walls of blood vessels, especially arterial vessels. *Oxytalan fibres* have also been identified. It should be remembered that elastic fibres contain elastin and fibrillin, whereas oxytalan fibres contain only fibrillin.

C. Of the *glycosaminoglycans* present in the mature pulp, 60% is *hyaluronic acid*, 20% *dermatan sulphate*, 12% *chondroitin sulphate* and the remainder *heparan sulphate*. However, during development, the majority of glycosaminoglycan is chondroitin sulphate.

D. The extracellular matrix contains other molecules:
   1. Typical of connective tissue, such as *fibronectin*, *tenascin* or *integrin*.
   2. Some are specific to bone tissue, such as *bone sialoprotein (BSP)*, *osteopontin (OPN)* and *bone morphogenetic protein (BMP)*. *Osteonectin* can also be found in the immature pulp of the tooth germ. The presence of these proteins may be justified by the similarities between dentine and bone.
   3. *Metalloproteinases* and *growth factors (TGF-β)* are also found.

17.5 Pulp Vascularisation

A. Blood vessels enter and leave the pulp through the apical foramen and accessory foramina.
   1. Normally, one or two *arterioles* enter through the apical foramen. The largest arterioles may be 150 μm in diameter.
   2. Other smaller vessels enter through the accessory foramina.
   3. The *veins* leaving the pulp run parallel to the arterioles.

B. Once in the pulp, the arterioles increase in diameter and decrease in wall thickness.
   1. In the root portion of the pulp, they run through the central zone and emit some smaller lateral branches that go towards the subodontoblastic zone, where they branch out.

2. In the crown, the arterioles emit more abundant branches that also lead to the subodontoblastic area, where a dense network of capillaries is produced, forming the *subodontoblastic capillary plexus*:
   a. It has mostly *continuous capillaries* and a small number of *fenestrated capillaries* (4–5%) with a continuous basal lamina.
   b. All of them have *pericytes*, and *dendritic cells* are frequent around them.
   c. Capillaries are found under and between the odontoblasts. They are sometimes seen in the predentine.

C. The capillaries empty their blood into *postcapillary venules*, which converge into *central venules*.

D. Central venules merge into other venules that exit through the apical foramen.
   1. Venules are usually similar in diameter to arterioles but have a thinner wall and a wider lumen.
   2. Smooth muscle cells form a thin, discontinuous layer.

E. *Arteriovenous anastomoses* are common in the dental pulp. They are the size of arterioles and have relatively high endothelial cells.

F. The blood vessels maintain a constriction tone regulated by the *sympathetic system* endings.
   1. Blood flow regulation is important, as the increase in vessel diameter (vasodilation) accompanied by an increase in permeability leads to an increase in the volume of tissue fluid (*oedema*), which is part of the inflammatory process.
   2. As the pulp is restricted in the pulp cavity, inflammation stimulates sensory nerve endings that produce *pain* sensation.

G. Today, dentists determine the vitality of the pulp using techniques that measure blood flow.

H. Although difficult to observe, *lymphatic vessels* have also been found in the subodontoblastic area and, to a greater extent, in the central area of the pulp.
   1. These vessels form as blind vessels in the peripheral regions of the crown and merge into vessels in the central area.
   2. One or two larger vessels leave the pulp through the apical foramen.

17.6 Pulp Innervation

A. The pulp is richly innervated. It has been estimated that a tooth can be penetrated by 2500 axons aggregated in *nerves* with little perineurium and endoneurium.

B. Nerves enter through the apical foramen, accompanying the blood vessels, forming the *neurovascular bundle*. If the accessory foramina are large, some nerves may also enter.
   1. At the root, the nerves accompany the blood vessels through the central area and give off some branches to the peripheral areas. Most of the fibres are directed to the coronary area, where they branch extensively.

a. Many of the branches are directed to the periphery, and a nerve plexus is formed in the subodontoblastic area (*Plexus of Raschkow*).
b. Some nerve endings penetrate between the odontoblasts and may form a *marginal plexus (plexus of Bradlaw)* between the odontoblasts and the predentine, while other nerve endings penetrate the dentinal tubules.

2. The nerves contain both *myelinated* and *amyelinated axons*, but the myelin is being lost, so there are fewer myelinated axons in the coronary zone. Those that reach Plexus of Raschkow are usually myelinated axons.

C. There are sensory and vegetative nerve endings in the dental pulp.

1. Sensory endings:
   a. There is increasing evidence that sensory fibres act in functions of regulating the metabolism of the dental pulp by releasing *neuropeptides* or other substances, such as *calcitonin gene-related peptide (CGRP)*, *enkephalin*, *somatostatin* and *endorphin*.
   b. These fibres come from *cranial nerve V (trigeminal nerve)*. This nerve produces three branches (*ophthalmic*, *maxillary* and *mandibular*):
      (1) The pulp of the maxillary teeth is innervated by fibres of the *maxillary branch*.
      (2) The pulp of the mandibular teeth is innervated by fibres of the *mandibular branch*.
   c. Sensitive fibres can be of two types:
      (1) *Type A myelinated fibres*, responsible for dentine sensitivity.
         (a) Type A myelinated fibres make up 25% of the fibres that penetrate a tooth.
         (b) They reach the peripheral zone, the basal zone of Weil, where they often lose myelin and form the *subodontoblastic nerve plexus of Raschkow*.
         (c) Many axons terminate here, but some axon terminals enter between the odontoblasts and into the dentinal tubules. These endings may expand in a brush or fan shape.
         (d) In these areas, they contact the body of the odontoblast or its process. Although no true synapses have been described, some molecules typical of synapses have been detected in the dentinal tubules.
         (e) They are responsible for a *sharp (acute), sort, fast, localised pain* sensation, which corresponds to *dentine sensitivity*.
         (f) There are collateral branches of these fibres that are directed towards the blood vessels, so that if the sensory termination in the periphery receives a stimulus, it would propagate to the nerve centres but also to the collateral branch, via the so-called *axonal reflex*, and produce vasodilation.
         (g) Factors that can cause dentine pain include:
            1. Many external agents, such as cold water or cold air.
            2. Actions by the dentist, such as mechanical contact with probes or drills and dehydration with cotton wool.

(2) *Type C amyelinated fibres*, which are responsible for the sensitivity and therefore the sensation of pain in the pulp.
   (a) Type C amyelinated fibres terminate in the central area of the pulp.
   (b) They capture dental pulpal pain, which manifests as *slow, dull, diffuse and prolonged ache*.
   (c) There are also collaterals that can innervate blood vessels and would perform an *axonal reflex* as we have seen in myelinated fibres.

2. Endings of the vegetative system:
   a. It consists of *type C amyelinated axons* with vasomotor function that innervate the smooth muscle of blood vessels.
   b. Most of these fibres are *sympathetic system* fibres that originated from the *superior cervical ganglion* and release *noradrenaline* and, in some cases, *neuropeptide Y*, causing vasoconstriction.
   c. Although not clearly demonstrated, it seems that there are also fibres of the *parasympathetic system*:
      (1) They produce *acetylcholine* or *vasoactive intestinal polypeptide (VIP)*, which stimulates the endothelial cells of the blood vessel to produce *nitric oxide (NO)*, which is transmitted to the muscle cells of the arteriole and causes vasodilation.
      (2) However, since the presence of parasympathetic system endings is not well demonstrated, it could be that vasodilation occurs:
         (a) By reflex mechanisms of the sensory axons.
         (b) By the production of substances in the tissue itself. For example, NO can be produced by *hypoxia* (lack of oxygen in the tissues) or by tissue damage.
         (c) In addition, as noted above, sensory nerve endings in the pulp release *CGRP*, a neuropeptide that causes vasodilation.

17.7 Pulp Activity and Functions

A. *Induction* of enamel production, during odontogenesis, as explained in the corresponding chapter.
B. *Dentine formation*, which also occurs during odontogenesis. In the adult tooth, the synthesis of a small amount of dentine is maintained.
C. *Maintenance* of dentine by providing substances that pass from the tissue fluid to the dentinal tubules or are secreted from the odontoblast.
D. *Uptake of painful stimuli*, as nerve fibres in the pulp are responsible for sharp (acute), sort, fast, localised dentine pain and slow, dull, diffuse and prolonged pulpal ache.
E. *Defence of the tooth*, which generally translates into the ability to *repair* the tooth through the production of dentine. The dentine that is produced can be of two types, which we will study in the corresponding chapter:

1. *Tertiary or reparative dentine*, which is produced by new odontoblasts in response to an aggression.
2. *Peritubular dentine*, which forms in the dentinal tubules and can blind them as a defence against, for example, caries.

17.8 Pulp Disorders

A. Over time, the pulp often undergoes various changes.
B. All alterations affect the ability of the pulp to respond to possible damage and are related to the ability to produce new odontoblasts and repair dentine.
C. Pulp disorders may include:
   1. *Volume reduction*. One of the most important changes is the reduction of the volume of the pulp cavity and, consequently, of the volume of the dental pulp.
      a. This reduction in volume is due to the continuous deposition of dentine by the odontoblasts.
      b. The reduction in volume narrows the root canal, which can become very thin. This can affect the innervation and irrigation of the pulp. Sometimes the root canal can become blocked.
   2. *Modification of tissue characteristics*. They may be due to several causes:
      a. Reduction in the number of cells.
         (1) The number and density of pulp cells decrease with age.
         (2) It has been estimated that from the age of 20 to 70, cell density decreases by half.
      b. Increase in the number of collagen fibres. With age, collagen fibre bundles also appear, leading to the transformation of the tissue into *semi-dense connective tissue*.
   3. *Nerve degeneration*. With age, degeneration of the nerve fibres also occurs, which leads to a loss of sensitivity.
   4. *Calcification*. One of the most striking alterations of the pulp is the appearance of calcified regions. These calcifications can be (Fig. 17.4):
      a. *Pulp stones* or *denticles*. Mineral deposits occur on a collagen centre. These stones are usually spherical or oval, although they can also have irregular shapes. They can be classified according to several criteria:
         (a) Existence or non-existence of dentinal tubules:
            1. *True denticles*, when *dentinal tubules* are visible. Odontoblasts are usually seen at their periphery, with processes penetrating through the tubules.
            2. *False denticles*, if they have only concentric layers of calcified material.
         (b) Location in the pulp:
            1. *Free*, if they do not contact dentine.
            2. *Attached*, if in contact with dentine.
            3. *Embedded*, when they become included in the dentine. A denticle may form as free, grow and become attached or embedded.

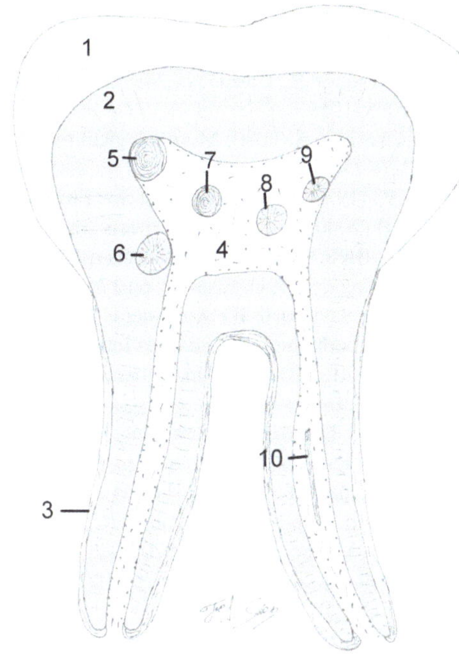

Fig. 17.4 Different types of dental pulp calcifications. (1) enamel; (2) dentine; (3) cementum; (4) dental pulp; (5) false attached denticle; (6) true embedded denticle; (7) false free denticle; (8) true free denticle; (9) true attached denticle; (10) calcified needle

b. *Diffuse calcifications*. They are frequent in the pulp of the root. They have an elongated appearance, which is why they are called *calcified needles*. They are formed following the path of collagen fibre bundles or surrounding blood vessels.

Take-Home Message

Several zones are usually distinguished in the dental pulp. The supraodontoblastic zone contains plexus of Bradlaw and dendritic cells. The odontoblastic zone is formed by odontoblasts, contains capillaries, and may be traversed by von Korff's reticular fibres and nerve endings. The basal zone of Weil is oligocellular, contains subodontoblastic Höhl cells, the nerve plexus of Raschkow, the subodontoblastic capillary plexus, von Korff's fibres and dendritic cells. The cell-rich zone has many of the connective tissue cell types in greater abundance than in the central zone. Between these two zones is the parietal nerve plexus. In the pulp, there are multipotent dental pulp stem cells that can produce cells from many types of tissues, including tissues not present in the tooth, and are being used in tissue engineering. There are motor and sensory nerve endings in the dental pulp. Type A (myelinated) sensory nerve fibres reach the dentine and are responsible for registering dentine pain, which is sharp, short, fast and localised. Type C (amyelinated) sensory fibres detect pulpal pain, which is slow, dull, diffuse and prolonged. With age, the pulp can become altered, including the formation of pulp stones.

Further Reading

Alvarez-Vásquez JL, Castañeda-Alvarado CP (2022) Dental pulp fibroblast: a star cell. J Endodont 48:1005–1019

Bae YC, Yoshida A (2020) Morphological foundations of pain processing in dental pulp. J Oral Sci 62:126–130

Barker BC (1980) Pulpal pain. Aust Soc Prosthodontists Bull 10:3–7

Bender IB (2000) Pulpal pain diagnosis – a review. J Endodont 27:175–179

Berkovitz BKB, Holland GR, Moxham BJ (2018) Dental pulp. In: Oral anatomy, histology & embryology, 5th edn. Elsevier, pp 176–194

Caruntu ID, Savinescu SD, Amalinei C (2014) Morphometric approach to pulp fibroblast development in tooth germ. Biomed Res Int 2014:836583

Chiego DJ (2023) Dental pulp. In: Essentials of oral histology and embryology, 6th edn. Elsevier, pp 110–130

Dammaschke T (2010) The formation of reparative dentine and Hohl cells in the dental pulp. Endo-Endodontic Pract Today 4:255–261

Dega V, Mogoanta L, Manolea H et al (2006) Radiological and microscopic aspects of the denticles. Romanian J Morphol Embryol 47:263–268

Fehrenbach MJ, Popowics T (2020) Dentin and pulp. In: Illustrated dental embryology, histology, and anatomy, 5th edn. Elsevier, pp 160–174

Feiglin B (1994) Aspects of dentinal and pulpal pain. Diagnosing dental pain. Ann R Australas Coll Dent Surg 12:143–152

Gartner LP (2017) Digestive system: oral cavity. In Textbook of histology, 4th edn. Elsevier, pp 419–434

Goga R, Chandler NP, Oginni AO (2008) Pulp stones: a review. Int Endod J 41:457–468

Goldberg M, Smith AJ (2004) Cells and extracellular matrices of dentin and pulp: a biological basis for repair and tissue engineering. Crit Rev Oral Biol Med 15:13–27

Gómez de Ferraris ME, Campos Muñoz A (2009) Complejo dentino-pulpar I: pulpa dental. In: Histología, Embriología e Ingeniería Tisular Bucodental, 3rd edn. Médica Panamericana, pp 231–254

Jain N, Grupta A, Meena N (2013) An insight into neurophysiology of pulpal pain: facts and hypotheses. Korean J Pain 26:347–355

Koling A, Raskandersen H (1983) Membrane junctions in the subodontoblastic region – a freeze-fracture study of the human dental-pulp. Acta Odontol Scand 41:99–109

Koling A, Raskandersen H, Baggersjoback D (1981) Membrane junctions on odontoblasts – a freeze-fracture study of the human dental-pulp. Acta Odontol Scand 39:355–360

Morse DR (1991) Age-related-changes of the dental-pulp complex and their relationship to systemic aging. Oral Surg Oral Med Oral Pathol Oral Radiol Endodont 72:721–745

Nair PNR (1995) Neural elements in dental pulp and dentin. Oral Surg Oral Med Oral Pathol Oral Radiol Endodont 80:710–719

Ngassapa D (1996) Correlation of clinical pain symptoms with histopathological changes of the dental pulp: a review. East Afr Med J 73:779–781

O'Dowd G, Bell S, Wright S (2023) Oral tissues. In: Wheater's functional histology. Elsevier, pp 241–252

Stenvik A, Mjor IA (1970) Epithelial remnants and denticle formation in human dental pulp. Acta Odontol Scand 28:721

Tamgadge S, Tamgadge A, Agre BP et al (2023) Odontoblasts in health and disease with an additional note on its three-dimensional histological perspective. J Sci Soc 50:4–9
Warshawsky H (1988) In: Weiss L (ed) The teeth. In cell and tissue biology – a textbook of histology, 6th edn. Urban & Schwarzenberg, pp 594–640
Wisniewska K, Rybak Z, Szymonowicz M et al (2021) Review on the lymphatic vessels in the dental pulp. Biology-Basel 10:1257
Zhang WB, Yellick PC (2021) Tooth repair and regeneration: potential of dental stem cells. Trends Mol Med 27:501–511

Dentine

18

What You Will Learn in This Chapter

In this chapter, we show you how dentine is, what its characteristics are, its composition, the types of dentine that can be distinguished, and its histological organisation. You will see that dentine is made up of dentinal tubules and an intertubular matrix, which are its basic structures. In addition, other secondary structures can be seen in dentine, produced by variations in the basic structures. You will be able to study how the differentiation of the odontoblast from an undifferentiated mesenchymal cell takes place. We will also show you how dentine is synthesised and mineralised. Finally, you will see the theories that try to explain how dentinal pain is captured.

18.1 General Characteristics

A. *Dentine* or *ivory* forms the main mass of the tooth, which is covered by enamel in the crown and cementum in the root. Dentine forms the wall of the pulp cavity in which the dental pulp is housed.

B. Dentine is a *yellowish*, *semi-transparent* substance when fresh.

C. From a mineral point of view, it is *harder* than compact bone, with which it shows certain similarities in its molecular organisation.

D. The dentine is made up of:

   1. A *mineralised extracellular matrix*, which is the major component.
   2. *Dentinal tubules*, tubular cavities, usually parallel to each other, which run through the entire thickness of the dentine. In these tubules are found:
      a. *Dentinal fluid*, which is found in the dentinal tubules.
      b. Complete cells are not observed, but in the dentinal tubules, immersed in the dentinal fluid, they are lodged:
         (1) *Prolongations of odontoblasts*.
         (2) *Nerve endings*.

F. J. Sáez, I. Badiola, *Essential Oral Histology*, Learning Materials in Biosciences,
https://doi.org/10.1007/978-3-032-03165-5_18

E. At the inner boundary of the dentine, in the area of contact with the pulp, the dentine is lined with a layer of *odontoblasts*, the cells that produce and maintain dentine and whose processes are housed in the dentinal tubules.
F. Between the dentine and the odontoblasts, there is usually a thin layer of non-mineralised dentine called *predentine*.
G. As dentine and odontoblasts of the dental pulp are closely related, the group formed by dentine and dental pulp is often known as the *dentine-pulp complex*.
H. The physical properties of dentine are very important for the dentist.
   1. Its yellowish colour, although variable, is responsible for the colour of the tooth.
   2. It is harder than bone and cementum, but less hard than enamel, due to the mineral content of its extracellular matrix.[1]
   3. For this reason, its degree of *radiopacity* is greater than that of bone and less than that of enamel, which is why it appears denser than bone but lighter than enamel on X-rays.[2]
   4. It is relatively *elastic*, which allows it to compensate for the hardness of the enamel and thus resist masticatory forces.
   5. It is more *permeable* than enamel due to the presence of dentinal tubules.
I. Different *varieties of dentine* in a tooth have been described according to different criteria. These varieties are described below.

18.2 Classification of Dentine According to Its Position in the Tooth (Topographical Classification)

A. Depending on its position in the tooth, dentine can be classified into three types (Fig. 18.1):
   1. *Mantle dentine*. It is the first to be produced and therefore the outermost.
      a. It is in contact with the enamel in the crown or the cementum in the root.
      b. It is about 20–150 μm thick.
      c. In their matrix, the *thick collagen fibres* (0.1–0.2 μm) are arranged in an orderly pattern, perpendicular to the dentine surface.
      d. Its degree of mineralisation is lower than in the rest of the dentine.
      e. As the dentinal tubules branch here, the number of tubules observed is greater than in the other areas of the dentine.
      f. In the root, it is usually seen with a granular appearance, so that some authors consider that in the root there is no mantle dentine but a layer of dentine called *granular layer of Tomes*.

[1] On the hardness of human mineralised tissues, it is useful to review the hardness table at the beginning of chapter 6.

[2] It should be noted that in an X-ray the denser is whiter and the less dense is blacker.

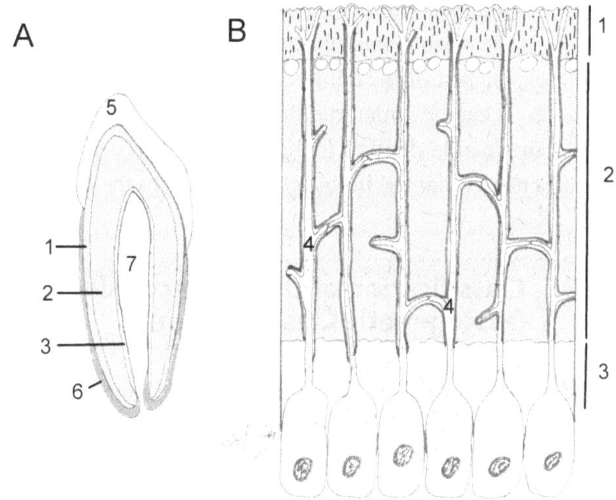

Fig. 18.1 Topographical classification of dentine: section of a tooth (**A**), and magnified view of dentine (**B**). (1) mantle dentine; (2) circumpulpal dentine; (3) predentine; (4) dentinal tubules with odontoblast processes; (5) enamel; (6) cementum; (7) pulp

2. *Circumpulpal dentine*. The remainder of the mineralised dentine that forms the main mass of the tooth.
   a. It is thick, but the thickness depends on the type of tooth and the area of the tooth observed.
   b. Its collagen fibres are thinner (50–200 nm) and irregularly arranged, intertwined in a very tight reticulum.
3. *Predentine*. A thin layer of non-mineralised dentine that is located in the innermost part, at the boundary with the pulp cavity and in direct contact with the odontoblasts.
   a. It is about 20–30 μm wide.
   b. The thickness of the predentine in a tooth usually remains stable, because as new predentine is synthesised on the inner side of the tooth, the outermost predentine mineralises and becomes dentine. However:
      (1) In periods with more active dentinogenesis, the thickness may be greater.
      (2) At older ages, it tends to become thinner.
   c. Its matrix is similar to that of *osteoid*, with 90% collagen and 10% non-collagenous proteins, with many sulphur compounds.
   d. In sections stained with haematoxylin and eosin, the dentine appears pinkish, but the predentine appears lighter, with less intensity of staining.

B. The dentine varieties described above are clearly visible in the crown. However, some peculiarities are observed in the root:
   1. Some authors consider mantle dentine to be exclusive to the crown, while in the root the outermost layer of dentine would be the *granular layer* mentioned above.
   2. An outer layer about 10–20 μm thick, at the boundary between cementum and dentine, called the *hyaline layer* or *Hopewell-Smith layer*, has been described.

a. It appears to be *atubular* and the collagen fibres are arranged perpendicular to the surface and oblique, which makes it possible to distinguish this layer from the rest of the dentine.
b. It can be a material that allows cementum to be attached to dentine.

C. As the mantle dentine is thin or absent in certain areas, most of the data on dentine structure explained in this chapter refer to the circumpulpal dentine.

18.3 Classification of Dentine According to Its Origin (Histogenetic Classification)

A. Dentine is synthesised and deposited throughout the life of the tooth. Dentine deposition is high during odontogenesis, but after tooth eruption, dentine synthesis slows down.

B. With this in mind, three varieties of dentine have been distinguished according to the time of its formation (Fig. 18.2):
   1. *Primary dentine*. This is the dentine that is produced during odontogenesis, from the time the crown begins to form until root formation is complete, i.e. until the tooth eruption is complete.
      a. It constitutes the major part of the dentine of the tooth.
      b. In relation to the topographical classification, primary dentine consists of mantle dentine and most of the circumpulpal dentine.

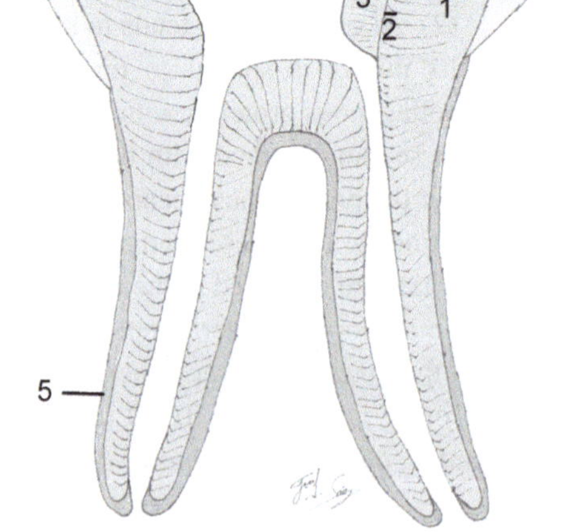

Fig. 18.2 Histogenetic classification of dentine. (1) primary dentine; (2) secondary dentine; (3) tertiary dentine; (4) enamel; (5) cementum

2. *Secondary dentine*, also called *adventitious, regular* or *physiological dentine*. It is the dentine that is produced after the completion of root formation.
   a. It constitutes the innermost part of the circumpulpal dentine.
   b. Its synthesis proceeds much more slowly than that of primary dentine throughout the life of the tooth, which causes the thickness of the dentine to increase slowly, while the volume of the pulp cavity decreases.
   c. The boundary between primary and secondary dentine is usually detected by a slight curvature of the dentinal tubules, which when viewed under the microscope is detected as a dark line.
   d. The thickness of secondary dentine differs in each region of the tooth.
      (1) In multi-rooted teeth, it is usually thicker at the roof and floor of the pulp chamber, but thinner at the pulp horns.
      (2) Secondary dentine deposition is high in the roots.
3. *Tertiary dentine*. Also known as *reparative, reactive, irregular* or *pathological dentine*. Dentine that is produced in response to a *noxa* (plural noxae, from the Latin noxa, 'damage') or localised stimulus.
   a. It only occurs in the affected area.
   b. As it grows inwards from the edge of secondary dentine, it deforms the pulp cavity.
   c. The microscopic structure of tertiary dentine is highly variable and can be distinguished (Fig. 18.3):
      (1) Normal dentine, the structure of which is like that of primary and secondary dentine that will be discussed later.
      (2) Dentine with very few dentinal tubules.
      (3) *Osteodentine*, with odontoblasts trapped in the matrix, forming a structure reminiscent of bone.

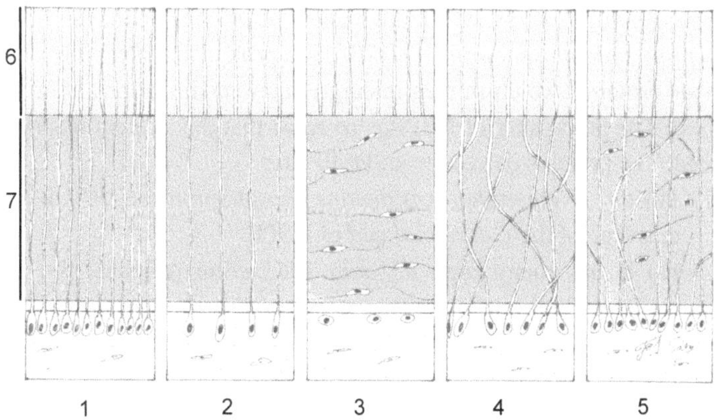

Fig. 18.3 Different forms of histological organisation of tertiary dentine. (1) normal dentine; (2) dentine with very few dentinal tubules; (3) osteodentine; (4) irregular dentine; (5) dentine showing a combination of several types of organisation; (6) primary or secondary dentine; (7) tertiary dentine

(4) Irregular dentine, with the dentinal tubules oriented in multiple directions, crossing over each other.
(5) Dentine showing a combination of several or all of the types described above.

d. Some authors distinguish between two types of tertiary dentine:
   (1) *Reactive* or *reactionary dentine*. Produced by the same odontoblasts that produce primary or secondary dentine. It is the *primitive odontoblasts* that react to the stimulus.
   (2) *Reparative dentine*. Produced by new odontoblasts formed from reserve cells because primitive odontoblasts have been damaged by the noxa.

18.4 Dentine Extracellular Matrix

A. Dentine contains no cells, only *odontoblast processes* and *nerve endings*, and most of its volume is made up of extracellular matrix. Therefore, in order to understand dentine, it is essential to study the organisation of its extracellular matrix.
B. *Water* may constitute 10% of the matrix. In this liquid medium, organic components (20%) and inorganic components (70%) are arranged, although this varies according to the region of the dentine.
C. Dentine is first produced as *predentine*, non-mineralised, and then mineralisation occurs. Therefore, when studying the dentine matrix, a distinction is usually made between the composition of the *organic matrix* and the *inorganic* or *mineral matrix*. However, it should always be borne in mind that the two are closely related.
D. *Organic matrix*. It is made up of the following components:
   1. *Collagen* (90% of the organic matrix). It is mainly *type I collagen*, although there is 1–2% of *type III* and *V collagen*. Most of the mineral in the dentine matrix is accommodated on and between the collagen fibrils.
   2. *Noncollagenous proteins*, which are located between the collagen fibres.
      a. *Dentine sialophosphoprotein (DSPP)* is a protein that is rapidly cleaved to form other proteins, which in turn undergo many modifications after being synthesised. The processing of DSPP is so rapid that this protein has never been isolated. The proteins derived from DSPP are:
         (1) *Dentine phosphophorin* or *dentine phosphoprotein (DPP)*. It is produced by cleavage of the *C-terminal end* of DSPP.
            (a) It has many *aspartic acid* and *serine* residues, the latter usually phosphorylated.
            (b) It has many negative charges and is able to bind calcium (Ca^{2+}).
            (c) It also appears to bind to collagen and facilitate the formation of hydroxyapatite crystals.
         (2) *Dentine sialoprotein (DSP)*. It is produced by cleavage of the *N-terminal end* of DSPP. Its distribution in dentine and its function are not at all clear.

(a) According to some, DSP is mainly found around the dentinal tubules (peritubular dentine), where it inhibits mineralisation to prevent the tubules from filling with dentine.

(b) Other authors place it exclusively in the mantle dentine.

(c) Some books indicate that it is found exclusively in the predentine.

(3) *Dentine glycoprotein (DGP)*. It is produced by cleavage of the *central region* of DSPP.

b. *Dentine matrix protein-1 (DMP1)*. DMP1 is mainly found around the dentinal tubules (*peritubular dentine*). It is very similar to DSP, and it appears that they may act synergistically.

c. *Gamma-carboxyglutamate-containing proteins (Gla proteins)*. These are proteins that contain a sequence of amino acids that bind Ca^{2+}.

(1) Among these amino acids is *glutamate*, which is carboxylated by the action of *vitamin K*.

(2) At least two proteins of this family, present in small quantities, have been found in dentine:

(a) *Osteocalcin* or *bone Gla protein (BGP)*.

1. It is present in osteoblasts, odontoblasts and cementoblasts.
2. Reversibly binds to hydroxyapatite crystals.
3. It may be involved in mineralisation processes.

(b) *Matrix Gla Protein (MGP)*. It is a protein that is also present in bone, where it is thought to be involved in calcification processes.

d. *Osteonectin*.

e. *Bone sialoprotein (BSP)*.

f. *Osteopontin*.

g. *Matrix extracellular phosphoglycoprotein*.

h. Serum proteins. Including *albumin* and *immunoglobulins*.

i. *Proteoglycans*. For example, decorin, biglycan and perlecan.

j. *Growth factors*. These include TGF, BMP, FGF and IGF.

k. Proteolytic enzymes. *Enamelysin* and *matrix metalloproteinases (MMP)* have been localised.

3. *Lipids*. *Phospholipids* (phosphatidylcholine and phosphatidylethanolamine) and *cholesterol* have been detected.

E. *Inorganic matrix*.

1. It consists primarily of *hydroxyapatite crystals*, smaller than those found in enamel, but similar to those found in bone.
   a. The crystals are about 36 nm long, 25 nm wide and 10 nm high.
   b. These crystals are positioned parallel to the collagen fibres, and are arranged between the fibres and within them, between their molecules.
2. Other minerals are present in smaller quantities, such as *carbonates*, *sulphates*, *fluorine*, *copper*, *iron*, *zinc* and *magnesium*.

18.5 Histological Organisation of Dentine

A. When studying dentine under the microscope, two types of structures have been observed:
   1. *Basic structures of dentine*. They are the actual components of dentine. There are two of them:
      a. Cylindrical cavities running through the dentine from the pulp to the dentine surface. These cavities are the *dentinal tubules*.
      b. A matrix that forms the major part of the dentine volume. It is called *intertubular matrix* or *intertubular dentine*.
   2. *Secondary structures of dentine*. They are structures that can be seen under the microscope, but are actually produced by the arrangement of the basic structures, by their different degree of mineralisation or by the relationship of dentine with cementum and enamel.
   3. We will study both types of structures separately.

B. *Basic structures of dentine*. There are two, the dentinal tubules and the intertubular matrix.
   1. *Dentinal tubules*. These are cylindrical cavities in which the odontoblast process is housed.
      a. The dentinal tubules pass through the dentine from the odontoblastic layer of the pulp to the boundary with the enamel (DEJ) or cementum (CDJ).
      b. Occasionally, the tubule is observed to partially penetrate the enamel or cementum.
         (1) This is because, during odontogenesis, the odontoblast process is introduced between the ameloblasts before enamel formation.
         (2) Consequently, when enamel begins to deposit, it does so around the odontoblast process and leaves a hollow tubular region where the odontoblast process is housed and is a continuation of the dentinal tubule.
      c. The dentinal tubules do not have the same thickness along their entire length.
         (1) They are wider near the pulp (2.5 μm) than near the enamel or cementum (1 μm).
         (2) This is due to the deposition towards the centre of the tubule of mineralised matrix in the regions furthest away from the odontoblast.
      d. The amount of dentinal tubules is high.
         (1) In the vicinity of the pulp there are between 60,000 to 75,000 tubules per mm^2.
         (2) In the vicinity of the enamel or cementum, the density of tubules is lower and may be reduced by half.
         (3) This difference is due to the fact that during dentine growth, dentine is deposited towards the interior of the tooth and the odontoblasts are crowded together as the pulp space decreases, which means that the tubules are arranged in less space.

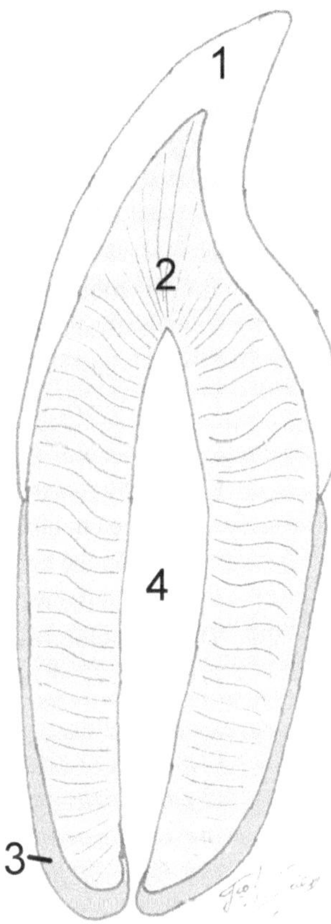

Fig. 18.4 Primary curvatures of the dentinal tubules. (1) enamel; (2) dentine; (3) cementum; (4) dental pulp

e. The dentinal tubules have two types of curvatures:
   (1) *Primary curvatures*. When a tooth is observed in a perpendicular section, from crown to apex, the dentinal tubules usually show a curved path from the DEJ or CDJ to the pulp (Fig. 18.4).
      (a) They are caused by the relative movements of the odontoblasts during dentinogenesis. These movements occur because the pulp chamber is decreasing in volume, and the odontoblasts are rearranging themselves in the available space.
      (b) These curvatures are usually in the form of an inclined (italic) 'S' shape.
      (c) This 'S' has a coronary convexity on the outer part of the dentine, and an apical convexity on the innermost part.
      (d) In the cusp or incisal regions, the curves are smaller, and the tubules may appear straight.
      (e) In the apical region, there may be a single curvature, with very little pronounced apical convexity, and they may even be rectilinear.
   (2) *Secondary curvatures*. These are small sinusoidal curves that appear relatively regularly along the entire course of a tubule.

   (a) They are included in the primary curvatures.
   (b) They may be a reflection of the spiral path taken by the odontoblasts as they move centripetally while synthesising dentine.

f. Dentinal tubules can branch. There are two types of branching:
   (1) *Collateral branches* or *secondary tubules*.
      (a) The collateral branches are very thin, less than 1 μm in diameter, and start at right angles to the main tubule, until they contact the nearby tubules.
      (b) Branches of the odontoblast process are introduced through these secondary tubules, which can communicate with the processes of neighbouring odontoblasts by *gap junctions*.
      (c) Branching is higher at the root than at the crown.
   (2) *Terminal branches*, present in the final course of the dentinal tubules, close to the enamel or cementum.
      (a) In the crown, these terminal branches are arborescent.
      (b) In the root, these endings are dichotomous.

g. The following structures can be found inside the tubule:
   (1) *Process of the odontoblast*. We have already studied it in the chapter on the pulp.
   (2) *Nerve endings*, naked. These are sensory endings coming from the pulp.
   (3) *Dendritic cell extensions*, whose cell somas are found in the dental pulp.
   (4) *Collagen fibres*, which form a very loose, non-mineralised matrix.
   (5) *Dentinal fluid* is a liquid medium, derived from the tisue fluid of the dental pulp, which fills the entire space of the tubule surrounding the odontoblast process (*periprocess space*) and the rest of the structures found in this space.

h. It is very difficult to determine the length of the odontoblast processes, the presence of nerve endings and dendritic cell extensions, so there is considerable controversy on these points.

i. Over time, the dentinal tubules develop a tubule wall (Fig. 18.5):
   (1) At the beginning of dentinogenesis, the newly formed dentinal tubules have the tubular cavity and its contents surrounded by the intertubular matrix, with no structure forming a true tubule wall.
   (2) After mineralisation of the intertubular dentine, a thin, sheath-like layer of material is deposited from the outside towards the centre of the tubule.
      (a) This sheath or wall is called *peritubular, tubular* or *intratubular matrix (peritubular, tubular* or *intratubular dentine).*
      (b) It is at the boundary with the intertubular dentine.
      (c) It is characterised by a high degree of mineralisation, very low collagen type III,[3] and an abundance of non-collagenous organic substances, such as DSP and DMP1 proteins.

[3] Some authors claim that there is no collagen in the peritubular dentine.

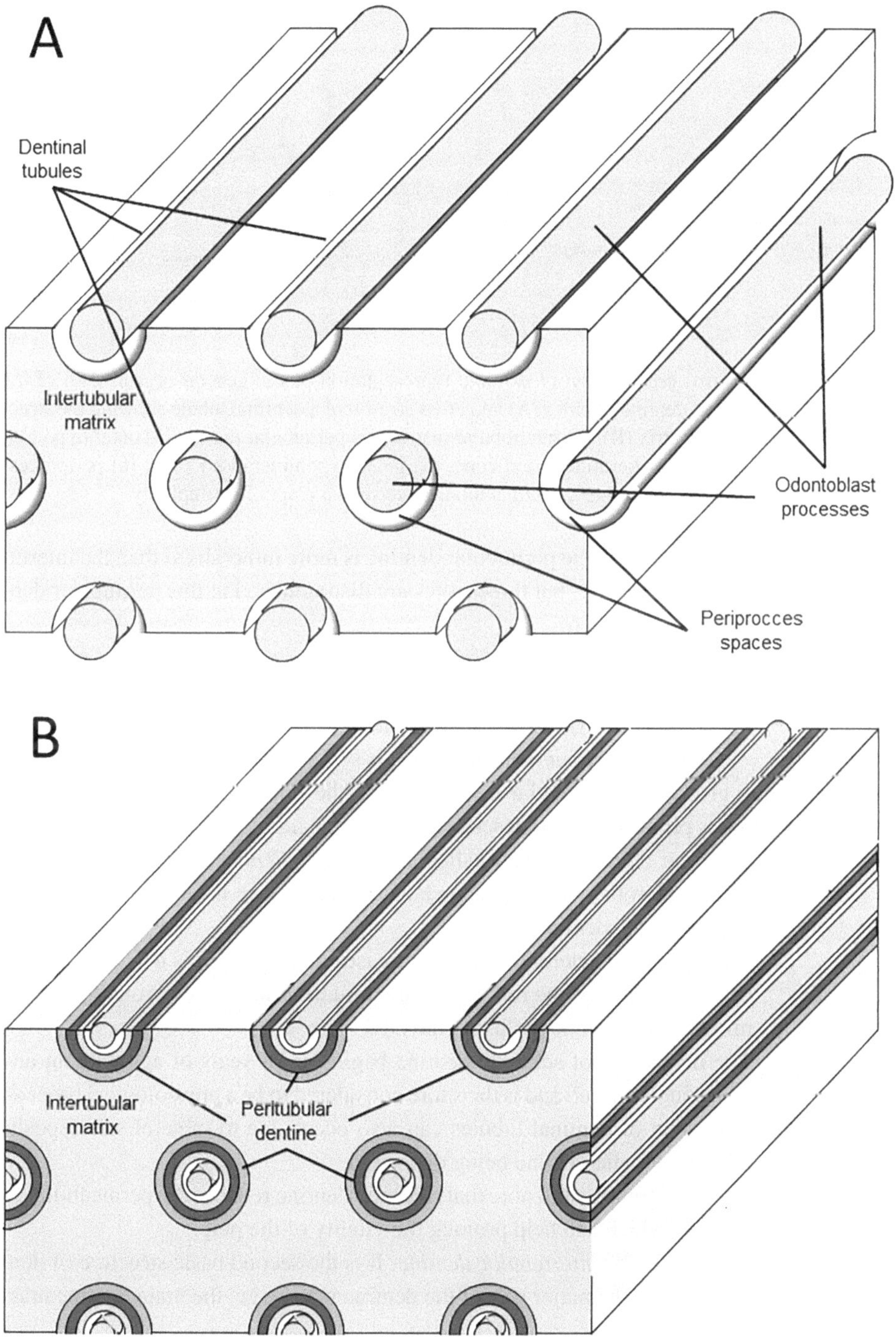

Fig. 18.5 Structure of newly formed dentinal tubules (**A**), which are cylindrical cavities through the intertubular matrix and lack a tubular wall. After some time (**B**), a mineralised dentine wall (peritubular dentine) is produced in the tubules

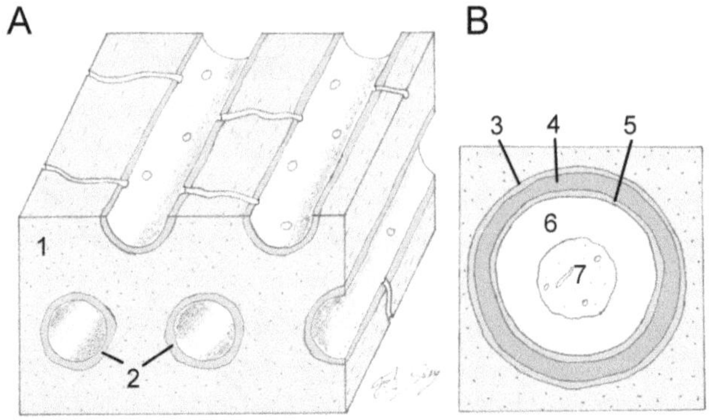

Fig. 18.6 Histological organisation of dentine: representation of the general organisation of the intertubular matrix and dentinal tubules (**A**) and cross section of a dentinal tubule showing the structure of the peritubular matrix (**B**). (1) intertubular matrix; (2) peritubular matrix; (3) outer hypomineralised zone; (4) medial hypermineralised zone; (5) inner hypomineralised zone; (6) periprocess space; (7) odontoblast process. In A, dentinal tubules are shown empty for simplicity

(d) Altogether, the peritubular dentine is more mineralised than the intertubular dentine, but three zones are distinguished in this peritubular dentine (Fig. 18.6):
   1. *Outer hypomineralised zone*. It is very thin and is the first to be deposited.
   2. *Medial hypermineralised zone*. It is the thickest.
   3. *Inner hypomineralised zone*. It is the most recently formed.

j. If the growth of peritubular dentine blinds the lumen of the tubule, *sclerotic dentine* is produced and is described as sclerotic *tubules*.
   (1) Sclerotic dentine is also called *translucent dentine*, because in regions where sclerotic dentine is abundant, its appearance is clearer than in other regions of the dentine.
   (2) The amount of sclerotic dentine increases with age, and is more frequent in the apical third of the root and in the middle zone of the coronary dentine, midway between the pulp and the DEJ.
   (3) The deposition of sclerotic dentine begins at 18 years of age without any identifiable stimuli and is therefore considered to be a physiological process.
   (4) Occlusion of dentinal tubules can also occur due to mineral salt deposits without organic dentine being deposited.
   (5) It is very important to note that sclerotic dentine reduces the permeability of dentine, which can help prolong the vitality of the pulp.

2. *Intertubular matrix* or *intertubular dentine*. It is the second basic structure of dentine and constitutes the major part of the dentine matrix, i.e. the main dentine mass of the crown and root.

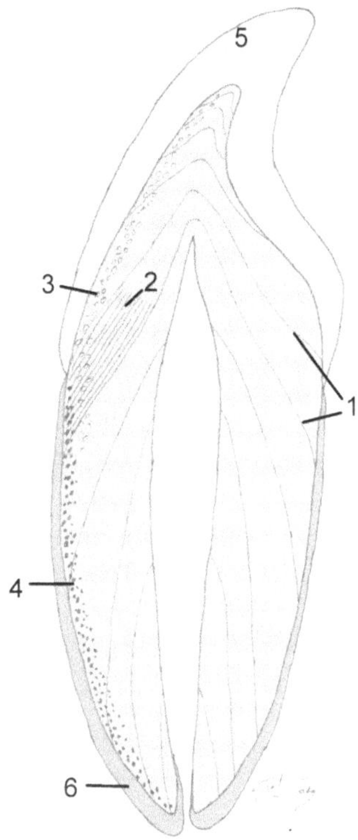

Fig. 18.7 Secondary structures of dentine. (1) contour lines of Owen; (2) minor incremental lines of von Ebner or imbrication lines; (3) interglobular dentine or spaces of Czermack; (4) granular layer of Tomes; (5) enamel; (6) cementum. Daily incremental lines and mineralisation lines are not shown

a. In the circumpulpal dentine, which is the largest part of the dentine, it consists of a mesh of type I collagenous fibrils, 50–200 nm thick, mineralised.
   (1) The degree of mineralisation is less than that of peritubular dentine.
   (2) The collagen mesh is formed because the fibrils are randomly arranged in a plane more or less perpendicular to the surface of the dentinal tubules.
b. In mantle dentine, the collagen fibres are thicker and arranged perpendicular to the dentine surface.

C. *Secondary structures of dentine.* These are structures that are observed under the microscope, usually in ground preparations, and are produced by variations in the mineralisation or organisation of the basic structures. Three are distinguished (fig. 18.7):
   1. *Incremental lines.* Dentine grows by *apposition*. Because the mineralisation process can vary over time, parallel lines are observed separating successively produced dentine layers.
      a. These lines are arranged more or less perpendicular to the dentinal tubules.
      b. There are several types of lines, which in primary dentine are[4]:

[4]The rate of deposition of secondary dentine is much slower, so the lines of increment follow a different and irregular pattern.

(1) *Contour lines of Owen*[5] . These are variably spaced and are thought to be due to alterations in mineralisation, as they are *hypomineralised lines*.
   (a) Longitudinal sections of the teeth show that in the crown they form cap-like structures, but in the root they are bands converging towards the pulp surface.
   (b) In cross-sections, they look like concentric rings.
   (c) The most obvious line of Owen is the *neonatal line*, which occurs at birth.
   (d) Other lines may be due to periods of malnutrition or illness.
(2) *Minor incremental lines of von Ebner (imbrication lines).*
   (a) These lines are separated by an average of 20 µm.
   (b) The dentine deposited between two lines corresponds to an activity period of 5 days.
(3) *Daily incremental lines.* Located between the lines of von Ebner.
   (a) They are about 4 µm apart, on average, but can vary from 8 µm at the crown to 3 µm at the root.
   (b) They delimit day-long mineralisation processes, as collagen fibrils change orientation from one day to the next.
(4) *Mineralisation line.* These are lines that appear between the daily incremental lines.
   (a) They are 1.7–2 µm apart.
   (b) They are formed every 12 hours.

2. *Interglobular dentine* or *spaces of Czermack.* They occur mainly in the coronary dentine, rarely in the root dentine, in the contact region between mantle dentine and circumpulpal dentine (or the granular layer and the circumpulpal dentine in the root), and mark the beginning of the latter.
   a. They are seen as *hypomineralised zones* surrounding spherical, globular mineralised zones, hence they are called interglobular.
   b. They are caused by a defect in the fusion of the *calcospherites*, which will be described later.
   c. They are abundant in people with *vitamin D* defects during odontogenesis.
   d. In these regions, the distribution and course of dentinal tubules are not affected, although they may lack peritubular dentine.
3. *Granular layer of Tomes.* A zone located in the peripheral part of the root dentine, close to the CDJ, observable in ground sections.
   a. It consists of granular structures that increase in number from the cervical to the apical region.
   b. This area is less mineralised than the circumpulpal dentine of the root.

[5]There is some confusion with this term. Owen described lines in the dentine that appear to be due to secondary curvatures of the tubules. Subsequently, other lines have been observed which have the same arrangement but are due to mineralisation problems. The latter are now called lines of Owen.

c. It is not known exactly what they are, as they are indistinguishable in neither decalcified and stained preparations nor under the electron microscope. There are two plausible hypotheses.
   (1) These are regions where the terminal branches of the dentinal tubules are intertwined.
      (a) The interlocking is due to movements of the odontoblasts in various directions in the early stages of root dentine formation.
      (b) These movements would occur because the activity of the root odontoblasts would not be as regulated as in the formation of crown dentine.
   (2) Collagenous and non-collagenous proteins are organised in a special way in this region close to the cementum.

18.6 Differentiation of the Odontoblast

A. Dentine formation or *dentinogenesis* begins in the late bell phase. The synthesis and deposition of dentine begins in the incisal or cusp regions and progresses as a wave towards the cervical loop regions of the enamel organ.
B. When the synthesis of the coronary dentine is complete, the formation of the root dentine begins.
C. The progression of dentine synthesis from the occlusal or incisal regions to the apical regions is reflected in the structure of the dentine-synthesising pulp cells, i.e. *odontoblasts*.
   1. Odontoblasts mature following a wave that starts in the incisal and occlusal regions and ends in the apical regions.
   2. Some authors speak of a life cycle of the odontoblast, but this is not a very correct term, because at the end of its life the odontoblast does not return to the beginning or generate new odontoblasts. It would therefore be more correct to speak of a *process of differentiation of the odontoblast*.
D. During the process of differentiation of the odontoblast, the odontoblast goes through several phases, many of which are visible in the same developing tooth by looking at different regions of the tooth germ (Fig. 18.8). These phases can be summarised as follows:
   1. *Undifferentiated mesenchymal cell*. These cells of the dental papilla are gradually located in the periphery, close to the enamel organ.
      a. They are small, stellate cells with a large nucleus, little cytoplasm with few organelles, which synthesise *type I* and *type III collagens*.
      b. Between these cells and the inner epithelium of the enamel organ, there is an acellular, amorphous layer of fine collagen fibrils.
      c. As the cells of the *inner epithelium of the enamel organ* mature, as evidenced by a change in their polarity—which will be explained in the corresponding chapter—they send signals to the mesenchymal cells for development.

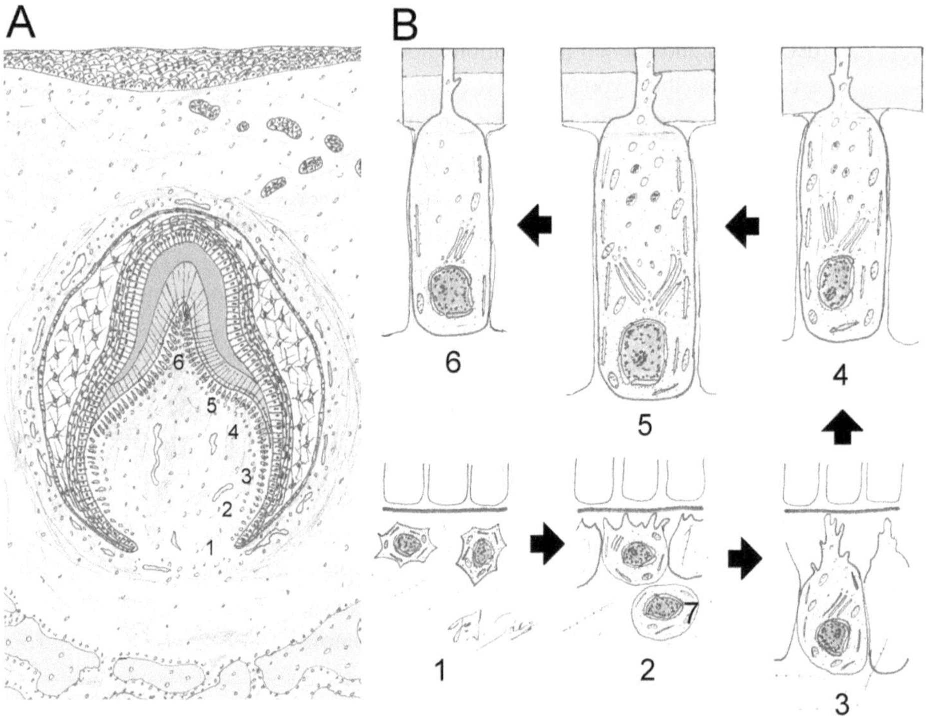

Fig. 18.8 Tooth germ at the appositional stage (**A**), in which most of the phases of the odontoblast differentiation process can be found (**B**). (1) undifferentiated mesenchymal cell; (2) preodontoblast; (3) young odontoblast; (4) secretory odontoblast; (5) mature secretory odontoblast; (6) transitional odontoblast; (7) subodontoblastic Höhl cell. In **A**, the numbers indicate the relative positions where the different stages of the process shown in **B** can be observed

2. *Preodontoblast*. They are low cuboidal or columnar cells.
   a. They are *polarised* cells, with a *proximal pole* (directed towards the enamel organ) and a *distal pole* in contact with the dental papilla.
   b. The cytoplasm of these cells has *cytoplasmic prolongations* on the proximal side.
   c. The cells are joined together to form an epithelial-like structure.
   d. The acellular zone between them and the basal lamina of the enamel organ is progressively reduced and disappears.
   e. These cells are still able to divide by *mitosis*. Some of the cells produced in these divisions are located at the distal pole, towards the papilla, forming the *subodontoblastic Höhl cells*.
3. *Young odontoblast*. These cells are taller than the previous ones and are organised as a columnar epithelium.
   a. They are already incapable of divide.

b. Cells are much more polarised, which is demonstrated by looking at the distribution of organelles in the cytoplasm.
   (1) The rough endoplasmic reticulum is abundant and lies parallel to the proximal-distal axis of the cell.
   (2) The Golgi apparatus is located between the nucleus and the proximal pole of the cell.
   (3) The cytoskeleton predominates at the proximal pole.
   (4) At this pole, one of the initial extensions becomes larger than the others and forms an *odontoblast process* that elongates perpendicular to the basal lamina of the enamel organ.

4. *Secretory odontoblast.* It starts to synthesise dentine.
   a. Dentine synthesis begins with the deposition of *predentine* between the cells and the basal lamina of the inner epithelium of the enamel organ.
   b. Type III collagen is no longer synthesised, and *type I collagen* and some other varieties are synthesised in much smaller quantities (type V and type VI collagens).
   c. Other molecules involved in dentine formation, such as *metalloproteinases* and some *proteoglycans*, are also synthesised. Some proteoglycans initially inhibit mineralisation.
5. *Mature secretory odontoblast.*
   a. They are characterised to be tall cells with a basal nucleus and basophilic cytoplasm with a region of the Golgi apparatus very evident.
   b. Under the electron microscope, the organelles indicated when describing the odontoblast in the chapter on dental pulp are observed (fig. 17.3).
   c. When dentine is synthesised, the *odontoblast process* remains inside a tubular cavity, which will be the *dentinal tubule.*
   d. The cell continues to synthesise organic matrix and starts mineralisation of the matrix further away from the odontoblast. It appears that the DMP1 protein inhibits mineralisation by blocking small calcium phosphate particles, and when this protein is broken down, mineralisation begins.
   e. Dentine secretion continues until mantle dentine and circumpulpal dentine are formed.
6. *Transitional odontoblast.* In this phase, odontoblasts are smaller cells.
   a. They occur after primary dentine has formed, when dentine synthesis slows down and secondary dentine begins to form.
   b. Its synthetic activity is lower, because *secondary dentine* synthesis is slow.
   c. They can recover the appearance of mature odontoblasts to form tertiary dentine.

18.7 Dentinogenesis

A. As mentioned above, crown dentine is produced first, and when the crown is already formed, root dentine is produced. In both cases, mantle dentine is produced first, followed by circumpulpal dentin.[6] After the eruption of the tooth, secondary dentine is produced in smaller quantities and at a slower rate.
B. Dentine formation occurs in several stages:
   1. *Synthesis and deposition of organic matrix.* Collagen fibrils and other dentine matrix molecules are synthesised and deposited on the proximal pole of the odontoblasts.
   2. *Mineralisation of the matrix*, by precipitation of salts, mainly calcium, as mentioned above.
C. Mineralisation starts in the outermost regions of the dentine, away from the odontoblast, and progresses to the innermost areas, close to the odontoblast.
D. There is one exception, the *hyaline layer*, the first dentine produced in the root.
   1. In this layer, mineralisation starts at an intermediate point and progresses towards the odontoblasts and towards the surface.
   2. This allows the surface of this layer to take time to mineralise and allows the dentine matrix to intermix with the cementum matrix, which favours the adhesion of both materials once mineralised.
E. There is some difference in the way mantle dentine and circumpulpal dentine are synthesised. To understand them properly, it is necessary to explain how mineralisation occurs. Then we will give some details about the formation of dentine.
F. Mineralisation mechanisms.
   1. Two mechanisms of mineralisation have been described (Fig. 18.9):
      a. *Mineral nucleation on collagen fibrils.*
         (1) Minerals are deposited on type I collagen fibrils, all under the control of the *DPP* protein. DPP is anionic and can take up calcium.

Fig. 18.9 (continued) produced and mineralisation occurs directly on the collagen fibres. Calcium ions (Ca^{2+}) can reach the dentine from the dental pulp by paracellular transport (between cells), but can also enter the odontoblast, which stores and releases it at the opposite pole. Collagen and proteoglycans are released in the proximity of the odontoblast body. At positions further away from the odontoblast body, close to the mineralising front, the odontoblast process releases Ca^{2+}, DPP protein, more proteoglycans and Gla protein. (1) dentine; (2) predentine; (3) odontoblasts; (4) linear mineralising front; (5) calcospherites; (6) matrix vesicles; (7) exocytosis vesicles. The arrows indicate the direction of transport of ions and organic molecules

[6] It should be remembered that the existence of mantle dentine in the root is disputed.

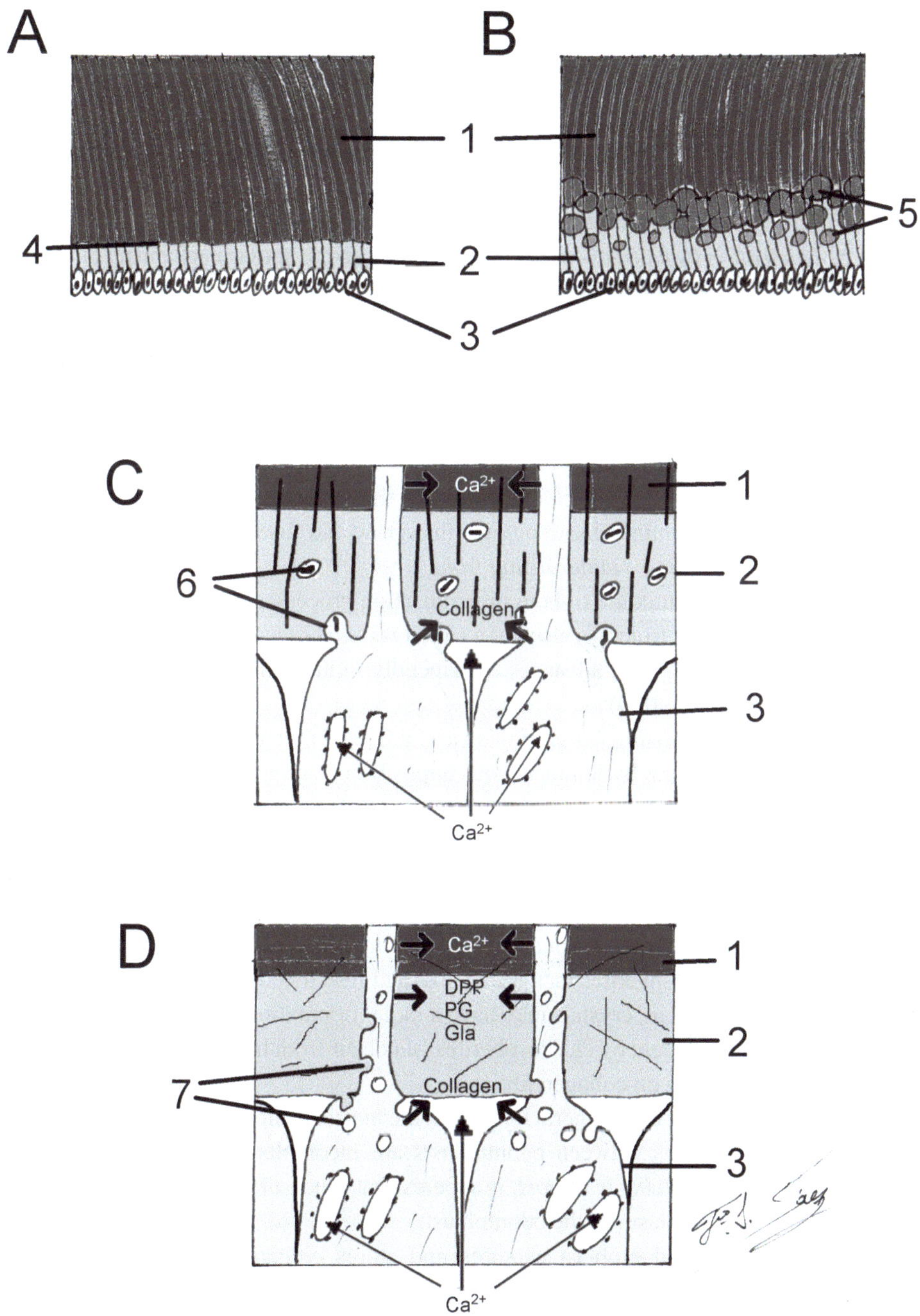

Fig. 18.9 Mineralisation of dentine. The dentine mineralisation front advances centripetally, either as a linear mineralising front (**A**) or as a globular mineralisation front (by calcospherites) (**B**). Mineralisation of mantle dentine (**C**) occurs primarily through the formation of matrix vesicles released by the odontoblast, probably by budding. In the circumpulpal dentine (**D**) no matrix vesicles are

(2) If DPP levels are greatly increased, mineralisation is inhibited. If the odontoblast reduces DPP production to lower levels, calcium deposition in the matrix occurs.
(3) Apparently, DPP locates the ideal mineralisation points and stabilises and orients the crystals.
(4) It appears that DPP is released in odontoblast processes at some distance from the cell soma, so there is no DPP in the predentine.

b. *Nucleation by matrix vesicles*. Simultaneous to the above mechanism, or instead of it, mineralisation may occur in *matrix vesicles* arising from the surface of the odontoblast by *budding*.
(1) Hydroxyapatite crystals grow inside these vesicles, apparently without the intervention of DPP.
(2) The process is similar to that which occurs in bone and which was described in the corresponding chapter.
(3) As indicated in Chap. 7, crystals exit the vesicle, are deposited in the matrix and form mineralisation cores that grow and fuse with neighbouring crystals, forming a *mineralising front*.

2. In some cases, a mixture of both mineralisation processes may occur.
3. The mineralising front has also been called *metadentine*.
4. This mineralising front advances centripetally in the dentine layer. It can advance in two ways (Fig. 18.9):
   a. *Linear mineralising front*. In this case, a line of mineralisation is seen progressively advancing by homogeneous apposition of mineral over the already mineralised zones.
   b. *Globular* or *calcospheric mineralising front*. What happens in this case is that heterogeneous mineralisation is observed, as there are spherical mineralised zones separated by non-mineralised zones.
      (1) These mineralised zones are called *calcospherites*.
      (2) The calcospherites increase in size and fuse together.
      (3) Within them, crystals may have a radial orientation.
      (4) Advancement by chalcospherites can occur from matrix vesicles, but also in nucleation on collagen fibrils.
5. It is not very clear how mineral ions reach the matrix from the dental pulp. Although the tight junctions between odontoblasts are incomplete, which is supposed to facilitate *paracellular transport*, it appears that much of the transport takes place through the cytoplasm of the odontoblast.
   a. The mature odontoblast captures and stores *calcium ions*. *Calcium pumps (ATPases)* in the distal membrane, which bring calcium into the cell, and *calcium channels* in the proximal membrane of the odontoblast, which release the ions into the dentine, have been demonstrated.

b. There are also *calcium pumps* that store this ion in the cisternae of the endoplasmic reticulum.
c. In addition, the odontoblast raises *phosphate ion* concentrations through *alkaline phosphatase*.

6. As the mineralisation process is delayed in relation to the synthesis of new predentine, there is always a strip of unmineralised predentine in the innermost part of the dentine, in contact with the odontoblasts.

G. Dentine synthesis.

1. During dentine synthesis, collagen and proteoglycans are deposited in the region closest to the body of the odontoblast.
2. In areas further away from the odontoblast, right at the border of the mineralised areas, i.e. at the so-called mineralising front, the odontoblast process releases *ions*, *Gla proteins* and new proteoglycans. The *DPP* protein, which seems to be responsible for the initiation of mineralisation, is also found here (Fig. 18.9).
3. *Mantle dentine synthesis*
   a. The first dentine to form is mantle dentine.
   b. The first sign of dentine synthesis is the appearance of *von Korff's fibres* between the odontoblasts. These fibres are of *type III collagen*, are thick, 0.1–0.2 mm in diameter, and are initially associated with *fibronectin*.
   c. Then they begin the synthesis of *type I collagen*. These are thick collagen fibres surrounded by a lot of amorphous matrix. The collagen fibres are arranged parallel to each other and perpendicular to the basal lamina of the inner epithelium of the enamel organ.
   d. Mineralisation by *matrix vesicle* predominates.
   e. The metadentine or mineralising front is produced by *calcospherites*.
4. *Circumpulpal dentine synthesis.*
   a. Mantle dentine can be 20–150 μm thick. When this thickness is reached, the odontoblasts start to produce the *circumpulpal dentine* that, as mentioned above, constitutes the largest part of the dentine.
   b. In circumpulpal dentine, the collagen fibres are thinner and irregularly arranged.
   c. Mineralisation occurs exclusively by *nucleation on the fibrils*, as no matrix vesicles have been described.
   d. In the metadentine, a *linear mineralising front* or a *calcospheric mineralising front* can be observed.
   e. At the border between mantle dentine and circumpulpal dentine of the crown, the final fusion process of the calcospherites is defective and hypomineralised regions occur between the adjoining globular regions, forming *interglobular dentine* or *spaces of Czermack*, which have been described as one of the secondary structures of dentine.

18.8 Dentine Sensitivity

A. Dentine and pulp are extremely sensitive structures, which cannot be explained from an evolutionary point of view.
B. Two types of *sensory nerve endings* reach the dental pulp (Chap. 17):
   1. Amyelinated type C fibres that terminate in the pulp. Type C fibres are responsible for *pulpal pain*, a slow, dull, diffuse and prolonged pain.
   2. Myelinated type A fibres which, stripped of their sheath, enter between the odontoblasts and can be identified within the dentinal tubules. They are responsible for a sensation of sharp (acute), sort, fast, localised *dentinal pain.*
C. Three theories have tried to explain how the sensation of sharp pain in dentine is captured (Fig. 18.10):
   1. *Theory of direct innervation.* The *nerve endings* that penetrate the dentinal tubules would be responsible for directly capturing the painful stimuli.

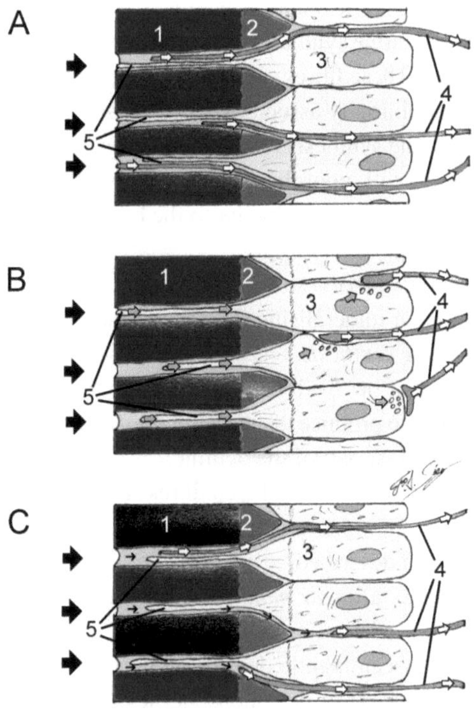

Fig. 18.10 The three theories proposed to explain dentine sensitivity. Painful stimuli are represented by the large black arrows. According to the theory of direct innervation (**A**), the nerve endings pick up the stimulus, which is transmitted by the nerve fibre (white arrows). According to the theory of transduction (**B**), painful stimuli are taken up by the odontoblast processes and transmitted by this cell (grey arrows), until it stimulates the nerve endings, which transmit the sensation of pain (white arrows). According to the hydrodynamic theory (**C**), painful stimuli cause a displacement of the dentinal fluid (small black arrows). This movement and the associated pressure changes stimulate the nerve endings, which transmit the stimulus (white arrows). (1) dentine; (2) predentine; (3) odontoblasts; (4) nerve endings; (5) odontoblast processes

   a. This option seems the most obvious, as it is logical to think that the nerve endings are the ones that capture painful stimuli.
   b. However, not all tubules are innervated and, when they are, the nerve endings do not usually reach the outermost part of the dentine.
   c. These and other data cast doubt on whether the sensation of pain is perceived directly by the nerve ending.
2. *Theory of transduction*. According to this theory, the *odontoblast process* is responsible for capturing the pain stimulus and transmitting it via synapses to the nerve endings.
   a. Neither the ability of the odontoblast process to receive the stimulus nor the synaptic junction of the cell with the axon has been demonstrated.
   b. *Gap junctions* between the odontoblast and axons have been seen, which could function as electrical synapses.
3. *Hydrodynamic theory*. According to this theory, painful stimuli would cause movements in the *dentinal fluid* of the tubules, with consequent *pressure changes*.
   a. These movements and pressure variations would stimulate the nerve endings.
   b. This theory is supported by quite a number of experimental results.

D. Although the third theory seems the most probable one, it cannot be ruled out that some of the dentine pain may be due to direct innervation or transmission through the odontoblast.

E. If the dentine is exposed, *hypersensitivity* to certain pain-causing stimuli occurs. In this case, the dentist must fill and plug the dentinal tubules in the exposed area to eliminate or reduce this hypersensitivity.

Take-Home Message

Dentine is the mineralised tissue that covers the dental pulp. It is an acellular tissue, although it has prolongations of odontoblasts and dendritic cells and nerve endings. Depending on its position in the tooth, three types of dentine are considered from the outside inwards: mantle dentine, circumpulpal dentine and predentine. Depending on the time of their formation, three types of dentine are distinguished: primary (synthesised before eruption), secondary (synthesised after eruption), and tertiary (synthesised in response to a noxa). The extracellular matrix of dentine contains type I and some type III collagen and other dentine proteins, and is mineralised with hydroxyapatite crystals similar to those present in the bone matrix. Dentine is made up of dentinal tubules and an intertubular matrix, which are its basic structures. In addition, other secondary structures can be seen in dentine, produced by variations in the basic structures. The differentiation process of an odontoblast begins with undifferentiated mesenchymal cells of the dental pulp close to the basal lamina of the enamel organ that develop into secretory odontoblasts. Odontoblasts synthesise the organic matrix of dentine and subsequently control the deposition of minerals, which is similar to that of bone. Two mineralisation fronts can be distinguished, linear mineralising front and calcospheric mineralising front. There are three theories to explain dentine pain uptake: the theory of direct innervation, the theory of transduction and the hydrodynamic theory.

Further Reading

Almeida AW, Almeida MLP, Dechichi P (1998) The mineralization process in mantle and circumpulpar dentin. J Dent Res 77:1153–1153
Berkovitz BKB, Holland GR, Moxham BJ (2018a) Dentine. In: Oral anatomy, histology & embryology, 5th edn. Elsevier, pp 152–175
Berkovitz BKB, Holland GR, Moxham BJ (2018b) Dentinogenesis. In: Oral anatomy, histology & embryology, 5th edn. Elsevier, pp 387–386
Chen S, Xie H, Zhao SL et al (2022) The genes involved in dentinogenesis. Organogenesis 18:1–19
Chiego DJ (2023) Dentin. In: Essentials of oral histology and embryology, 6th edn. Elsevier, pp 97–109
Chung G, Jung SJ, Oh SB (2013) Cellular and molecular mechanisms of dental nociception. J Dent Res 92:948–955
Fehrenbach MJ, Popowics T (2020) Dentin and pulp. In: Illustrated dental embryology, histology, and anatomy, 5th edn. Elsevier, pp 160–174
Goldberg M (2014) Dentin, pulp, and tooth pain. In: Hand AR, Frank ME (eds) Fundamentals of oral histology and physiology. Wiley Blackwell, pp 85–112
Gómez de Ferraris ME, Campos Muñoz A (2009) Complejo dentino-pulpar II: dentina. In: Histología, Embriología e Ingeniería Tisular Bucodental, 3rd edn. Médica Panamericana, pp 255–290
Johansen E (1964) Microstructure of enamel and dentin. J Dent Res 43(Suppl):1007-1020
Kuhnisch J, Dietz W, Heinrich-Weltzien R et al (2003) The intra-tubular dentine structure in permanent teeth—a SEM-investigation. Microsc Microanal 9:440–441
Magloire H, Couble ML, Thivichon-Prince B et al (2009) Odontoblast: a mechano-sensory cell. J Exp Zool Part B Mol Develop Evol 312B:416–424
Magloire H, Maurin JC, Couble ML et al (2010) Topical review. Dental pain and odontoblasts: facts and hypotheses. J Orofac Pain 24:335–349
Maurin JC, Couble ML, Thivichon-Prince B, Magloire H (2013) Odontoblast: a key cell involved in the perception of dentinal pain. M S Med Sci 29:293–299
Mjör IA, Smith MR, Ferrari M, Mannocci F (2001) The structure of dentine in the apical region of human teeth. Int Endod J 34:346–353
Moss ML (1974) Studies on dentin. 1. Mantle dentin. Acta Anatomica 87:481–507
Nagahama S (1982) A histological study on the nature of human interglobular dentin. Okajims Folia Anatomica Japonica 58:383–390
Nanci A (2013) Dentin-pulp complex. In: Ten Cate's Oral histology, 8th edn. Elsevier, pp 165–204
O'Dowd G, Bell S, Wright S (2023) Oral tissues. In: Wheater's functional histology. Elsevier, pp 241–252
Sato H, Kagayama M, Sasano Y, Mayanagi H (2000) Distribution of interglobular dentine in human tooth roots. Cell Tissues Organs 166:40–47
Scott JN, Weber DF (1976) Microscopy of primary-secondary dentin interface. J Dent Res 55:B120–B120
Sun YT, Sanders AM, Pashley DH et al (2025) Beyond hydrodynamics: the role of ion channels in dentine hypersensitivity. J Dent 157:105745
Symons NB (1965) Interglobular dentine and the calcospherite pattern. Arch Oral Biol 10:1009–1010
Taintor JF, Biesterfeld RC, Langeland K (1981) Irritational or reparative dentin—a challenge of nomenclature. Oral Surg Oral Med Oral Pathol Oral Radiol Endodontol 51:442–449
Warshawsky H (1988) The teeth. In: Weiss L (ed) Cell and tissue biology—a textbook of histology, 6th edn. Urban & Schwarzenberg, pp 594–640

Enamel

19

What You Will Learn in This Chapter

In this chapter, you will learn that enamel, the tissue that covers the dentine in the crown, is the hardest and most mineralised tissue in the body. You will see that it is an acellular tissue whose matrix has a high mineral content, a low content of organic matter, and that it lacks collagen. You will see that the basic structure of enamel is the enamel rod, and that under the microscope, secondary structures can be seen due to variations in the shape, organisation and composition of the rods. You will study the characteristics of the enamel surface. Finally, you will see how amelogenesis occurs and the process of ameloblast differentiation.

19.1 General Characteristics

A. *Enamel* (from Anglo-French enamailler, from en- 'in' and amailler 'to glaze'), *adamantine tissue* (from the Ancient Greek ἀδαμάντινος adamántinos, 'diamond-like') or *adamantine substance*, is a mineralised structure that coats the crown of the tooth.

B. Origin:
   1. It is formed during odontogenesis, while the *tooth germ* is embedded in the *mesenchyme* or *connective of the oral mucosa*, prior to eruption, by the activity of the cells of the *enamel organ*.
   2. During tooth eruption, the enamel organ disappears, and the enamel is exposed in the oral cavity.
   3. After eruption, although it may undergo modifications, no more enamel can be synthesised.

C. It is an *acellular* tissue, as it lacks cells or cellular extensions in its interior. For this reason, although it can undergo modifications, if enamel is lost, it cannot regenerate.

D. Enamel can be lost through wear due to tooth-to-tooth contact, masticatory activity, heavy mechanical brushing and the action of acidic substances such as carbonated soft drinks, which promote demineralisation.

F. J. Sáez, I. Badiola, *Essential Oral Histology*, Learning Materials in Biosciences,
https://doi.org/10.1007/978-3-032-03165-5_19

E. The enamel surface is coated with organic materials that form a *film* attached to the enamel.
F. Enamel is the most mineralised tissue in the human body, more than bone or cementum and more than dentine.
G. As a highly mineralised tissue, it is a very *hard* tissue from a mineralogical point of view, but it is *fragile.*[1]
   1. Its strength is improved by the elasticity of the dentine on which it rests. If the dentine is damaged or destroyed by caries or by a bad intervention by the dentist, the enamel is easily fragmented.
   2. Although enamel is softer than geological materials with a similar composition, it has a higher strength due to the organisation of the crystals into *prisms* or *rods*, which we will explain later.
H. It is a *translucent* tissue, so the colour we see in the tooth usually corresponds to that of the underlying tissues.
I. It is very poorly permeable.
J. Due to its high mineral content, it is the most *radiopaque* substance in our body, which is why it appears very white on X-rays.
K. The area of contact with dentine forms the *dentinoenamel junction* or *DEJ*.
L. The distance from the DEJ to the enamel surface, i.e. the thickness of the enamel, varies from tooth to tooth, but also in a single tooth.
   1. Normally, the enamel is thicker in the incisal and occlusal regions (up to 2.5 mm) than in the lateral regions (1.3 mm), while it is very thin and thinning in the cervical regions.
   2. It is thicker in the vestibular (buccal or labial) regions than in the lingual or palatal regions.
   3. The enamel is thinner in the grooves and pits between the cusps.

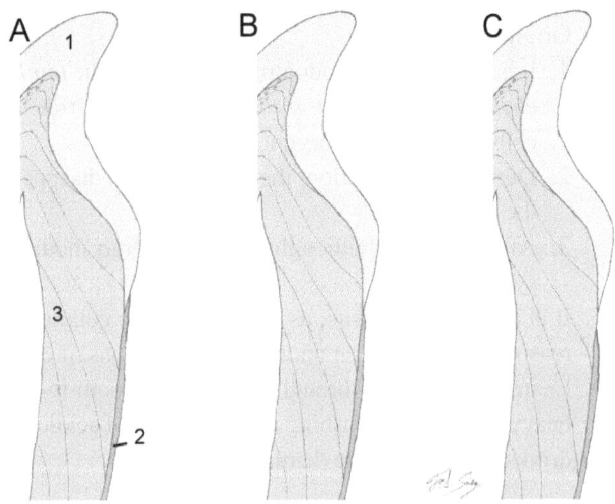

Fig 19.1 Relationship between enamel and cementum: cementum covers enamel (**A**, overlapping), cementum and enamel touch (**B**, contact) and cementum and enamel do not touch and dentine is exposed (**C**, separation). (1) enamel; (2) cementum; (3) dentine

[1] It is useful to review the table of tissue hardnesses, available at the beginning of Chap. 6.

M. In the cervical region, the enamel thins and eventually ceases to cover the dentine at the root. In the cervical region, the enamel is related to the *cementum* covering the root dentine. This occurs in three possible ways (Fig. 19.1):
   1. *Overlapping*. The cementum covers the enamel. This is the most frequent and occurs in 60% of cases.
   2. *Contact*. Cementum and enamel touch, but the enamel is not covered. The dentine is always protected by one of the two tissues. It is less frequent than the previous one, present in approximately 30% of teeth.
   3. *Separation*. The dentine is exposed in a strip because enamel and cementum are separated by a distance. This is the least frequent.

19.2 Enamel Composition

A. Enamel is an *acellular* and highly *mineralised tissue*.
   1. In mature enamel, the *mineral fraction* constitutes 95–96% of the total weight and 88-90% of the volume.
   2. *Organic molecules* account for only 1–2% by weight.
   3. The rest, around 2% by weight (5–10% of the volume), is *water*.

B. The *inorganic matrix* consists mainly of *hydroxyapatite crystals*, with the formula $Ca_{10}(PO_4)_6(OH)_2$, i.e. as in bone or dentine.
   1. In contrast to other mineralised tissues, enamel does not appear to contain amorphous calcium phosphate.
   2. Other minerals, such as *carbonates* and *sulphates*, may be present, but in much smaller quantities.
   3. Some *trace elements* such as *magnesium*, *iron* or *fluoride* may be present in small amounts. The latter is very important, as *fluoride ions* can replace some hydroxyl groups to produce *fluorhydroxyapatite*, which is more resistant to the demineralising action of acids and therefore to caries.
   4. Hydroxyapatite crystals in enamel are larger than those in bone or dentine. They can be between 100 and 1000 nm long, 30 and 70 nm wide and 10 and 40 nm high, although it should be noted that in each source very different dimensions are given (Fig. 19.2A).
   5. The crystals have traditionally been described as elongated and *hexagonal* in outline, but some authors believe that the hexagonal shape is due to the projection of a *parallelepiped* on a plane, which would be their true outline (Fig. 19.2B, C).
   6. Within the crystals, the chemical elements are molecularly organised in structures called *unit cells* (Fig. 19.2D).
      a. These unit cells are hexagonal-shaped structures, with *calcium ions* at the vertices, *phosphate groups* on the sides, and a *hydroxyl group (OH)* in the centre. Three other calcium atoms are located around the central OH.

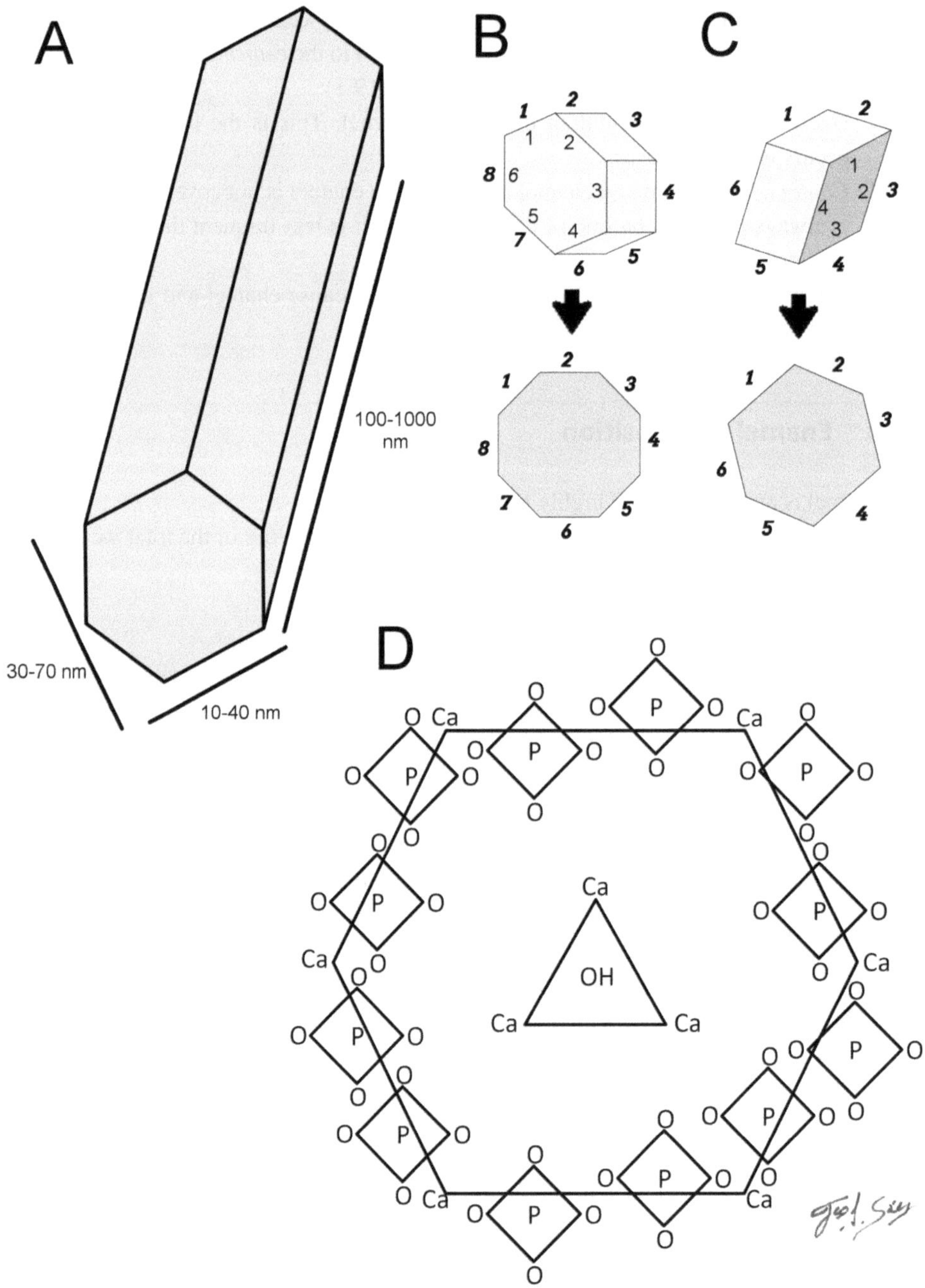

Fig. 19.2 The probable structure of the hydroxyapatite crystal (**A**) is that of an elongated structure with a hexagonal profile, the dimensions of which are variable. Some authors propose an alternative structure (**B**, **C**). According to them, if the crystals were prisms with a hexagonal base, they should be seen in microscopic sections with an octagonal outline (**B**). In contrast, the hexagonal appearance would be produced by the projection on a plane of a hexahedron (**C**). In **B** and **C**, the round numbers indicate the faces of the crystal profile, while the bold italic numbers indicate the faces of the crystal that are projected on the microscopic section. The structure of the enamel unit cell is also shown (**D**)

b. This arrangement is repeated, with unit cells next to each other and in different planes, to form the enamel crystal.
c. The basic structure of the cell may be modified in some places.
(1) Phosphates and, to a lesser extent, hydroxyls can be replaced by *carbonates*.
(2) Calcium ions can be replaced by *magnesium*.
(3) Hydroxyls can be substituted with *fluoride*.

7. The crystals are tightly packed together, with some *gaps* or *pores* of less than 5 nm between them, which may be occupied by water molecules or organic molecules.

C. The *organic matrix* is scarce where crystals are highly ordered, and is higher where crystals are more disorganised.

1. It consists mainly of proteins, among which collagen is not found, as enamel lacks this protein.
2. From the beginning to the end of enamel synthesis, there are many changes in composition.
a. Some proteins can only be found at certain stages of enamel development.
b. At the beginning of enamel maturation, proteins may constitute 23–30% of the enamel matrix, but in mature enamel, it may be 1%.
3. Some of the best-characterised proteins of immature enamel are:
a. *Amelogenins*. They are the most abundant during enamel formation (90%).
(1) The ameloblasts secrete several *isoforms* of the same protein.
(2) They have a tendency to self-aggregate, forming *nanospheres* of about 20 nm in diameter that cluster into chains and ribbons.
(3) They appear to prevent lateral growth of hydroxyapatite crystals.
b. *Ameloblastin* (formerly called *amelin*). It is less abundant than the previous one (5%).
(1) It is found in greater quantities in newly formed enamel (closest to the surface) than in older enamel (closest to the dentine).
(2) It breaks down into peptide fragments, some of which bind calcium.
c. *Enamelin*. Present in very small quantities.
(1) It is a glycoprotein and is phosphorylated.
(2) It fragments and some fragments are found on the growth surface of the enamel, but not in deeper regions.
(3) Other smaller fragments are found in deeper regions of the enamel and inhibit crystal growth.
(4) It is believed to regulate crystal growth.
d. *Enamelysin (matrix metalloproteinase-20, MMP-20)*.
(1) While the former proteins seem to be involved in enamel growth, enamelysin is produced later and is involved in the final *maturation* of the enamel.
(2) It may be involved in the fragmentation of the proteins mentioned above.
e. *Enamel matrix serine protease* (*kallikrein 4*).
(1) It is secreted when all the enamel has been synthesised.
(2) It appears to be involved in degrading other proteins.

f. *Amelotonin.*
   (1) It is associated with the *basal lamina* of ameloblasts for a short period at the beginning of the enamel maturation phase.
   (2) It is believed to be involved in the formation of the final *aprismatic enamel* layer.
g. *Odontogenic ameloblast-associated protein (ODAM).* It is synthesised at the same time as the amelotonin and is also associated with the basal lamina.
h. *Tuftelin* or *tuft protein.*
   (1) It is located at the dentinoenamel junction when enamel begins to form.
   (2) Actually, the tuftelins are several similar proteins.
   (3) It is found in many other tissues, so its function is unclear, but it is thought to be a signal molecule involved in induction processes between ameloblasts and odontoblasts.
   (4) Derivatives of tuftelins have been detected in mature enamel.
4. As we can see, many of the proteins fragment into smaller molecules once secreted. The whole protein is usually found in the forming enamel in regions close to the ameloblasts, while the fragments are located in older regions of the enamel, away from the ameloblasts.
5. There is a small amount of *glycosaminoglycans* in the enamel. A small amount of *lipids* has also been identified.

D. *Water* may be in the pores between crystals, surrounding organic materials, filling defects in crystals and forming a hydrated layer around crystals. The presence of water is important as *fluorine* diffuses through water.

19.3 Basic Structural Unit of the Enamel

A. When the histological organisation of the enamel is studied, as in the case of dentine, two types of structures are described:
   1. *Basic structures* that constitute the enamel.
   2. *Secondary structures* that are visible under the microscope as a consequence of the different organisation of the basic units.
B. The basic structural unit of enamel, known by its acronym *BSUE*, is the *enamel prism* or *enamel rod*. It is important not to confuse this structure with the much smaller hydroxyapatite crystals found within these BSUEs.
C. The enamel rod was described as hexagonal in shape, hence the name enamel prism, but in reality the shape of the rods is not regular, so the term prism is undesirable (Fig. 19.3).
D. Enamel rods are elongated structures, about 5–6 μm thick and up to 2.5 mm long, extending from the DEJ to the enamel surface.

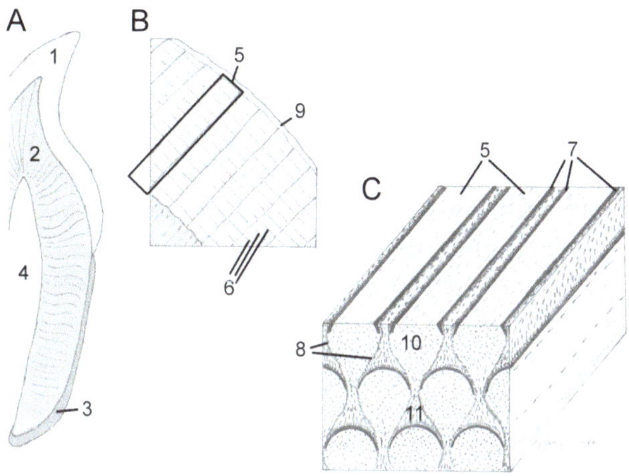

Fig. 19.3 Basic structural unit of enamel (BSUE): general view of the arrangement in the tooth (**A**), orientation of the BSUEs in the enamel (**B**) and magnified detail of the organisation of the BSUEs (**C**). (1) enamel; (2) dentine; (3) cementum; (4) dental pulp; (5) enamel prism or rod; (6) cross-striations; (7) rod sheath; (8) interrod enamel or interprismatic substance; (9) aprismatic enamel; (10) rod head; (11) rod tail

E. Enamel rods are formed by the packing of numerous *hydroxyapatite crystals* (several million). These crystals are generally oriented with their major axis in the longitudinal direction of the enamel rod.
F. The rods or BSUEs are held together by a small amount of *interprismatic substance* or *interrod enamel*.
G. In the interrod enamel we find many hydroxyapatite crystals with a different orientation to those in the rod.
H. Between the enamel rod and the interrod enamel, a small region, the *rod sheath*, has been described where the *organic matrix* is concentrated.
    1. In a small region of the rod, there may be no sheath, and the hydroxyapatite crystals of the rod meet those of the interrod enamel.
    2. Some authors believe that there is no true rod sheath, but that there is a higher concentration of organic materials and water at the edges of the rod. This higher concentration is due to the fact that they occupy the microscopic spaces produced by the different organisation of the rod crystals and the interrod enamel.
    3. The dentist should note that the rod sheath is more acid resistant than the rod. This is important for sealing activities performed to prevent caries, which require acid etching to improve adhesion.
I. As there are many variations regarding the organisation of the rods, they are considered to be cylindrical structures embedded in the interrod enamel. However, when looking at the enamel rods cut transversely, three *patterns of organisation* can be observed (Fig. 19.4):

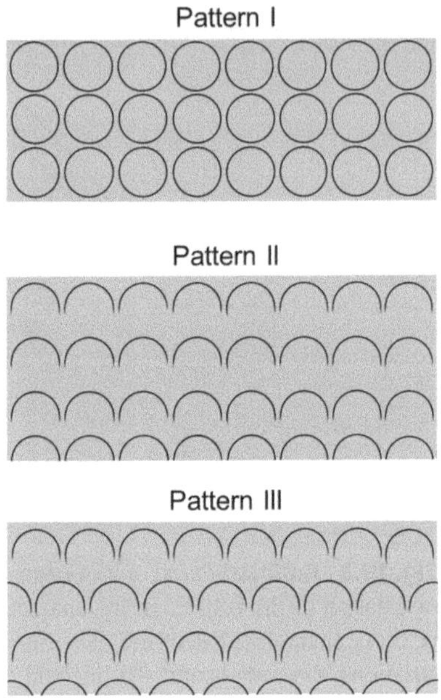

Fig. 19.4 The three patterns seen in cross-sections of the enamel rods. Pattern I shows circular rods with well-defined boundaries. Pattern II and pattern III show the rods with some poorly defined edges. Pattern II shows the rods in parallel rows. In pattern III, the rods are in staggered rows

1. *Pattern I.* The rods show a perfectly delimited circular profile, as the crystals of the rod and those of the interrod enamel have different orientations. It predominates in the areas close to the dentine and enamel surface.
2. *Pattern II.* No clear boundary is observed on some sides of the rod, as its crystals and those of the interrod enamel have similar orientations. The rods are arranged in parallel rows. It is rare.
3. *Pattern III.* No clear boundary is observed at some edges of the rod. The rods are arranged in staggered rows. This is the predominant pattern in humans.

J. In sections transverse to the longitudinal axis of the rods, the most frequent arrangement of the rods and the interrod enamel, which is pattern III, takes the form of an *old keyhole* or, sometimes, of *fish scales* (Figs. 19.3 and 19.4).
   1. Due to the old keyhole shape, a *head*, a *neck* and a *tail* are usually distinguishable in cross-sections across the longitudinal axis of the rod.
   2. The head is about 5 μm wide, the tail about 1 μm wide.
   3. From head to tail, the rod measures about 9 μm.

K. Normally, at the head of the rod, the crystals follow the longitudinal axis of the rod, but at the neck, the crystals adopt more oblique orientations, until at the tail, they are arranged perpendicular to the longitudinal axis.

19.4 Aprismatic Enamel

A. There are regions of enamel that lack the typical rods and form the *aprismatic (non-prismatic) enamel*, which usually appears on the outer surface of the enamel.
B. It usually occurs in *primary teeth*, where it is 20–100 μm thick, and in some permanent teeth. In the latter case, it is found in the cervical regions and in the coronary sulci, where its thickness is usually between 20 and 70 μm.
C. It has also been described that the first enamel to form, in contact with dentine, is a thin layer of aprismatic enamel of very few microns.
D. In aprismatic enamel, the crystals are arranged parallel to each other and perpendicular to the tooth surface.
E. This material is more mineralised due to the absence of organic molecules that are usually found between the rods.
F. It is believed that when this enamel is synthesised, rods are not formed because at the beginning and end of amelogenesis the ameloblast lacks the *Tomes' process*.

19.5 Secondary Structures of the Enamel

A. When looking at the enamel with certain microscopic techniques, one can see formations that we call *secondary structures*.
   1. As in the case of dentine, these elements are not constituents of the enamel but are structures due to the organisation of the real constituents of the enamel, the rods.
   2. They may be due to several factors.
B. *Cross-striations*.
   1. The enamel rods show lines or *cross-striations* transverse to their longitudinal axis every 3–6 μm (Fig. 19.3).
   2. Normally, the distance between these lines is smaller near the dentine than near the surface.
   3. They seem to mark the secretion of enamel produced by ameloblasts every 24 hours.
   4. It is not known whether these are areas with different amounts of organic material or slight variations in crystal orientation.
C. *Striae of Retzius, perikymata and imbrication lines of Pickerill* (Fig. 19.5).
   1. *Striae of Retzius* are growth lines delimiting bands arranged at intervals of 20–40 μm.
      a. On the cusps and incisal edges, they are arranged from DEJ to DEJ from one side to the other, like caps covering the tooth.
      b. On the lateral surfaces of the crown, there are oblique bands from the outer surface to the DEJ.
      c. Logically, in cross-sections of the tooth, they are seen as concentric rings.
      d. In enamel formed before birth, striae are hardly visible or absent.

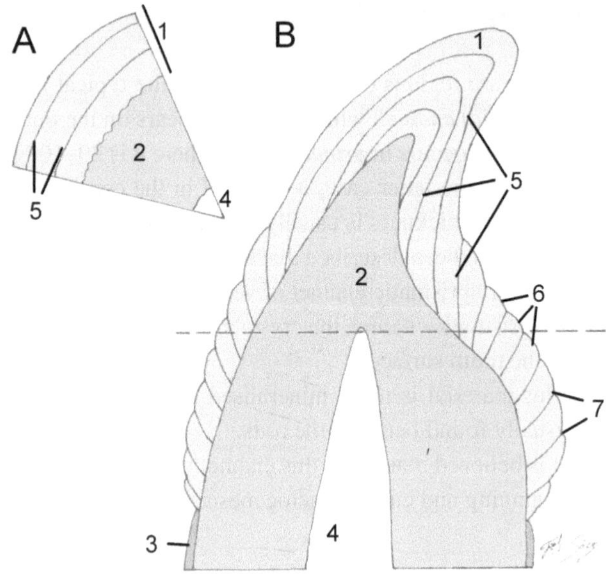

Fig. 19.5 Some secondary structures of enamel shown in cross-section of the tooth (**A**) and incisoapical section (**B**). The dashed line in B indicates the location of the cut section A. (1) enamel; (2) dentine; (3) cementum; (4) dental pulp; (5) Striae of Retzius; (6) perikymata; (7) imbrication lines of Pickerill

e. Although the actual cause of striae of Retzius is controversial, what is clear is that they are *growth lines.*
f. It is not sure whether these are signs of rhythmic enamel production leading to slight changes in the composition, shape or thickness of the rods, or whether they signal phases of incorporation into enamel production of new cohorts of cells (ameloblasts) as the crown grows.
g. There is one line that is thicker than the others and that is the *neonatal line*, which occurs at birth.
h. More marked lines may occur due to periods of illness or malnutrition. These lines, such as the neonatal line, are thick (20–40 μm) and represent a stress situation of several days, where the rods show a variation in thickness and a change in direction.
i. In the human tooth, it has been shown that there are between 7 and 9 transverse striations between each stria of Retzius.
   (1) As transverse striae are thought to form every 24 hours, each stria of Retzius signals a period of enamel synthesis of 7 days.
   (2) This data has been used to analyse the growth of hominid *fossil teeth.*

2. At the points where the striae of Retzius contact the lateral surface of the enamel, grooves are formed, called *imbrication lines of Pickerill.* The imbrication lines are 15–20 μm apart in the cervical region and up to 100 μm apart in areas further away from the cervical region of the tooth.
3. Between the imbrication lines, the enamel surface forms smooth, slightly pronounced ridges, which are called *perikymata* (Fig. 19.5). Like the striae of Retzius, each perichyma may be thought to indicate a period of synthesis of 7 days.

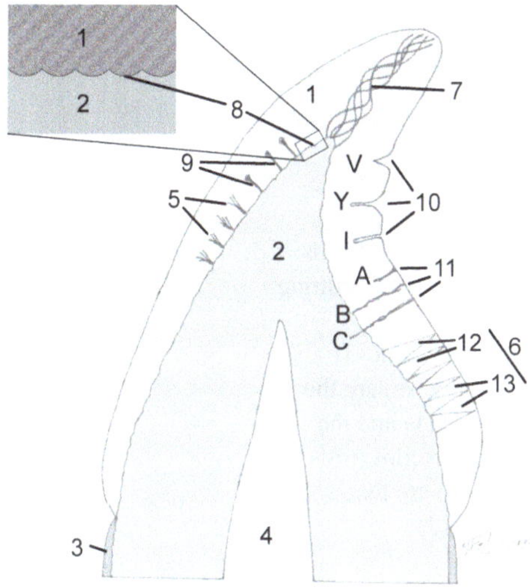

Fig. 19.6 Some secondary structures of enamel. (1) enamel; (2) dentine; (3) cementum; (4) pulp; (5) enamel tufts of Lindererer; (6) Hunter–Schreger bands; (7) gnarled enamel; (8) DEJ; (9) enamel spindles; (10) occlusal fissures (type V, Y and I); (11) enamel lamellae (type A, B and C); (12) diazones; (13) parazones

D. *Enamel tufts of Lindererer*.
   1. *Enamel tufts of Linderer* appear as faults or fractures in the enamel that have a branching shape and resemble a bush or a tuft of grass (Fig. 19.6).
   2. They arise from the DEJ and extend along the innermost third of the enamel, in the same direction as the enamel rods.
   3. They are abundant and usually spaced at intervals of about 100 μm. They are several rods thick and are *hypomineralised* areas.
   4. They seem to originate from variations in the orientation of clusters of rods. It has also been suggested that they may be due to mineralisation defects.
   5. What is clear is that in tufts, the organic matrix is more abundant than in other areas of the enamel. This is where the protein *tuftelin* is usually found.

E. *Hunter–Schreger bands*.
   1. In longitudinal ground sections of the tooth using polarised light, light and dark bands can be seen in the inner four-fifths of the enamel, which constitute the *Hunter–Schreger bands* (Fig. 19.6).
      a. The bands are about 50 μm thick.
      b. Clear bands are called *parazones*.
      c. The dark bands are the *diazones*.
   2. Scanning electron microscopic studies suggest that these bands are due to the winding course of the rods.
      a. When the tooth is cut, some groups of rods are cut at one angle and others at a different angle, and each group reflects the light in a different way.
      b. The rods that are cut longitudinally are the parazones, and the rods that are cut transversely are the diazones (Fig. 19.7).
      c. In the superficial third of the enamel, the rods are all oriented in the same direction, which is why Hunter–Schreger bands are not observed there.

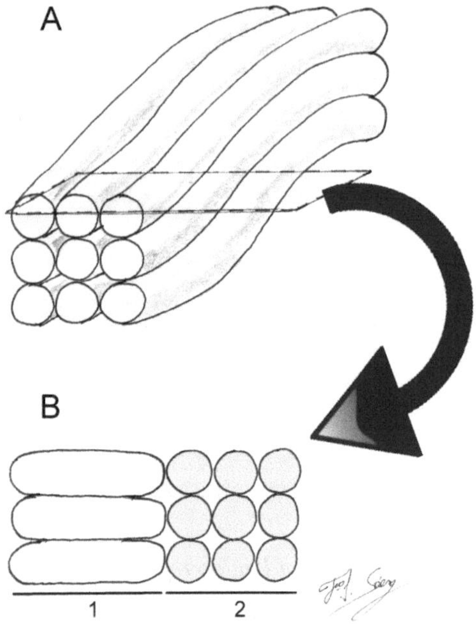

Fig. 19.7 Hunter–Schreger bands are due to the sinuous course of the rods in the inner part of the enamel. When a section is made in one plane (**A**), it passes through groups of rods showing different cutting profiles (**B**). The elongated profiles (rods cut lengthwise) are the parazones (1) and the rounded profiles (rods cut crosswise) are the diazones (2)

F. *Gnarled enamel.*
   1. *Gnarled enamel* is a structure visible on the cusps of teeth that is formed by the coiling and intertwining of the rods with each other (Fig. 19.6).
   2. This is thought to increase the strength of the enamel in these regions most exposed to the forces of mastication.

G. *Dentinoenamel junction (DEJ).*
   1. The *dentinoenamel junction* or *DEJ* is the plane of contact between enamel and dentine (Fig. 19.6).
   2. It allows the enamel to adhere to the dentine because it is not a straight border, but has a scalloped profile showing a succession of convex enamel regions of about 25 to 100 μm, which fit into concavities in the dentine.
   3. This scalloped profile is more evident in the cusp or incisal area, while the DEJ is smoother in the lateral areas of the tooth.
   4. In more detailed studies, it has been observed that there is another, smaller scalloping of 2–5 μm.
   5. In nanostructural studies, collagen fibrils in dentine have been shown to mix with enamel crystals at the boundary between the two materials.
   6. The DEJ structure seems to be set up to fulfil two functions:
      a. To increase adhesion between dentine and enamel and to prevent enamel detachment caused by mastication forces.

b. Limit the propagation of enamel fractures and prevent them from spreading to dentine or at least slow down their propagation[2] .

H. *Enamel spindles.*
1. The *enamel spindles* are club-shaped tubules arising perpendicularly from the DEJ, with a diameter of 0.5–1.5 μm and a length of 10 to 25 μm, sometimes more (Fig. 19.6).
2. They are tubular structures formed at the beginning of enamel synthesis:
   a. Some odontoblast processes lodge between the ameloblasts for a period of time (*remnant odontoblast processes*), so that no enamel is deposited at these sites.
   b. When the odontoblast process disappears, the region becomes hollow and filled with *dentinal fluid.*
3. Because they are a continuation of a dentinal tubule, enamel spindles have been called *penetrating dentinal tubules*, although they are thicker than dentinal tubules.
4. Enamel spindles are more frequent in the cusp regions.
5. After eruption, they usually no longer contain the extension of the odontoblast, although it is debated whether they may contain *collagen.*

I. *Occlusal fissures.*
1. *Occlusal fissures* are regions of the enamel surface of premolars and molars where it shows inward valleys.
2. Three types have been described. V-fissures, I-fissures and Y-fissures (Fig. 19.6).
   a. *V-fissures* are wider at the surface and slowly narrow as they deepen, so that in cuts they have a 'v' shape.
   b. The *I's* are the narrowest, and in the cuts they have an 'i' shape.
   c. The *Y*'s are shaped like this letter, wide at the beginning and very narrow at the bottom.

J. *Enamel lamellae.*
1. *Lamellae* are structures visible to the naked eye, arising from the surface inwards in a straight line and sometimes extending into the DEJ.
2. Some authors consider two types:
   a. *Primary lamellae*:
      (1) Produced by failure of mineralisation during amelogenesis.
      (2) Usually contain organic material, either components of the enamel itself or remnants of the enamel organ or the connective tissue surrounding the tooth germ during eruption.
   b. *Secondary lamellae:*
      (1) These are cracks produced after eruption of the tooth as a result of tensile forces, trauma or temperature changes.
      (2) If they contain organic material, it is from *saliva.*
3. Three types of lamellae can be distinguished (Fig. 19.6):

---

[2] Most enamel fractures stop at the DEJ, and the few that penetrate the dentine do so by a few microns.

a. *A* (do not extend beyond the outer third of the enamel). They are primary.
b. *B* (may go through the DEJ). They are primary,
c. *C* (may go through to the dentine). They are secondary.

19.6 Enamel Surface

A. From the point of view of the *dentist's clinical activity*, the enamel surface is very important. Here, enamel comes into contact with *food*, *restorations* are applied, *orthodontic brackets* are attached, *whitening* takes place, *remineralisation* takes place, *caries* begins, etc.
B. The enamel surface is different from the rest of the enamel.
  1. Physically, the surface enamel is harder, less porous, more radiopaque, and less soluble than the rest of the enamel.
  2. The composition is also different, as it is more mineralised, has a higher concentration of certain elements, such as *fluoride*, but has less carbonate. Because it is more mineralised, it is more resistant to caries.
  3. As indicated above, the surface enamel is usually yellowish.
C. The surface of the enamel is not smooth. It may show:
  1. *Lamellae*. They are observed as lines on the surface.
  2. *Perikymata*.
     a. Perikymata become smoother over time due to wear and abrasion.
     b. The wear and abrasion are less in the cervical areas.
  3. *Small pits*.
     a. Found in the cervical areas, where reduced enamel epithelium may persist for some time after eruption.
     b. They are impressions of the end of the ameloblasts.
     c. They are 1–1.5 μm deep.
  4. *Enamel caps*. They are small elevations 10–15 μm wide, frequently found in the lateral areas of the tooth.
  5. *Focal holes*. Also present in lateral areas of the tooth.
  6. *Brochs* (from Scottish brough, 'fort'). Surface elevations of 30–50 μm in diameter.
D. Enamel can be lost through three processes:
  1. *Attrition*, due to contact between the teeth, either in the occlusal plane or in a proximal plane. This causes the contact surface between the teeth to increase with age.
  2. *Abrasion*, due to contact with materials of external origin. Excessive tooth brushing can cause abrasion.
  3. *Erosion*, due to the action of acidic substances, such as certain drinks or gastric acids
E. The dentist should be aware that the cementum used to secure the *brackets* can cause demineralisation of the enamel.

F. Finally, the enamel is covered by layers that are formed by the accumulation of materials of different origins. These layers can be:
   1. *Primary enamel cuticle*. It is a material produced by the ameloblasts when the formation of the crown ends. It is therefore the last secretion product of these cells.
      a. Covers the tooth when eruption occurs.
      b. It is about 30 μm thick, consists of *proteins* and may contain *immunoglobulins* and cell debris from the *reduced enamel epithelium*.
      c. In the cervical region, it may contain other molecules (*proteoglycans* and *glycosaminoglycans*) provided by gingival tissues.
      d. It gradually disappears after the eruption.
      e. The enamel cuticle and the remains of the *reduced enamel epithelium* that may erupt form the *Nasmyth's membrane*.
   2. *Secondary, exogenous* or *acquired pellicle*. It is formed by the deposition of *salivary proteins* and others from the *gingival crevice*[3] that form a film 0.5–1 μm thick. It is initially acellular and devoid of bacteria. However, *bacteria* easily adhere to its surface.
   3. *Dental plaque* or *biofilm*.[4] It is caused by the growth of *bacteria* on the secondary pellicle.
      a. When bacterial deposition is high, a layer forms and coats the secondary pellicle.
      b. In physiological situations, they are usually *Gram-positive, facultative, anaerobic cocci and filaments*.
      c. In chronic inflammation of the periodontium, the bacterial flora is more varied and dominated by *Gram-negative, anaerobic bacteria*, including *cocci, rods, filaments* and *motile bacteria (spirochaetes)*.
      d. The growth of plaque is prevented by brushing the teeth.
      e. Bacteria can produce acidic substances that can demineralise enamel and lead to *caries*.

19.7 Amelogenesis

A. *Amelogenesis* occurs during *odontogenesis* from the *enamel organ*. As the latter disappears with eruption, no new enamel can form on the mature tooth.
B. Amelogenesis is a continuous process, but for the purpose of study, it is usually divided into phases:
   1. First, an *organic matrix* is produced which is slightly mineralised (30% mineral matter). It is likely that the mission of the organic matrix is to produce and organise the mineralisation.

[3] The gingival crevice is described in Chap. 20.

[4] Communities of microbes growing on a surface are often referred to as biofilm.

2. In a second phase, the removal of most of the organic matrix and water and the growth of hydroxyapatite crystals takes place.

C. The process of enamel mineralisation is not well understood.
   1. Calcium passes into the enamel from the enamel organ, probably via the ameloblast, using active transport by *calcium pumps*.
   2. No *matrix vesicles*, as in dentine or bone, have been described. It is possible that crystallisation of minerals occurs more or less spontaneously.
   3. It has been proposed that mineralisation is triggered by *amelogenins*. The amelogenins assemble into *nanospheres of* about 20–100 nm in diameter, in or around which crystals form.

D. The apposition of enamel onto dentine begins in the *appositional stage*, the last phase of crown formation.

E. Enamel synthesis is due to cells of the inner epithelium of the enamel organ, which differentiate into *ameloblasts* in the so-called *ameloblast differentiation process*[5], where they undergo phenotypic changes leading to different activities at different times.

F. The process of ameloblast differentiation is continuous, but in order to study it better, it has been divided into several *phases*. These phases can be visible in the same enamel organ from the cusp regions towards the cervical regions. Some authors have grouped these phases into three *stages*.

G. The steps of the differentiation process (phases and stages) are shown in Fig. 19.8 and described below:
   1. *Presecretory stage.*
      a. *Morphogenetic phase.*
         (1) The cells become *preameloblasts*.
         (2) They are low columnar or cuboidal cells, with a central nucleus and little Golgi apparatus towards the distal zone, where there is a junctional complex.
         (3) Mitochondria and other organelles are scattered in the cytoplasm.
         (4) These cells are actively dividing.
      b. *Histodifferentiation phase.*
         (1) The preameloblast differentiates into a *young ameloblast*.
         (2) It increases in volume and becomes more columnar.
         (3) The nucleus and organelles are concentrated at the distal pole.
         (4) The Golgi apparatus develops and moves to a proximal position, and the number of cisternae of the endoplasmic reticulum increases, so that the cytoplasm becomes *basophilic*.
         (5) Another junctional complex develops in the proximal region. Thus, two junctional complexes surround the cell, one at each pole.
         (6) In the cytoplasm, at the level of these complexes, actin filaments are seen forming a *terminal web* at each pole.

[5] As with odontoblast development, some books refer to the ameloblast life cycle, but here we prefer to avoid this term.

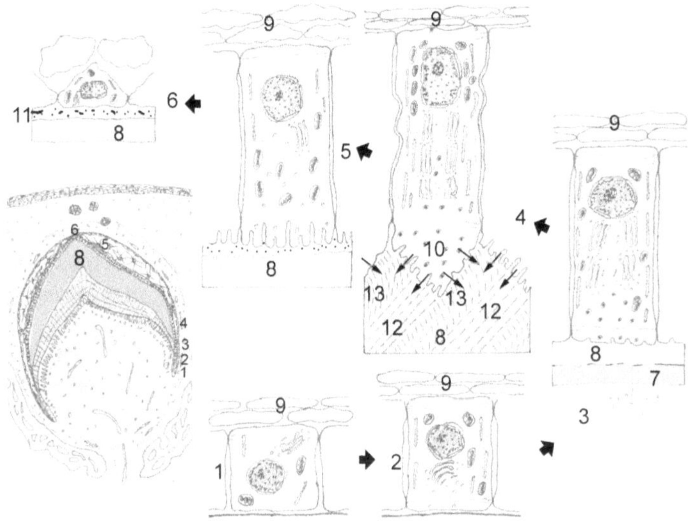

Fig. 19.8 Process of ameloblast differentiation. In the illustration depicting a tooth germ at the appositional stage, the relative positions of the cell types indicated by each number are indicated by small numbers. Big numbers: (1) morphogenetic phase (preameloblast); (2) histodifferentiation phase (young ameloblast); (3) initial secretory phase (ameloblast); (4) secretory phase (mature secretory ameloblast); (5) maturation phase (ruffle-ended ameloblast, smooth-ended ameloblast not shown); (6) protective phase (protective ameloblast); (7) predentine; (8) enamel; (9) stratum intermedium; (10) Tomes' process; (11) primary cuticle; (12) rod head; (13) rod tail. The transitional phase (between 4 and 5) and the desmolytic phase (after 6) are not shown. The thick arrows indicate the progression of ameloblast development, the thin arrows indicate the direction of enamel deposition by ameloblasts

(7) The reorientation of organelles within the cell is called *reversal of polarity*.
(8) This cell can still divide and begins to synthesise some enamel proteins.
(9) For a short period of time, it synthesises *dentine sialoprotein*, a protein also produced by odontoblasts.
(10) At this phase, the basal lamina of the inner dental epithelium begins to degrade.

2. *Secretory stage.*
   a. *Initial secretion phase.*
      (1) At this phase, the secretion is already intense.
      (2) The cell (*ameloblast*) is taller and has more organelles, but the Tomes' process characteristic of the mature ameloblast has not yet developed.
      (3) The ameloblasts are attached to the cells of the *stratum intermedium* by *desmosomes*.
      (4) The first enamel to be synthesised does not form rods (BSUEs), but is instead a *aprismatic enamel*.
   b. *Secretory phase.*

(1) At this stage, the *Tomes' process* appears and enamel is actively synthesised. The Tomes' process is a conical extension of the cytoplasm that occurs at the *proximal pole*, i.e. the pole facing the DEJ.
(2) In these cells, the Golgi apparatus is highly developed and forms a cylindrical structure surrounded by cisternae of the rough endoplasmic reticulum. Both organelles are located proximal to the distal nucleus.
(3) There are numerous *secretory vesicles* that migrate to the Tomes' process and are released by *exocytosis* without being stored in the cytoplasm. These vesicles have been called *ameloblastic bodies* or *adamantine bodies*.
(4) The enamel that is synthesised is already *prismatic enamel.*
(5) Along with the organic matrix that is deposited, the first *hydroxyapatite crystals* are incorporated and even intermingle with the crystals of the neighbouring dentine in the DEJ.
(6) As secretion progresses, the position of the ameloblast nucleus may vary, from close to the distal pole to closer to the proximal pole.
(7) Each pole of the cell has a *terminal web* associated with junctions between cells.
   (a) At the proximal pole of ameloblasts, the junctions between cells surround the entire cell perimeter (zonulae), thus sealing the intercellular space.
   (b) At the distal pole, the intercellular junctions are not continuous and do not seal the intercellular space.
(8) The Tomes' process has two enamel-forming zones:
   (a) A secretory zone with an *irregular surface*, with numerous *invaginations*, which forms the head of the rods.
   (b) A *smooth-surfaced* secretory zone that is responsible for the formation of the rod tails.
(9) The regions of the Tomes' process closest to the cusps of the process form the enamel rods. The interrod enamel is produced from the regions of the Tomes' process furthest from the cusps, i.e. those closest to the junctional complexes.
(10) Four ameloblasts are involved in the formation of a BSUE, one of which forms the head, two others form the neck, and the last one produces the tail. Each of these ameloblasts is involved in the formation of certain parts of four rods (Fig. 19.9).

3. *Maturation stage.*
   a. This stage is characterised by hardening of the enamel. This is not due to the deposition of new hydroxyapatite crystals, but to the fact that the existing crystals increase in size.
   b. The process is slow and can take several years from crown formation to tooth eruption.
   c. Several phases can be distinguished at this stage:

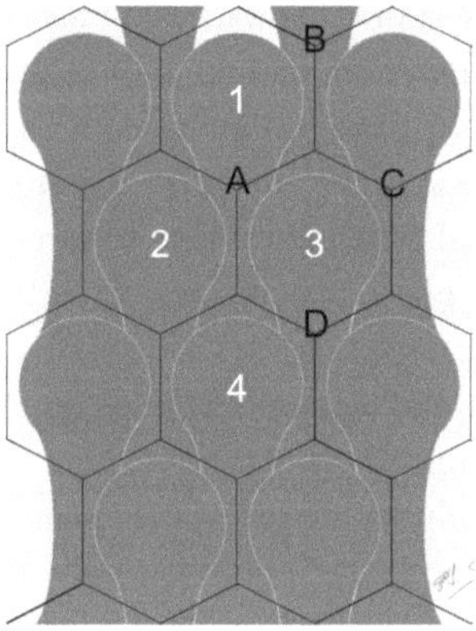

Fig. 19.9 Formation of the enamel rods. The profiles of the rods (dark) and the ameloblasts (light hexagons) have been superimposed. Some rods have been numbered with letters (A to D) and some ameloblasts have been numbered (1 to 4). Each rod is produced by materials provided by four ameloblasts; for example, rod A is made up of materials synthesised by ameloblasts 1, 2, 3 and 4. In turn, each ameloblast contributes materials to four rods; for example, in a simplified way, it can be seen that ameloblast 3 contributes materials to the head of rod D, to the necks of rods A and C, and to the tail of rod B

(1) *Transitional phase.*
   (a) This is a short phase in which the morphology of the ameloblast is modified.
   (b) The cell decreases in volume, and the number of organelles is reduced.
   (c) In rodents, some cells, up to 25% of them, die by *apoptosis*.
   (d) In this phase, invaginations of the outer enamel epithelium occur, which accommodate *blood vessels*. Although these do not penetrate into the enamel organ, the blood supply is facilitated to approach the ameloblasts.
(2) *Maturation phase.*
   (a) During this stage, there is a reduction of water and organic material from the enamel.
      1. Before maturation, the enamel consists of 65% water, 20% organic material and 15% hydroxyapatite.
      2. In the final composition of the enamel, hydroxyapatite makes up 96% by weight. The elimination of water and organic molecules is due to the activity of the ameloblast.
   (b) During this period, more ameloblasts die by *apoptosis* (in rodents another 25% of the initial number).
   (c) In addition, the proximal surface of the ameloblast, which has lost the Tomes' process, undergoes cyclical variations consisting of the formation of a *striated* surface, which becomes *smooth*, becomes striated again, becomes smooth again, and so on in a cyclical manner.

1. This process of creating, losing and recreating a striated surface is called *modulation*, and up to three cycles occur daily.
2. These cycles manifest themselves as a *wave* that runs from the less mature areas of the enamel to the more mature areas, i.e. starting in the cervical area and moving towards the incisal area.

(d) During this phase, therefore, two forms of ameloblast can be distinguished:

1. *Smooth-ended ameloblast*. Each ameloblast spends 20% of its time in this form.
   a. In this phase, the *pH* of the enamel approaches neutrality.
   b. The smooth ameloblast has tight junctions that are labile (permeable) proximally and very tight distally.
   c. Smooth endoplasmic reticulum cisternae can be seen inside.
   d. The smooth ameloblast has little endocytic activity and does not pump calcium, which can move exclusively by diffusion. As calcium is highly concentrated in the enamel, it is difficult for new calcium to reach the enamel at this stage.
2. *Ruffle-ended ameloblast*. Each ameloblast spends 80% of its time in striated form.
   a. The pH of enamel is slightly acidic. This acidity is thought to be due to the release of protons (*hydrogen ions*) produced by crystal growth.
   b. At this stage, the ruffle-ended ameloblast has tight junctions proximally and weak junctions distally.
   c. The ruffle-ended ameloblast does a lot of *endocytosis*.
   d. It can also regulate *calcium* transport by pumping calcium into the matrix.
   e. Lysosomes can be seen in the cytoplasm.

(e) In summary, enamel crystal growth occurs primarily during the ruffle-ended ameloblast phases.

1. In this phase, calcium is actively pumped into the striated border, and the pH is lowered by crystal growth.
2. Crystal growth releases protons which are neutralised by the release of *bicarbonate*, produced by *carbonic anhydrase* in the ameloblasts.

(f) At this phase of maturation, the ameloblast has the capacity to absorb water and to degrade and absorb organic substances (*absorptive ameloblast*). For this, the ameloblast produces *lysosomal enzymes*, *alkaline phosphatase* and a calcium-dependent membrane ATPase (*calcium pump*).

1. Enamel proteins are degraded by enzymes that are released into the matrix by exocytosis of the lysosomes. In this way, smaller peptide

fragments are produced which can cross the labile junction between the smooth-ended ameloblasts, pass between the cells and are absorbed there.

2. In the ruffled-ended ameloblast, endocytosis occurs at the striated border.

(g) During the maturation stage, a material is deposited between the enamel and the surface of the ameloblast.

1. This material resembles a basal lamina, but lacks its typical proteins and has a different form of laminin (*laminin 332*).
2. The ameloblast is attached to this structure by *hemidesmosomes*.

(3) *Protective phase*.

(a) Regression of the ameloblast (*protective ameloblast*) occurs.

(b) The cells flatten, except in areas where fissures occur, no longer form a well-defined layer and are indistinguishable from those of the *stratum* intermedium, with which they merge.

(c) They also fuse with the *outer epithelium* of the enamel organ. Together, they form the so-called *reduced enamel epithelium* or *reduced dental epithelium*.

(d) At this stage, the ameloblast produces the *primary cuticle*.

(4) *Desmolytic phase*.

(a) Once the reduced enamel epithelium is formed, it secretes enzymes that can destroy the connective tissue that covers it and separates it from the oral epithelium (*desmolysis*).

(b) This facilitates the eruption of the tooth.

Take-Home Message

Enamel is the hardest mineralised tissue in our body. It is an acellular tissue. The enamel matrix is mostly mineral, consisting of large crystals of hydroxyapatite. In immature enamel, there are organic substances, mainly proteins that regulate mineralisation, but in mature enamel, these are scarce. There is no collagen. The basic structural unit of enamel is the enamel rod or enamel prism. In some areas, there may be aprismatic enamel. Depending on variations in the organisation and composition of the rods, several types of secondary structural units can be observed. The enamel surface is not smooth, and various formations can be distinguished. Enamel can be lost on the surface by attrition, abrasion and erosion. In addition, the enamel may be covered by layers of organic material on which micro-organisms can grow and form a biofilm. Amelogenesis occurs during odontogenesis. A slightly mineralised organic matrix is first synthesised, followed by removal of organic matrix and growth of hydroxyapatite crystals. To produce enamel, the ameloblasts undergo a differentiation process that goes through several phases. At the end of amelogenesis, the enamel organ becomes the reduced enamel epithelium.

Further Reading

Bei M (2009) Molecular genetics of ameloblast cell lineage. J Exp Zool Part B Mol Develop Evol 312B:437–444
Berkovitz BKB, Holland GR, Moxham BJ (2018a) Amelogenesis. In: Oral anatomy, histology & embryology, 5th edn. Elsevier, pp 369–386
Berkovitz BKB, Holland GR, Moxham BJ (2018b) Enamel. In: Oral Anatomy, histology & embryology, 5th edn. Elsevier, pp 123–143
Berkovitz BKB, Holland GR, Moxham BJ (2018c) Investing organic layers on enamel surfaces. In: Oral anatomy, histology & embryology, 5th edn. Elsevier, pp 144–151
Chiego DJ (2023a) Biofilms. In: Essentials of oral histology and embryology, 6th edn. Elsevier, pp 195–202
Chiego DJ (2023b) Enamel. In: Essentials of oral histology and embryology, 6th edn. Elsevier, pp 88–96
Deutsch D, CatalanoSherman J, Dafni L et al (1995) Enamel matrix proteins and ameloblast biology. Connect Tissue Res 32:92–107
Duverger O, Lee JS (2024) The intricacies of tooth enamel: embryonic origin, development and human genetics. J Struct Biol 216:108135
Fan Y, Zhou YC, Zou XD et al (2015) Epigenetic control of gene function in enamel development. Curr Stem Cell Res Ther 10:405–411
Fehrenbach MJ, Popowics T (2020) Enamel. In: Illustrated dental embryology, histology, and anatomy, 5th edn. Elsevier, pp 149–159
Fernandes CP, Chevitarese O (1991) The orientation and direction of rods in dental enamel. J Prosthet Dent 65:793–800
Fukumoto S, Yamada A, Nonaka K, Yamada Y (2005) Essential roles of ameloblastin in maintaining ameloblast differentiation and enamel formation. Cell Tissues Organs 181:189–195
Gil-Bona A, Bidlack FB (2020) Tooth enamel and its dynamic protein matrix. Int J Mol Sci 21:4458
Gómez de Ferraris ME, Campos Muñoz A (2009) Esmalte. In: Histología, Embriología e Ingeniería Tisular Bucodental, 3rd edn. Médica Panamericana, pp 291–332
Hand AR (2014) Enamel. In: Hand AR, Frank ME (eds) Fundamentals of Oral histology and physiology. Wiley Blackwell, pp 63–84
Ida-Yonemochi H (2025) Role of glucose metabolism in amelogenesis. J Oral Biosci 67:100667
Lacruz RS, Habelitz S, Wright JT, Paine ML (2017) Dental enamel formation and implications for oral health and disease. Physiol Rev 97:939–993
Moradian-Oldak J (2012) Protein-mediated enamel mineralization. Front Biosci Landmark 17:1996–2023
Nanci A (2013) Enamel: composition, formation and structure. In: Ten Cate's oral histology, 8th edn. Elsevier, pp 122–164
O'Dowd G, Bell S, Wright S (2023) Oral tissues. In: Wheater's functional histology. Elsevier, pp 241–252
Robinson C (2015) Enamel maturation: a brief background with implications for some enamel dysplasias. Front Physiol 5:338
Romaniuk K, Shroff FR (1965) Directional variation in terminal portions of enamel rods in human deciduous teeth. J Dent Res 44:1208
Simmer JP, Papagerakis P, Smith CE et al (2010) Regulation of dental enamel shape and hardness. J Dent Res 89:1024–1038

Skobe Z, Stern Z (1980) The pathway of enamel rods at the base of cusps of human teeth. J Dent Res 59:1026–1032

Warshawsky H (1988) The teeth. In: Weiss L (ed) Cell and tissue biology—a textbook of histology, 6th edn. Urban & Schwarzenberg, pp 594–640

Wu K, Li XC, Bai YY et al (2024) The circadian clock in enamel development. Int J Oral Sci 16:56

Yin KF, Paine ML (2017) Bicarbonate transport during enamel maturation. Calcif Tissue Int 101:457–464

Protective Periodontium 20

What You Will Learn in This Chapter

This chapter explains the histological organisation of the protective periodontium, consisting of the gingiva and the dentogingival junction. You will learn that you can distinguish an attached gingiva, bonded to the alveolar bone, a free gingiva, not bonded to the alveolar bone, and you will see how the mucosa of both gingivae is organised. It is explained that the dentogingival junction is formed by the gingival sulcus, the sulcular epithelium and the junctional epithelium, which are supported by a lamina propria. You will learn the peculiar characteristics of the junctional epithelium. You will also see the structure of the interdental gingiva. The organisation of the gingival ligament is explained. Finally, you will study the origin of the protective periodontium, its innervation and vascularisation.

20.1 Components of the Periodontium

A. *Periodontium* (from the classical Greek περι peri 'around' and ὀδούς, ὀδόντος odoús, odontos 'tooth') refers to the tissues and structures that are associated with the teeth and provide them with protection and support.

B. The periodontium can be divided into two different but related parts:

   1. *Protective periodontium.*
      - This part of the periodontium separates the inner tissues of the periodontium and isolates them from the environment of the oral cavity by a seal in the cervical region of the dental crown.
      - It is made up of the *gingiva* (from the Latin gingīva 'gum') and the *dentogingival junction*. The latter is the part of the gingiva that is in intimate contact with the tooth.
      - These structures are specialised regions of the oral mucosa.

F. J. Sáez, I. Badiola, *Essential Oral Histology*, Learning Materials in Biosciences,
https://doi.org/10.1007/978-3-032-03165-5_20

2. *Insertion periodontium (attachment periodontium).*
   a. Formed by the internal tissues, not in contact with the oral cavity, which surround the dental root and support the tooth.
   b. The components of the attachment periodontium are *cementum*, *periodontal ligament* and *alveolar bone*.
   c. Cementum is the tissue that surrounds the dentine of the tooth root, to which it is attached. It can be considered as a dental tissue, but it is formed at the same time and with the same origin as the periodontal ligament and the alveolar bone, so these three tissues form a functional unit and are often studied together.

20.2 Components of the Protective Periodontium

A. The components of the protective periodontium are the gingiva and the dentogingival junction (Fig. 20.1). Both are made up of the *gingival mucosa*, which is a specialised part of the oral mucosa.
B. Apical to the gingiva is a region of the oral mucosa that is loosely attached to the underside of the alveolar bones and is called the *alveolar mucosa* (Fig. 20.1).
   1. The alveolar mucosa consists of a thin, *non-keratinised epithelium*, a *lamina propria* with few *connective tissue papillae* and a broad *submucosa* with many *minor salivary glands* of mucous secretion.

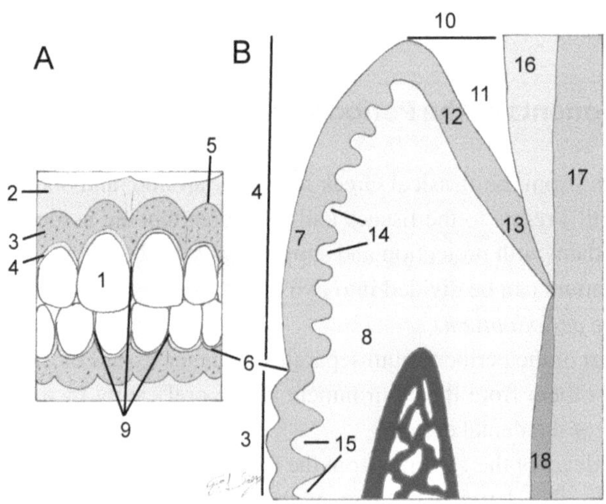

Fig. 20.1 Frontal view of the dentition (**A**) and section of the vestibular side of the cervical region of the tooth (**B**), showing the protective periodontium. (1) tooth; (2) alveolar mucosa; (3) attached gingiva; (4) free or marginal gingiva; (5) mucogingival junction; (6) free gingival groove; (7) gingival epithelium; (8) lamina propria; (9) interdental gingiva or interdental papilla; (10) dentogingival junction; (11) gingival sulcus or gingival crevice; (12) crevicular or sulcular epithelium; (13) junctional epithelium; (14) adelomorphous papillae; (15) delomorphous papillae; (16) enamel; (17) dentine; (18) cementum

C. The boundary between the gingival mucosa and the alveolar mucosa is called the *mucogingival junction* and is clearly distinguishable as a *mucogingival groove* on the mucosal surface.
D. On the palatal surface of the maxillary teeth, there is no alveolar mucosa and the gingival mucosa of the attached gingiva directly contacts the mucosa of the hard palate, without a clear boundary.
E. The gingiva (Fig. 20.1) is the part of the *oral mucosa* that lines the outside of the alveolar bones and attaches to the cervical regions of the teeth.
F. The mucosa of the gingiva is called the *gingival mucosa*. Like all the oral mucosa studied in Chap. 14, it is composed of:
   1. An *epithelium*.
   2. A *lamina propria* of connective tissue.
G. Two regions can be distinguished in the gingiva:
   1. *Attached gingiva*. It is the closest to the mucogingival junction, and it is located covering the edges or alveolar crests of the *alveolar bones*.
   2. *Free* or *marginal gingiva*. This is the gingiva closest to the neck region of the teeth. It is in a coronal position with respect to the attached gingiva. It is not in contact with the alveolar bone.
H. The point where the two (free and attached) gingivae meet is easy to see with the naked eye, as it forms a small groove called the *free gingival groove*.
   1. This free gingival groove is approximately at the level of the junction between enamel and cementum.
   2. In some individuals, this groove is absent, and the boundary between the two gingivae shows a line of irregularly aligned dots.
I. The *free gingiva* is a ridge-shaped structure with two sides:
   1. The visible side, which is oriented towards the vestibular (buccal or labial) region or towards the lingual or palatal region, depending on the jaw, is what we actually call the free gingiva.
   2. The side facing the tooth is the *dentogingival junction*.
J. The dentogingival junction consists of several structures:
   1. *Gingival sulcus* or *gingival crevice*. The small space, sometimes a virtual space, that separates the gingival side of the gingiva from the tooth enamel.
   2. *Crevicular* or *sulcular epithelium*. This is the epithelium lining the dental side of the gingiva in the area of the gingival sulcus.
   3. *Junctional epithelium*. This is the part of the epithelium that covers the dental side of the gingiva in the area closest to the neck of the tooth. In this area, unlike in the area of the gingival sulcus, the epithelium is intimately bound to the enamel of the dental crown.
   4. *Lamina propria*. It is connective tissue without clear boundaries with the lamina propria of the free gingiva.

K. The region of the free gingiva between two neighbouring teeth is called the *interdental gingiva* or *interdental papilla*.
   1. Between the anterior teeth, this interdental gingiva is usually pyramidal in shape.
   2. In molars, on the other hand, it is flattened in the vestibulo-lingual direction, and even shows a concave depression. In this case, it is called *interdental col*, and in vestibulo-lingual sections, it is saddle-shaped.

20.3 Histology of the Gingiva

A. The gingiva is separated from the alveolar mucosa by the *mucogingival junction*. A *mucogingival groove* is visible on the mucosal surface.
   1. On one side of the groove, the mucosa is bright pink (alveolar mucosa).
   2. On the gingival side, the mucosa is pale pink (gingival mucosa).
   3. The microscope shows that at this junction there is an abrupt transition in the mucosa, both in the type of epithelium and in the organisation of the lamina propria:
      a. Type of epithelium.
         (1) The gingival epithelium is *parakeratinised*.
         (2) The epithelium of the alveolar mucosa is somewhat thicker and not keratinised.
      b. Variety of lamina propria.
         (1) The lamina propria of the attached gingiva has many bundles of collagen fibres attached to the periosteum.
         (2) The lamina propria of the alveolar mucosa is a loose connective tissue with many elastic fibres extending into a thick *submucosa*.

B. The gingiva is divided into two regions, the *attached gingiva* and the *free gingiva*, separated by the *free gingival groove*.
   1. The attached gingiva has a pale pink colour, with a rough surface, usually (but not always) with a dotted appearance, often reminiscent of an 'orange peel'. This gingiva has a firm consistency.
   2. The free gingiva usually has a bright coral-pink appearance, a smooth surface and a softer consistency.
   3. When the histological structure of the two gingival areas is observed under the microscope, there is not a great difference between the two, so that in the area of the free gingival groove there is no abrupt transition, as in the mucogingival junction.

C. *Gingival epithelium.*
   1. Both gingival regions are dominated by *parakeratinised epithelium.*
   2. In the attached gingiva, there are orthokeratinised regions, but the majority is parakeratinised epithelium (75%), and there may be some non-keratinised areas (10%).

D. *Lamina propria.*

1. Connective tissue of the lamina propria:
   a. The lamina propria of the attached gingiva is *dense* and adheres to the periosteum (*mucoperiosteum*).
   b. The lamina propria of the free gingiva is *semi-dense*. It is above the crest of the alveolar bone and therefore does not contact any periosteum.
2. In the region of contact between epithelium and lamina propria, *connective tissue papillae* can sometimes be observed (connective tissue papillae were explained in the chapter on the oral mucosa).
   a. In attached gingivae, the papillae are long and narrow, push on the epithelium and cause elevations of the free surface of the epithelium, which gives rise to its rough appearance (orange peel). They are therefore said to be *delomorphous papillae* (from the Ancient Greek δῆλος dêlos, 'visible' and μορφο morpho, μορφος morphos, 'shape').
   b. In the free gingiva, the papillae are short, numerous and *adelomorphous*, which implies that the surface of the epithelium is smooth.
3. In the lamina propria of both gingivae, relatively thick bundles of *type I collagen fibres* intermingle, especially in the area of the free gingiva, and form the so-called *gingival* or *supracrestal ligament*.
   a. In addition to collagen fibres, *elaunin fibres* have been described in the gingival ligament.
   b. The main function of the fibres is to support the free gingiva, to attach the attached gingiva to the bone and the tooth and to attach neighbouring teeth.
   c. These fibres form several groups, which are called by different names according to their arrangement. Although many groups of fibres have been described, the most important ones are listed here (Fig. 20.2):
      (1) *Dentogingival group*, consisting of fibres that run from the root surface of the tooth (cementum) to the free gingiva.
      (2) *Alveologingival group*, formed by fibres that start from the periosteum of the alveolar crest and go towards the free gingiva.
      (3) *Dentoperiosteal group*, which has fibres that start from the dental cementum and reach the periosteum of the external side of the alveolar crest.
      (4) *Circular group*, with fibres forming a ring around the neck of the tooth
      (5) *Transseptal* or *dentodental group*, formed by fibres that go from the cementum of one tooth to the cementum of another tooth, above the alveolar crest, crossing the interdental gingiva.
   d. Some authors consider the gingival ligament to be part of the *periodontal ligament*. In fact, the fibres of both intermingle in the cervical region.

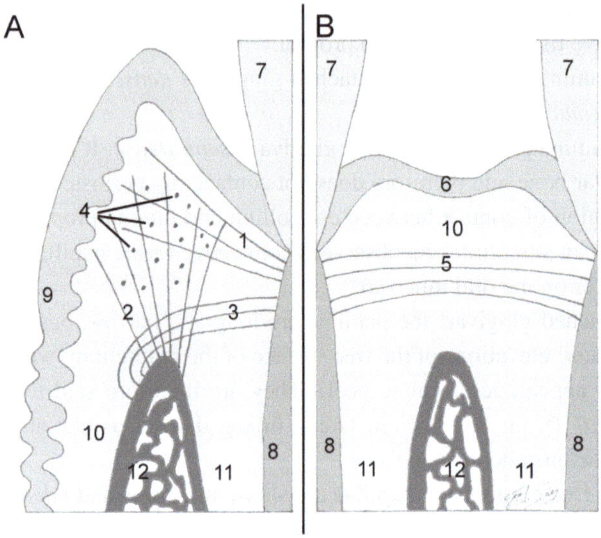

Fig. 20.2 Vestibulolingual (**A**) and mesiodistal (**B**) section of the cervical region of teeth showing the organisation of the gingival ligament. (1) dentogingival group; (2) alveologingival group; (3) Dentoperiosteal group; (4) circular group; (5) transeptal or dentodental group; (6) interdental col; (7) enamel; (8) cementum; (9) gingival epithelium; (10) lamina propria; (11) periodontal ligament; (12) alveolar bone

4. Although the gingival lamina propria is in continuation with the periodontal ligament, with no clear boundary between the two, it is important to note that there are important differences, which are outlined in this table:

	Gingiva	Periodontal ligament
Type III collagen	Little	A lot
Hyaluronan	Little	A lot
Collagen production	Low	High
Collagen turnover rate	Slow	Fast
Alkaline phosphatase	No	Yes
Origin of fibroblasts	Mesenchyme (mesoderm)	Neural crests (ectomesenchyme)
Contractile proteins in fibroblasts	No	Yes
Fibroblast proliferation	Low	High
Prostaglandin production in response to histamine	No	Yes

5. The gingival lamina propria is highly *vascularised*. In each connective tissue papilla, there is an ascending arteriole and a descending venule, connected by capillaries.

20.4 Histology of the Dentogingival Junction

A. The *dentogingival junction* corresponds to the inner or dental side of the free gingiva. It consists of the *crevicular (sulcular) epithelium*, which is separated from the dental enamel by the *gingival crevice (gingival sulcus)*, the *junctional epithelium* and the *lamina propria* beneath both epithelia.

B. At the dentogingival junction, the junctional epithelium seals the periodontal tissues and separates them from the oral cavity.

C. The *gingival crevice* is 0.5–3 mm deep.
   1. When it is larger than 3 mm, it is considered a pathological situation and is referred to as a *periodontal pocket*.
   2. In this gingival crevice, there is liquid, the *gingival crevicular fluid*, which passes from the lamina propria through the junctional epithelium, *desquamated cells* of the junctional and crevicular epithelia and *inflammatory cells*, mainly *neutrophils*.
      a. It is interesting to note that in the other epithelia of the oral mucosa, the passage of fluids and cells as through the junctional epithelium has not been detected.
      b. Gingival crevicular fluid consists of *inorganic molecules* (sodium, potassium, calcium and magnesium), *proteins* (immunoglobulins, cytokines, fibrinogen, proteases), *cells* (desquamated epithelial cells, neutrophils and bacteria) and *cellular debris*.
      c. Its composition can vary in pathological states and can be used as an indicator of the severity of a possible infection.

D. The *crevicular epithelium* is found on the dental side of the free gingiva, lining the gingival side of the gingival crevice.
   1. This epithelium is a continuation of the epithelium of the free gingiva, with which it contacts at the gingival crest, and therefore has the characteristics of the epithelium of the oral mucosa.
   2. It is a *non-keratinised stratified squamous epithelium. Cytokeratin 4*, typical of non-keratinised epithelia in the oral mucosa, is produced in its superficial layers.

E. The *junctional epithelium* or *attachment epithelium* is located at the floor of the gingival crevice (Fig. 20.3).
   1. It forms a kind of belt or band around the cervical margin of the tooth.
   2. It is partly derived from the *reduced enamel epithelium*, as during eruption, this epithelium fuses with the epithelium of the oral mucosa and is crossed by the dental crown (dental eruption will be explained in another chapter).
   3. This epithelium is attached to the neck of the clinical crown, so that it fixes the gingiva to the *enamel* surface[1] and seals and protects the tissues of the periodontium.

---

[1] In reality, the junctional epithelium is attached to the pellicle overlying the enamel, but neither this pellicle nor the enamel are visible in the decalcified microscope samples commonly used.

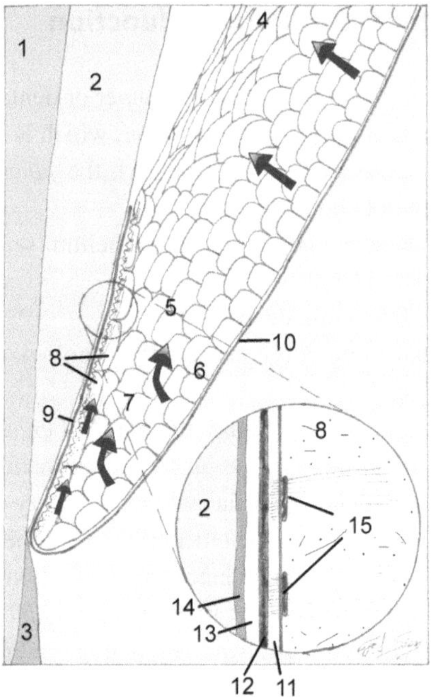

Fig. 20.3 Histological organisation of the junctional epithelium. The circled area is enlarged to show the structure of the internal basal lamina. (1) dentine; (2) enamel; (3) cementum; (4) crevicular epithelium; (5) junctional epithelium; (6) basal cells; (7) suprabasal cells; (8) DAT cells (directly attached to the tooth); (9) internal basal lamina; (10) external basal lamina; (11) lamina lucida; (12) lamina densa; (13) sublamina lucida; (14) cuticle; (15) hemidesmosomes. The arrows indicate the direction of ascent of the epithelial cells

4. The junctional epithelium is a non-keratinised epithelium and forms a lining mucosa with the underlying lamina propria. However, it is a special epithelium. It has two special features:
   a. The cells of this epithelium do not undergo the differentiation process that is usual in the stratified epithelia of the oral mucosa. This can be seen from several data:
      (1) They produce different cytokeratins:
         (a) The cells of this epithelium do not express cytokeratin 4.
         (b) *Cytokeratins 5, 14* and *19*, which are typical of the stratum germinativum, are expressed in all strata. They are also those expressed in the reduced enamel epithelium.
         (c) In addition, junctional epithelial cells express *cytokeratins 8* and *18*, which are commonly found in simple epithelia.
      (2) They produce *ODAM* (odontogenic ameloblast-associated protein), a protein also produced by ameloblasts. This protein has been found to be produced in certain neoplastic epithelia, whose cells remain permanently

undifferentiated, suggesting that it has some role in maintaining the undifferentiated state.

(3) *Tonofilaments* and *desmosomes* are less abundant than in the other epithelia of the oral mucosa.

(4) There is neither a *stratum granulosum* nor a *stratum superficial.*

(5) The cells show no *keratohyalin granules* or *lamellar bodies.*

b. It has two *basal laminae.*

(1) There is an *external (outer) basal lamina* separating the epithelium from the lamina propria. That is, it is at the basal pole of the epithelium.

(2) At the deepest end of the epithelium, the basal lamina surrounds it and forms a second basal lamina, the *internal (inner) basal lamina*, which covers the apical pole of the epithelium and is located between it and the tooth enamel. The internal basal lamina is peculiar:

(a) It adheres to mineralised tissue and not to non-mineralised connective tissue, as is the case with most basal laminae.

(b) Its composition is peculiar:

1. It lacks laminin and type IV collagen present in most basal laminae.
2. It has a special variety of laminin, *laminin 332*, which is also present in the basal lamina of the ameloblast during the maturation phase.
3. It contains ameloblast proteins such as *amelotonin* and *ODAM protein.*

(c) The transmission electron microscope shows three zones in the basal lamina, from the epithelium towards the tooth:

1. *Lamina lucida.* This is the lighter area in contact with the surface of the epithelial cells.
2. *Lamina densa.* A darker band following the previous one.
3. *Sublamina lucida.* A clear band between the lamina densa and the tooth surface.

5. Cells of the junctional epithelium:

a. The *basal cells*, which are more or less cuboidal, are located on the external basal lamina. As in the other epithelia of the oral mucosa, they divide to produce new cells that move up through the epithelium.

b. The suprabasal cells flatten as they move upwards in the epithelium and away from the external basal lamina.

c. There is a *stratum spinosum.*

d. There is no stratum granulosum or stratum superficial.

e. There are large spaces between the cells and fewer *desmosomes* than is usual in other epithelia of the oral mucosa. This facilitates the passage of fluid and cells from the lamina propria to the crevicular fluid.

f. *Neutrophils* are often found infiltrating between the epithelial cells.

g. The *turnover rate* of the junctional epithelium cells is the highest of all epithelia of the oral mucosa.

   h. The movement of the cells is peculiar, as they move upwards in a coronal direction, almost parallel to the tooth surface.
   i. Finally, there are cells that are in contact with the internal basal lamina and the enamel and are called *DAT cells* (from *D*irectly *A*ttached to the *T*ooth). DAT cells adhere via *hemidesmosomes* to the internal basal lamina.
6. When the junctional epithelium is damaged, it can be rapidly regenerated by the crevicular epithelium and the oral epithelium. The cells that regenerate this epithelium quickly begin to produce the *ODAM protein*.
7. As soon as the tooth appears in the oral cavity (*eruption*), the junction between the epithelium and the tooth descends.
   a. Immediately after eruption, the junctional epithelium is located in the cervical half of the anatomical crown.
   b. With age, the attachment point decreases in an apical direction. If a person suffers exposure of the tooth root (due to age or pathological changes), the attachment epithelium can proliferate in the apical direction and attach to the cementum.

F. The *lamina propria* of the crevicular epithelium and junctional epithelium is a *loose connective tissue*.
   1. It has few *fibroblasts* and *collagen fibres*.
   2. It has many *transient cells* (lymphocytes, neutrophils, monocytes) and macrophages, suggesting that it is in a state of permanent *inflammation*.
      a. It is possible that the permanently inflamed state of this tissue is responsible to emit signals to maintain the characteristics of the epithelium.
      b. In certain experiments, when inflammation is successfully reduced, *keratinisation* of the crevicular epithelium occurs.

G. There is a rich vascularisation beneath the junctional epithelium with many *postcapillary* venules. These vessels are the source of fluid, molecules and cells of the immune system that subsequently pass through the epithelium to form the *gingival crevicular fluid*.

20.5 Interdental Gingiva

A. The *interdental gingiva* is the part of the gingiva between two neighbouring teeth.
B. Shows attached gingiva and free gingiva.
C. Their shape varies depending on the form of contact between the teeth.
   1. Seen from the outside or inside of the mouth, it is usually wedge-shaped.
   2. When seen in buccolingual sections:
      a. The anterior teeth usually have the same shape.
      b. In posterior teeth, which have a larger contact surface, the interdental gingiva is usually concave and saddle-shaped. This structure is called an *interdental col*.

D. The interdental col has an epithelium similar to the junctional epithelium, is non-keratinised, thin, has the same origin and is maintained by cell divisions.
   1. It is a site sensitive to *infection* because food debris and *bacteria* tend to accumulate between the teeth. I*nflammatory cells* are often observed in the lamina propria.
   2. If the posterior teeth are wide apart, the interdental col is not visible, the interdental gingiva is flattened and may be keratinised.

20.6 Origin and Development of the Protective Periodontium

A. Three phases can be distinguished in the formation of the gingiva and the rest of the protective periodontium (Fig. 20.4):
   1. Before the eruption:
      a. Once the crown is formed, during the *ameloblast* protective phase, the *primary cuticle* is formed, the ameloblasts fuse with the rest of the cells of the enamel organ and form the *reduced enamel epithelium.*
      b. This epithelium consists of several layers of flattened cells, an *internal basal lamina* (contacts the tooth) and an *external basal lamina*, which separates the connective tissue from the lamina propria of the oral mucosa.
      c. The cells that are in contact with the enamel are attached by *hemidesmosomes* to the basal lamina that separates them from the tooth (internal basal lamina of the reduced enamel epithelium).

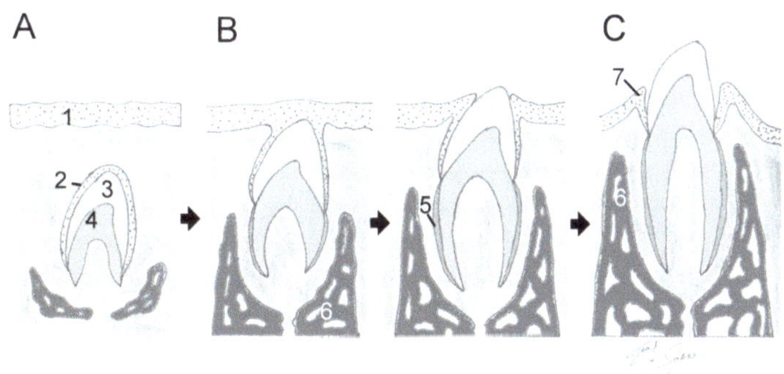

Fig. 20.4 Formation of the protective periodontium during tooth eruption (see also Chap. 22): Preeruptive stage (**A**), prefunctional eruptive stage (**B**) and functional eruptive or posteruptive stage (**C**). (1) oral epithelium; (2) reduced enamel epithelium; (3) enamel; (4) dentine; (5) cementum; (6) alveolar bone; (7) gingival epithelium. The arrows indicate the progress from one stage to another

2. During the eruption:
   a. The reduced enamel epithelium approaches the oral epithelium.
   b. The connective tissue between the two degenerates (*desmolysis*), and the epithelia contact and fuse.
   c. The external basal lamina of the reduced enamel epithelium fuses with the basal lamina of the oral epithelium.
   d. In the central area of the fused epithelia, *necrosis* (from the Late Latin necrōsis, from the Greek νεκρο- nekro-, 'dead' and -σις -sis, 'irregular state, disease') occurs. The cells die and a hole is formed through which the tooth can erupt.
3. After the eruption:
   a. When the tooth erupts, the oral mucosa invaginates around it, forming the *gingival crevice*, the *crevicular epithelium* and the *junctional epithelium.*
   b. The *junctional epithelium*:
      (1) It is formed mainly by cells from the reduced enamel epithelium and some cells from the oral epithelium.
      (2) Over time, most of the cells are replaced by cells from the oral epithelium.
   c. The rest of the *gingiva (free* and *attached)* is derived from the oral mucosa.

B. The gingiva forms with the eruption. If the tooth falls out, the gingiva disappears (Fig. 20.5).

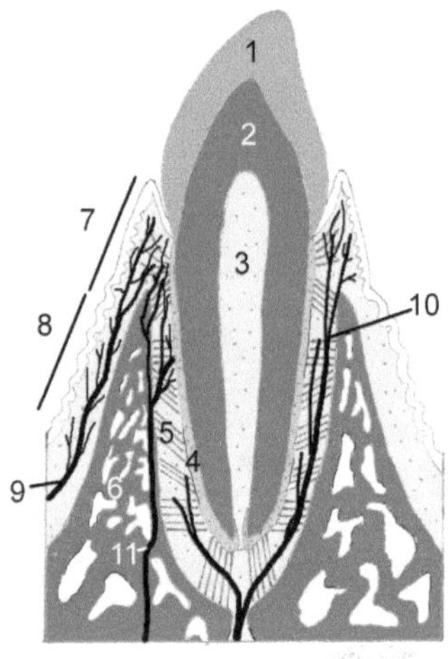

Fig. 20.5 Vascularisation of the protective periodontium. Three types of vessels supply the gingiva: supraperiosteal vessels, originating from the alveolar mucosa or palatal mucosa (9), vessels of the periodontal ligament, which reach the gingiva (10), and vessels of the alveolar bone, which produce branches supplying the periodontal ligament and the gingiva (11). For simplicity, each vessel type has been represented on one side of the tooth only. (1) enamel; (2) dentine; (3) dental pulp; (4) cementum; (5) periodontal ligament; (6) alveolar bone; (7) free gingiva; (8) attached gingiva

20.7 Vascularisation

A. The blood vessels supplying the gingiva have three origins:
   1. *Supraperiosteal vessels*: circulate through the alveolar mucosa and palate and enter the gingivae.
   2. *Vessels of the periodontal ligament*: reach the gingival lamina propria and anastomose with the supraperiosteal vessels.
   3. *Vessels of the alveolar bone*: they leave the alveolar bone and give branches towards the gingiva, where they anastomose with the other vessels. They also supply the periodontal ligament.

20.8 Innervation

A. The protective periodontium is innervated by divisions of the *trigeminal (V) nerve*. Specifically, the following divisions arrive:
   1. Superior labial of the infraorbital nerve.
   2. The superior alveolar nerve.
   3. The anterior palatine nerve.
   4. The sublingual nerve (terminal of the lingual nerve).
   5. The inferior alveolar nerve.
   6. The mental nerve.

B. In attached gingiva, the nerve fibres end in the lamina propria, only rarely are endings seen that reach the epithelium.

C. The dentogingival junction may be richly innervated with *free nerve endings* in the connective tissue and between the cells of the junctional epithelium. This rich innervation may be related to the inflammatory state of this area.

Take-Home Message

The periodontium consists of the structures associated with the tooth. A distinction is made between a protective periodontium and an insertion periodontium. The protective periodontium consists of the gingiva and the dentogingival junction. The gingiva can be divided into two areas, an attached gingiva, in a more apical position with respect to the tooth, attached to the alveolar bone, and a free gingiva, in a more cervical position, which does not contact the alveolar bone. In the lamina propria of both gingivae is the gingival ligament, which consists of several groups of collagenous fibres. Important differences have been found between the connective tissue of the gingiva and that of the periodontal ligament. The dentogingival junction is formed by the gingival sulcus, the sulcular epithelium and the junctional epithelium, the latter supported by its lamina propria. The junctional epithelium is attached to the tooth and has an internal and an external basal lamina. Between two teeth is the interdental gingiva, which between the posterior teeth forms the interdental col. The epithelia of the protective periodontium is formed during tooth eruption, by fusion of the reduced enamel epithelium and the oral epithelium.

Further Reading

Arnim SS, Hagerman DA (1953) The connective tissue fibers of the marginal gingiva. J Am Dent Assoc 47:271–281

Bartold PM, Walsh LJ, Narayanan S (2000) Molecular and cell biology of the gingiva. Periodontology 2000(24):28–55

Bosshardt DD, Lang NP (2005) The junctional epithelium: from health to disease. J Dent Res 84:9–20

Caimi M, Duarte G, Vicente I, Pedemonte R (1989) Elastic system in human gingiva in health and periodontal disease. An Fac Odontol 25:21–48

Delima AJ, Van Dyke TE (2003) Origin and function of the celular components in gingival crevice fluid. Periodontology 2000(31):55–76

Fehrenbach MJ, Popowics T (2020) Gingival and Dentogingival junctional tissue. In: Illustrated dental embryology, histology, and anatomy, 5th edn. Elsevier, pp 124–134

Garnick JJ, Ringle RD (1988) The dentogingival junction as seen with light-microscopy and scanning electron-microscopy. Scanning Microsc 2:1113–1122

Giannopolou C, Cimasoni G (1996) Functional characteristics of gingival and periodontal ligament fibroblasts. J Dent Res 75:895–902

Gómez de Ferraris ME, Campos Muñoz A (2009) Periodonto de protección: encía y unión dentogingival. In: Histología, Embriología e Ingeniería Tisular Bucodental, 3rd edn. Médica Panamericana, pp 333–354

Goodson JM (2003) Gingival crevice fluid flow. Periodontology 2000(31):43–54

Hokama MM, Foster RA, Archer SY et al (1987) Comparative characterizacion of gingival and periodontal-ligament cells, invitro. J Dent Res 66:357–357

Hou LT, Yaeger JA (1993) Cloning and characterization of human gingival and periodontal-ligament fibroblast. J Periodontol 64:1209–1218

Jany Z, Jakubovsky J, Kovar M et al (1984) Contribution to the knowledge on the ulstrastructure of marginal gingivae. Czech Med 7:135–144

Krstić RV (1991) Digestive apparatus. In: Human microscopic anatomy. An atlas for students of medicine and biology. Springer Verlag, pp 161–255

McKnight H, Kelsey WP, Hooper DA et al (2014) Proteomic analyses of human gingival and periodontal ligament fibroblasts. J Periodontol 85:810–818

Newman HN (1990) The tooth-gingiva interface: weak or strong? J West Soc Periodontol Periodontal Abstr 38:5–9

Nishio C, Wazen R, Moffat P, Nanci A (2013) Expression of odontogenic ameloblast-associated and amelotonin proteins in the junctional epithelium. Periodontology 2000(63):59–66

O'Dowd G, Bell S, Wright S (2023) Oral tissues. In: Wheater's functional histology. Elsevier, pp 241–252

Palmer RM, Lubbock MJ (1995) The soft connective tissues of the gingiva and periodontal ligament: are they unique? Oral Dis 1:230–237

Pöllänen MT, Salonen JI, Uitto VJ (2003) Structure and function of the tooth-epithelial interface in health and disease. Periodontology 2000(31):12–31

Schroeder HE, Listgarten MA (1997) The gingival tissues: the architecture of periodontal protection. Periodontology 2000(13):91–120

Schroeder HE, Listgarten MA (2003) The junctional epithelium: from strength to defense. J Dent Res 82:158–161

Shimono M, Ishikawa T, Enokiya Y et al (2003) Biological characteristics of the junctional epithelium. Microscopy 52:627–639

Somermanb MJ, Archer SY, Imm GR, Foster RA (1988) A comparative-study of human periodontal-ligament cells and gingival fibroblasts invitro. J Dent Res 67:66–70

Squier CA (1981) Keratinization of the sulcular epithelium—a pointless pursuit. J Periodontol 52:426–429

Stern IB (1981) Current concepts of the dentogingival junction—the epithelial and connective-tissue attachments to the tooth. J Periodontol 52:465–476

Subbarao KC, Nattuthurai GS, Sundararajan SK et al (2019) Gingival crevicular fluid: an overview. J Pharm Bioall Sci 11:S135–S139

Warshawsky H (1988) The teeth. In: Weiss L (ed) Cell and tissue biology—a textbook of histology, 6th edn. Urban & Schwarzenberg, pp 594–640

Insertion Periodontium

21

What You Will Learn in This Chapter

The embryonic origin and histological organisation of the insertion periodontium, formed by the cementum, the periodontal ligament and the alveolar bone, is explained here. You will see that the three tissues are formed during the development of the tooth root from the ectomesenchymal cells of the dental sac. You will learn that cementum is a mineralised tissue very similar to bone, although several types of cementum can be distinguished, in some of which there are no cells. You will see that, unlike bone, cementum does not usually remodel and is very resistant to resorption. You will learn that the periodontal ligament is a connective tissue with groups of collagen fibres running through it, from the alveolar bone to the cementum. Finally, you will study that the alveolar bone is odontodependent and you will see that different sides can be distinguished and the organisation of each one.

21.1 General Characteristics

A. The *insertion* or *attachment periodontium* consists of *cementum*, *periodontal ligament* and *alveolar bone*.

B. The three structures are closely related and have the same origin from the *dental sac*, which is derived from the *ectomesenchyme.*

C. The three structures of the attachment periodontium develop and evolve in a coordinated manner and in relation to the movements of the tooth. When the tooth is lost, all three structures disappear:

   1. The cementum and most of the periodontal ligament are lost with the tooth.
   2. The remaining parts of the ligament, as well as the alveolar bone, are lost through tissue *regression.*

D. The junction between the tooth and the alveolar bone is a joint, the *alveolodental joint*, which is considered a *gomphosis*. It holds the tooth in place and allows it to resist the forces of mastication.

F. J. Sáez, I. Badiola, *Essential Oral Histology*, Learning Materials in Biosciences,
https://doi.org/10.1007/978-3-032-03165-5_21

21.2 Origin and Development of the Insertion Periodontium

A. The insertion periodontium is formed with the *tooth root* (Fig. 21.1).
   1. As we have already seen in Chap. 13, after the formation of the dental crown, *Hertwig's epithelial root sheath* develops by the union of the inner and outer dental epithelia of the cervical loops of the enamel organ.
   2. This sheath grows apically. As it grows, it surrounds the tissues of the dental papilla and induces the formation of *odontoblasts*, which begin to synthesise root *dentine.*
   3. The first layer of dentine to form, about 10–20 μm thick, constitutes the *hyaline layer* found on the surface of the root dentine. It is believed that this layer also contains typical enamel substances produced by Hertwig's root sheath cells.
   4. As the root sheath grows apically, it fragments in the cervical regions, where dentine is being produced.
      a. Fragmentation occurs because some cells of the sheath die by *apoptosis*.
      b. The spaces that occur are invaded by tissues of the dental follicle.

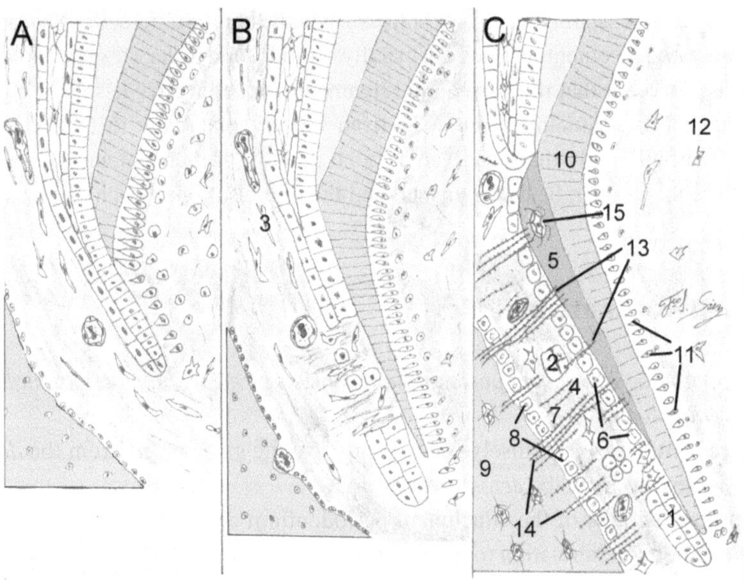

Fig. 21.1 Tooth root development: early stages of root dentine formation (**A, B**)—already seen in Fig. 13.12—and the beginning of the formation of the attachment periodontium (**C**). (1) Hertwig's epithelial root sheath; (2) epithelial cell rests of Malassez; (3) dental sac or follicle; (4) mesenchyme of the dental follicle invading the space between the epithelial cell rests of Malassez; (5) cementum; (6) cementoblasts; (7) periodontal membrane (periodontal ligament); (8) osteoblasts; (9) alveolar bone; (10) dentine; (11) odontoblasts; (12) dental pulp (dental papilla); (13) perforating fibres; (14) Sharpey's fibres; (15) cementocyte

5. From the dental sac that invades these spaces, the following are produced:
    a. *Cementoblasts*, close to the dentine, which begin to synthesise *cementum*. As the mineralisation of the hyaline layer of dentine is slow, a mixing of dentine and cementum collagen fibres can occur prior to mineralisation, which makes it easier for the cementum to become firmly attached to the dentine.
    b. *Fibroblasts*, which synthesise connective tissue with *collagen fibres*. This tissue has disordered fibres and is called *periodontal membrane*. Subsequently, the fibres of the tissue are organised into groups and form a very fibrous tissue, the *periodontal ligament*.
    c. *Osteoblasts*, from cells of the dental sac located slightly further away from the dentine, which synthesise *alveolar bone* by *intramembranous ossification* (see Chap. 7). This alveolar bone surrounds the periodontal ligament and root.
6. The apical end of Hertwig's sheath continues to grow and induces the formation of new dentine in more apical positions, while positions away from its apical end fragment and are invaded by new mesenchyme from the dental sac. The process repeats itself as the sheath grows.
7. The result of this process is the growth of the root (dentine) and the formation of the associated insertion periodontium.

B. In summary, all root and periodontal tissues are of *ectomesenchymal origin* from two different tissues:
    1. From the *dental papilla*: pulp and root dentine.
    2. From the *dental sac*: cementum, periodontal ligament and alveolar bone.

C. Some remnants of Hertwig's root sheath may remain in the periodontium. These *epithelial cell rests of Malassez* are the only tissues of the periodontium that originate from the *surface ectoderm*.

D. Although the previous explanation is the most widely accepted, there are other theories to explain the origin of the insertion periodontium.
    1. It has been proposed that Hertwig's epithelial root sheath cells may transform into cementoblasts in a process called *epithelial-mesenchymal transition*. This has not been demonstrated.
    2. It has been pointed out that acellular cementum and cellular cementum (two varieties of cementum explained below) have different origins:
        a. The *acellular cementum* would originate from Hertwig's sheath.
        b. The *cellular cementum* would be formed from cells from the dental sac.
    3. If these theories are accurate, cementum would have a dual origin from surface ectoderm and from ectomesenchyme.

E. Sometimes, remnants of Hertwig's epithelial sheath remain on the root surface producing small masses of an enamel-like material. These structures are called *enamel pearls*. They usually appear in the bifurcation areas of the root.

21.3 Cementum

A. *General characteristics.*

1. Cementum is the *mineralised tissue* that covers the dental root, from the neck to the apex. It is attached to the dentine on one side and to the periodontal ligament on the other. Its main function is to provide support and fixation to the fibres of the periodontal ligament.
2. Cementum is an *avascular tissue* that *lacks* innervation.
3. It is a tissue very similar to bone, due to its cells, the chemical composition of its matrix, and its histological organisation.
   a. However, lacking vessels and nerves, cementum differs from bone in that it lacks *Haversian canals* and *Volkmann canals*.
   b. Unlike bone, cementum *does not remodel* and is very *resistant to resorption*, which is essential for *orthodontic techniques*.
   c. Another difference with bone is that cementum can grow in thickness throughout the life of the tooth.
4. Cementum is deposited discontinuously, so it is deposited in layers separated by *imbrication* or *incremental lines of Salter* (actually resting lines), which are dark lines separating lighter layers of cementum.
5. The cementum is thicker in the apical regions (50–200 μm, although it may exceed 600 μm) than in the cervical regions (10–15 μm).
   a. Slow and continuous growth throughout the life of the tooth modifies the shape of the tooth root.
   b. The thickness increases with the age of the tooth.
6. In the apical region, in addition to covering the dentine, it may cover the final portion of the *root canal.*
   a. As a result, the root canal may become coated in its final part with cementum.
   b. This causes the root canal to have the shape of two cones facing each other at their apexes, i.e. with a constriction at the point where the canal begins to be lined with cementum. These two cones are:
      (1) The *dentine cone*.
      (2) The *cementary cone*. At the end of this is the *apical foramen*, the point where the dental pulp intermingles with the periodontal ligament and forms *Black's undifferentiated space* or *periapex*, in which there are *multipotent mesenchymal stem cells.*
7. There are three possible forms of relationship between cementum and enamel in the cervical region. These three forms of relationship are shown in Fig. 19.1.
   a. In 60% of teeth, cementum can be found overlying the end of the coronary enamel (*overlapping*).
   b. It can also be found in contact with the enamel, without covering it, which is observed in 30% of the pieces (*contact*).

c. In the rest of the teeth, there is a small gap between the cementum and the enamel, leaving a small strip of dentine uncovered by any mineralised tissue (*separation*).

B. *Physical properties.*
   1. Cementum is less mineralised than bone or dentine.
   2. It is yellowish and not shiny.
   3. Although it is lighter than dentine, it is difficult to differentiate from dentine in appearance, but it can be distinguished because it is softer.[1]
   4. Permeability depends on the type of cementum (cellular cementum is more permeable) and on age (decreases with age).
   5. As the tissue is somewhat soft and the cervical regions are thin, if the gingiva descends and exposes regions of cementum, these may be lost, exposing the dentine.

C. *Extracellular matrix.*
   1. The extracellular matrix of cementum consists of *organic matter* (20%), *inorganic matter* (45–50%) and *water* (32%):
   2. *Organic matter* is difficult to analyse due to the difficulty in obtaining sufficient sample to analyse. In it we find:
      a. *Type I collagen fibres*. They make up 90% of the matrix proteins. However, there may be small amounts of other forms of collagen, such as *type III collagen*. Fibres can be of two types:
         (1) *Intrinsic fibres*. They are so called because they are inside the cementum and do not exceed its limits. They are produced by the *cementoblasts* during cementum synthesis.
         (2) *Extrinsic fibres*. These are fibres of the periodontal ligament that penetrate the cementum matrix. Therefore, they are fibres produced by the *fibroblasts* of the periodontal ligament. They are also called *perforating fibres*. They are similar to Sharpey's fibres that penetrate the bone.
      b. *Proteoglycans*, such as lumican, versican, decorin, fibromodulin and biglycan. These substances are also found in bone. They are concentrated near the lacunae that house the cells and around the canaliculi that connect the lacunae.
      c. *Glycosaminoglycans*, mainly chondroitin sulphate, but also hyaluronan and dermatan sulphate. Chondroitin sulphate is concentrated around the cell-hosting lacunae.
      d. *Non-collagenous proteins*. Most of them are also present in bone. Among them, the existence of the following has been demonstrated:
         (1) *Cementum attachment protein (CAP)*. It is a protein that stimulates the attachment of mesenchymal cells to the matrix. It has been localised in the matrix and in cementoblasts. It is not found in bone.
         (2) *Alkaline phosphatase*.

[1] It is worth recalling the table comparing the hardness of tissues and minerals shown in Chap. 6.

(3) *Bone sialoprotein (BSP).*
(4) *Dentine matrix protein-1 (DMP1).*
(5) *Dentine sialoprotein (DSP).*
(6) *Fibronectin.*
(7) *Osteocalcin.*
(8) *Osteonectin.*
(9) *Osteopontin.* It is found in the incremental lines.
(10) *Tenascin.*
(11) *Growth factors*, although they are often found inactive or blocked in some way.

3. *Inorganic* or *mineral matter* is made up of:
   a. *Hydroxyapatite* (*calcium phosphate*). It is the major mineral component.
      (1) It forms smaller crystals than those of enamel or dentine, about 55 nm wide, 8 nm thick and variable in length.
      (2) The crystals lodge in and between the collagen fibres.
      (3) The hydroxyapatite in cementum is less crystalline than in other dental tissues and is therefore easier to demineralise.
   b. *Calcium carbonates* and some *trace elements* such as iron, sulphur and fluorine may also be present.
      (1) *Fluoride* levels are high, higher than in other mineralised tissues.
      (2) Fluoride levels are higher in acellular cementum than in cellular cementum.
      (3) These fluoride levels may be responsible for making the cementum *resistant to resorption.*

D. *Types of cementum.*
1. There are several types of cementum (Fig. 21.2):
   a. *Acellular extrinsic fibre cementum (AEFC)*, consisting only of a mineralised matrix in which no cells are included.
      (1) Called *primary cementum* because it is usually synthesised early in root formation.
      (2) It is also called *acellular cementum.* This name is inappropriate, as we shall see, because there are also other varieties of acellular cementum.
      (3) It has *extrinsic fibres*:
         (a) At the beginning of its formation, the cementoblasts produce fibres perpendicular to the surface of the dentine with which they contact.
         (b) Later, these fibres contact the principal fibres of the periodontal ligament produced by its fibroblasts.
         (c) At this point, the cementoblasts stop synthesising collagen fibres and only synthesise other matrix components.
         (d) Therefore, in mature cementum, the fibres penetrate from the periodontal ligament and are considered *extrinsic collagen fibres*, produced by fibroblasts.

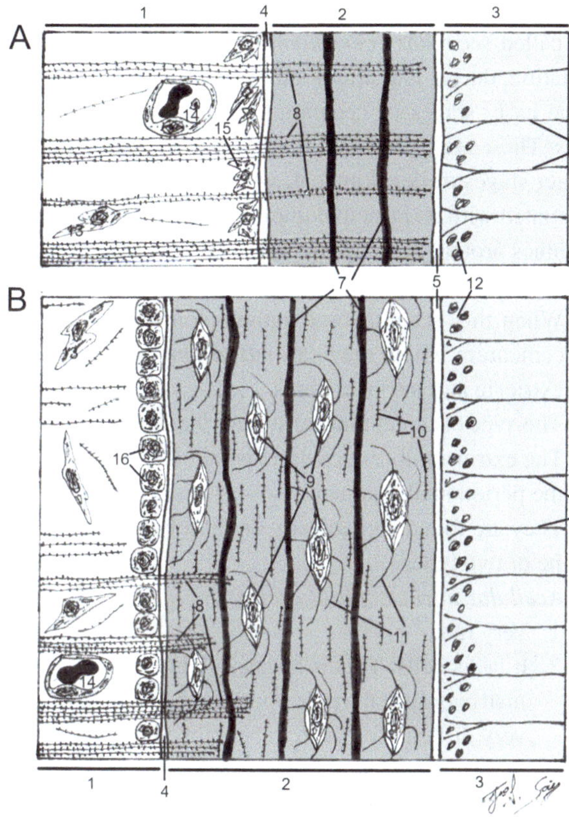

Fig. 21.2 Histological organisation of the two main varieties of cementum, acellular extrinsic fibre cementum (AEFC) and cellular intrinsic fibre cementum (CIFC). AEFC (**A**) lacks intrinsic collagen fibres and cementocytes included in the matrix, which is synthesised by fibroblast-like cementoblasts (15). Perforating fibres (extrinsic fibres, 8), synthesised by fibroblasts (13), penetrate from the periodontal ligament (1). The CIFC (**B**) has in its matrix intrinsic collagen fibres (10) and cementocytes (9) enclosed within lacunae. The matrix is synthesised by osteoblast-like cementoblasts (16). In early stages of development, CIFC lacks extrinsic fibres (top in **B**), but in later stages, fibres from the periodontal ligament are often entrapped in the cementum matrix as perforating or extrinsic fibres (8, bottom in **B**). (2) cementum; (3) dentine; (4) cementoid; (5) hyaline layer; (6) dentinal tubules; (7) imbrication or incremental lines of Salter; (11) calcophore ducts; (12) granular layer of Tomes; (14) blood capillaries

   (e) However, the deeper portion of these fibres must be considered to be produced by cementoblasts and could be considered intrinsic fibres.

(4) It lacks some of the glycosaminoglycans and proteins found in other types of cementum.

(5) It is usually found abundantly from the cervical region to the boundary with the apical third of the root. In front teeth, it may also cover apical regions of the root.

b. *Intrinsic fibre cementum*:
   (1) It is called *secondary cementum* because it is produced at the end of root formation, during and after tooth eruption.
   (2) In the early stages of cementum development, it contains only *intrinsic fibres*. These are collagen fibres arranged parallel to the dentine surface.
   (3) In later stages of development, when the periodontal ligament is well differentiated with its principal fibre groups well formed, cementum synthesis continues around these fibres which are integrated into the cementum surface and form *extrinsic fibres*:
      (a) When this occurs, some authors consider it to be another variety of cementum which they call *mixed fibre cementum*, as it contains both extrinsic and intrinsic fibres.
      (b) This type of cementum constitutes the majority of secondary cementum.
      (c) The extrinsic fibres penetrate perpendicularly into the cementum from the periodontal ligament.
      (d) They are found sporadically or densely arranged in parallel.
   (4) Can be of two types:
      (a) *Acellular intrinsic fibres cementum.*
         1. Very rare.
         2. It lacks cells. This is because the cementoblasts secrete cementum matrix only at the pole opposite the periodontal ligament. Thus, it is very difficult for them to be trapped in the cementum matrix.
      (b) *Cellular intrinsic fibre cell cementum (CIFC).*
         1. Contains cells called *cementocytes.*
         2. During their formation, cementoblasts secrete matrix around their perimeter, so they can be trapped by the secreted matrix. They thus become cementocytes.
         3. It is also called *cellular cementum.*
         4. It is abundant in the apical region of the root, and very abundant in the interradicular areas.
         5. This cementum is produced faster than AEFC.
         6. It is the only variety of cementum that can quickly repair a defect of the root caused by resorption. This is due to its high growth rate.

c. *Acellular and afibrillar cementum.* There may be small, thin areas of acellular cementum without collagenous fibrils.
   (1) Normally coats the enamel and dentine in the cervical region, in the area of the cementum-enamel junction.
   (2) Not involved in tooth attachment.
   (3) It is not clear how and why this type of cementum is produced, it is debated whether it can be produced by cells of *mesenchymal* origin or by epithelial cells originating from the *reduced enamel epithelium.*

   d. *Cellular mixed stratified cementum.* Appears in regions near the apex. It consists of alternating layers of AEFC and CIFC.
2. An *intermediate cementum*, synthesised by Hertwig's epithelial sheath cells on the dentine surface, has been described before cementum is produced by cementoblasts.
   a. This intermediate cementum is a hard, fibril-like, thin tissue, about 10 μm thick.
   b. It would be made up of enamel proteins.
   c. According to other authors, the cementum intermedium does not exist as such but is simply *acellular cementum* or a region where the collagenous fibres of dentine and cementum intertwine and form an *intermediate zone*.
   d. This layer is not observed on all teeth.
3. Cementum distribution:
   a. *Acellular cementum* is usually deposited more in the cervical region and without covering the apical third of the root, although a thin layer may also be deposited on the apical third.
   b. *Cellular cementum* is usually deposited in the apical region of the root, sometimes on a thin layer of acellular cementum.
   c. This arrangement is not unique, and many variations can be found.
      (1) In some roots, acellular cementum can be observed covering the cellular cementum.
      (2) In other roots, alternate layers of cellular and acellular cementum are seen.
4. There is controversy about the origin of the *cementoblasts* that form each type of cementum.
   a. Some authors believe that acellular cementum is synthesised by cementoblasts originating from *Hertwig's root sheath*, while cellular cementum is synthesised by cementoblasts of *ectomesenchymal origin*.
   b. This has not been proven with certainty, but it is true that the cementoblasts that produce both types of cementum differ phenotypically, as indicated in the following table.

Cementoblasts of acellular cementum	Cementoblasts of cellular cementum
Short life	Long life
Fibroblast-like	Osteoblast-like
Produce cytokeratins	They do not produce cytokeratins
They do not produce osteocalcin	They produce osteocalcin
They do not possess parathormone receptors.	Possess parathormone receptors
They express TGF-β and IGF growth factors.	They do not express TGF-β and IGF growth factors.

E. *Cells of the cementum.*
   1. Two cell types can be found in the cementum, *cementoblasts* and *cementocytes*.
   2. *Cementoclasts* or *odontoclasts* may sometimes be found in the close proximity of the cementum surface.

   a. They are cells very similar to *osteoclasts* and carry out *resorption* of dentine and cementum.
   b. These cells only appear in pathological situations (inflammation and excessive physical forces) and during the resorption of primary tooth roots (*rhizoclasia*).
3. *Cementoblasts.*
   a. Cementoblasts are found on the surface of the cementum, attached to it, at the boundary with the periodontal ligament and therefore partially included in it.
   b. They are organised as an epithelium.
   c. When active, they are cuboidal, highly basophilic cells.
      (1) Under the electron microscope, they have an excentric *nucleus* with one or two *nucleoli.*
      (2) They have many *mitochondria*, well-developed *rough endoplasmic reticulum* and *Golgi apparatus.*
      (3) There are usually many *glycogen granules*, *intermediate filaments* and *actin microfilaments*.
      (4) The cells are joined by *desmosome* and *gap* junctions.
   d. When inactive, they are flattened with a very heterochromatic nucleus.
   e. Between the cells and the mature cementum, there is usually a layer of unmineralised cementum called *cementoid* or *precementum.*
   f. These cells lack alkaline phosphatase activity, which osteoblasts have.
4. *Cementocytes.*
   a. Cementocytes are cementoblasts that are embedded in the mineralised cementum.
   b. Cementocytes are ovoid cells, with their major axis parallel to the longitudinal axis of the root, with small, dark, almost *pyknotic* nuclei and acidophilic cytoplasm.
   c. Electron microscopy shows very few organelles in the cytoplasm, with very few mitochondria and few dilated vesicles of the rough endoplasmic reticulum, indicating cells with little activity. In fact, there is no evidence that the cementocyte performs any activity in cementum maintenance.
   d. The cementocytes lodge in cavities in the matrix called *lacunae, cementocels* or *cementoplasts*.
   e. Empty cavities are sometimes observed due to the death of the cementocytes. This is more evident when the cementum is very thick, and is probably due to the fact that the cells die due to the difficulty of getting nutrients, as the cementum lacks blood vessels.
   f. From the lacunae arise *calcophore ducts* like those of osteocytes, so that the organisation of these cells and the organisation of the cementum is very reminiscent of the organisation of bone.
   g. The cementocytes have *cellular extensions* that lodge in the calcophore ducts and contact the extensions of neighbouring cementocytes.

(1) Most extensions and calcophore ducts, regardless of which side of the cell they are produced on, are usually directed towards the outer surface of the cementum: those coming from the inner side of the cementocyte make a turn and are directed towards the outer surface.
(2) This is because the nutrition of the cells comes from the vessels of the periodontal ligament.
(3) For this reason, cells located in deep regions of the cementum, where nutrients are more difficult to reach, may die.

F. *Resorption and repair.*
1. The cementum is very resistant to resorption. However, small areas of resorption may be found. The causes of this resorption are not known, but it is thought to be a response to minor trauma.
2. Resorption is carried out by cells called *cementoclasts* or *odontoclasts*, as they also resorb dentine.
3. Several theories have been proposed to explain the resistance of cementum to resorption:
   a. There is a *thin layer of non-mineralised matrix on the surface of the cementum*, with tightly packed collagen, which protects the cementum.
   b. The cementoblasts lining the cementum do not respond to hormonal stimuli (*parathormone*) like the osteoblasts lining the bone. Consequently, they do not detach from the surface of the cementum and prevent the action of the cementoclasts.
   c. In bone, bone-lining cells degrade the osteoid that coats the bone tissue as a preliminary step for the action of osteoclasts. There is no cell that performs this function in cementum.
4. If a pathological situation leads to resorption of cementum or dentine in the root, *repair* can occur through the synthesis and deposition of cementum by cementoblasts (Fig. 21.3), which are programmed to keep the root surface smooth.
   a. At the point where cementum resorption stops and new cementum is deposited, the so-called *inversion line* or *reversal line* is observed.
   b. If the cementum repair is successful, the *perforating fibres* of the periodontal ligament are integrated into the surface of the new cementum and the original functionality of the periodontium is restored.
   c. If the cementum repair is not correct, the cementum may fuse with the alveolar bone, with loss of the periodontal ligament that separates them, resulting in a loss of joint mobility or *ankylosis* (from the ancient Greek ἀγκύλωσις ankýlōsis, 'welding').
   d. With cementum resorption, some dentine may be lost, which may be replaced by new cementum when it is re-deposited.
   e. Only *CIFC* is produced during cementum repair.

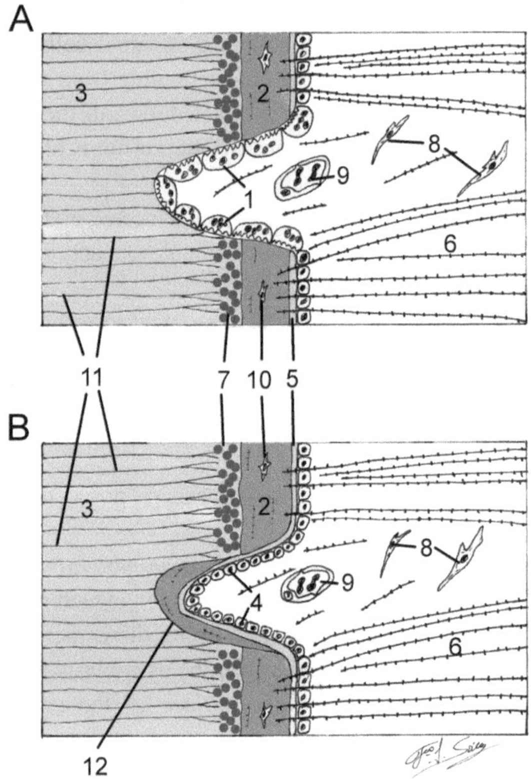

Fig. 21.3 Pathological resorption in the tooth root (**A**) and subsequent repair (**B**). Resorption is carried out by odontoclasts (1), which destroy cementum (2) and dentine (3). If conditions permit, the root can be repaired by the activity of new cementoblasts (4) that synthesise new cementum, which replaces the resorbed dentine and cementum. (5) cementoid; (6) periodontal ligament; (7) granular layer of Tomes; (8) fibroblasts; (9) blood vessels; (10) cementocytes; (11) dentinal tubules; (12) inversion or reversal line

5. With age, the cementum changes.
   a. Its smooth surface becomes rougher and rougher.
   b. The constant deposit of cementum can obstruct the *apical foramen.*
   c. The deeper lacunae may be empty, and only the more superficial ones are occupied by living cells.
   d. Although cementum is resistant to resorption, aged cementum can be partially resorbed. This process is cyclical, so that cementum is lost, resorption stops and cementum is redeposited in the resorbed area.

21.4 Periodontal Ligament

A. The periodontal ligament surrounding the root is a fibrous connective tissue. Its principal fibres contact the alveolar bone and cementum, into which they are inserted.
B. It has also been called *desmodont, alveolodental ligament, pericementum, periodontal fibre* and *dentoalveolar gomphosis.* Its functions are:
   1. *Clamping* the tooth. If the ligament is lost or fails, the tooth falls out.
   2. *Cushion* or *resist masticatory forces*, thanks to the fibres and the ground substance.

3. *Receive sensations* and send them to the brain for the correct functioning of mastication. This is done thanks to its *innervation*. However, this feature is poorly understood in detail. Stimulation of receptors in the ligament causes reflex opening of the jaw.
4. *Nourish* the ligament itself, the cementum and the alveolar bone, thanks to its *blood vessels*.
5. *Maintain* teeth and periodontium.
   a. Cells of the periodontal ligament (*cementoblasts* and *osteoblasts*) are responsible for *regenerating* cementum and alveolar bone if necessary.
   b. In addition, the periodontal ligament has *mesenchymal cells* capable of tissue regeneration.
6. *Maintain adequate thickness* throughout the life of the tooth, although thickness gradually decreases with age.
7. *Limit the mineralising activity* of osteoblasts and cementoblasts to the bone and tooth, respectively, in order to keep the periodontal ligament unmineralised, even though it is located between two mineralised tissues.
   a. There is evidence that fibroblasts in the periodontal ligament produce substances that regulate the activity of adjacent mineralised tissues to prevent them from proliferating and invading the space occupied by the ligament.
   b. When this regulation fails, the ligament mineralises and dental *ankylosis* occurs.
8. *Adapt to functional activity*.
   a. If the activity increases, the ligament and its fibre bundles thicken.
   b. If activity decreases, ligament and fibre bundle thickness decreases.

C. At the apex of the tooth, the periodontal ligament contacts the connective of the *dental pulp*, while in the cervical region of the tooth, it mixes with the gingival lamina propria (*gingival ligament*), allowing all these structures to form a functional system.

D. The width of the periodontal ligament varies for each individual and for each tooth:
1. It can be between 0.15 and 0.38 mm.
2. It usually decreases with age:
   a. In individuals aged 11–16 years, it averages 0.21 mm.
   b. From age 32 to 52, 0.18 mm.
   c. From 51 to 67 years old, 0.15 mm.

E. There is a narrower area of the ligament that serves as a support or lever for lateral movements and is approximately at the boundary between the middle third and the apical third of the root. This area is called the *fulcrum* (from Latin fulcrum, 'support').

F. *Cells of the periodontal ligament*.
1. As a connective tissue, the periodontal ligament may contain *resident* and *transient* cells, but it may also contain other cells.
2. *Fibroblasts*:
   a. Fibroblasts are the fundamental cells of the ligament, as they synthesise the matrix and renew and repair the tissue.
   b. They are of *ectomesenchymal origin*.

c. They usually have the characteristics of active fibroblasts with highly developed biosynthetic organelles, and *protein synthesis* has been found to be high.
   (1) As a consequence of this activity, the fibroblasts of the periodontal ligament perform a very rapid *renewal* of the matrix, especially of the collagen fibres, which are constantly being remodelled.
   (2) The differences between periodontal ligament fibroblasts and gingival fibroblasts were outlined in Chap. 20.
d. A peculiarity of fibroblasts in this tissue is that they show a *cilium* with an internal *9+0* structure, instead of the usual 9+2 structure, i.e. lacking the central microtubules.
e. Fibroblasts are connected to each other with many intercellular contacts, *desmosomes* and *gap junctions*, which is unusual in other tissues.
f. It has been proposed that periodontal ligament fibroblasts have contractile and migration capabilities that are important in participating in tooth movements during and after eruption.
   (1) Contractile activity has been demonstrated in cell culture, where cells can take on the appearance of *myofibroblasts*.
   (2) In tissues, they never appear as myofibroblasts, although they have been described as having *actin microfilaments*.
   (3) Fibroblasts have been shown to have some ability to move along collagen fibres in an apical-coronal direction.

3. *Undifferentiated mesenchymal cells*:
   a. In the periodontal ligament, there are undifferentiated mesenchymal cells in the vicinity of the *blood vessels* that are capable of producing fibroblasts, osteoblasts and cementoblasts.
   b. This replaces the cells that have died by *apoptosis*.
   c. What is not clear is whether there are different populations of mesenchymal cells, each capable of renewing a single cell type, or whether all mesenchymal cells are the same and can produce fibroblasts, osteoblasts and cementoblasts.
   d. Some of these cells may be *pluripotent stem cells*, capable of forming other tissues with the right stimuli.
      (1) It has been shown that these cells can transform into *adipocytes*, *cementoblasts*, *osteoblasts* and *chondroblasts*.
      (2) These cells are of interest as they can be used for the regeneration of tissues of the individual and are relatively easy to obtain.
4. *Cells from bordering tissues*:
   a. At the border of the ligament with neighbouring tissues, there are cells associated with these tissues.
      (1) *Cementoblasts* on the surface in contact with the cementum (Fig. 21.2).
      (2) *Osteoblasts* on the contact surface with the alveolar bone (Fig. 21.1C). In the absence of active bone synthesis, osteoblasts become *bone-lining cells*.
   b. If root resorption occurs, *cementoclasts* (*odontoclasts*) may be observed at the cementum boundary (Fig. 21.3).

5. *Epithelial cells*:
   a. Residual cells of Hertwig's epithelial sheath, which form the *epithelial cell rests of Malassez*, can be observed in the periodontal ligament.
      (1) These epithelial remnants are clusters of cuboidal cells, which in histological sections look like clusters, but have been shown to form intermingled parallel cords, which form a *reticulum* about 25 μm from the surface of the cementum.
      (2) The cell clusters are separated from the connective tissue by a *basal lamina*. The cells show *tonofilaments*, *desmosomes*, *hemidesmosomes* and *tight junctions*.
   b. These cells may play a role in the initial development of the cementum and may be involved in the maintenance of the mature ligament.
6. *Defence cells: Macrophages*, *mast cells*, *eosinophils* and *dendritic cells* are usually found in the periodontal ligament.

G. *Extracellular matrix of the periodontal ligament*.
   1. The extracellular matrix contains *type I collagen fibres* (70% of the total), but also some *type III* (20%).
   2. Type III fibres are found preferentially around blood vessels, nerves and Sharpey's fibres that penetrate the alveolar bone.
   3. *Type XII collagen* has also been described.
      a. Type XII collagen does not form fibres, but serves to bind other collagen fibres.
      b. It arises in the tooth ligament after eruption and appears to be involved in maintaining the integrity of the periodontal ligament.
      c. When orthodontic treatments are performed, type XII collagen is concentrated on the side of the ligament under pressure.
   4. Collagen fibres are organised in two ways:
      a. *Principal fibres*. These are fibres that are grouped in bundles about 5 μm thick, with different orientations in each area of the periodontal ligament.
      b. *Secondary fibres*. These are fibres that do not clump together and are apparently disordered.
   5. The principal fibres form different groups according to their orientation (Fig. 21.4):
      a. *Alveolar crest group* or *dentoalveolar crest fibres*.
         (1) These fibres mingle with those of the gingival ligament, as they are arranged in an upward direction from the alveolar crest (the top of the alveolar bone) to the cementum-enamel contact zone.
         (2) They prevent *extrusion movements* of the tooth.
      b. *Horizontal group*.
         (1) Perpendicular to the tooth root surface.
         (2) They are located in two places: one group below the crestoalveolar group and the other near the root apex.
         (3) They resist *horizontal forces relative to* the tooth and prevent lateral movements.

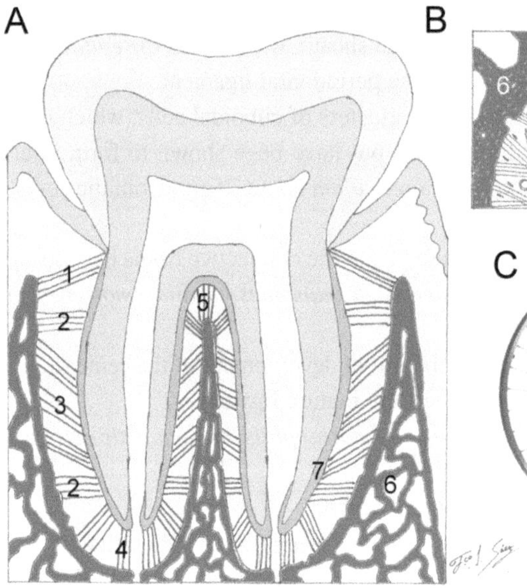

Fig. 21.4 Principal fibres of the periodontal ligament (**A**), cross section of a dental root to show the direction of the periodontal ligament fibres (**B**) and image of a bicycle wheel to show the arrangement of the spokes (**C**). (1) Alveolar crerst group (dentoalveolar crest fibres); (2) horizontal group; (3) oblique group; (4) apical group; (5) interradicular group; (6) alveolar bone; (7) cementum

c. *Oblique group.*
   (1) It is the largest group of major fibres.
   (2) The fibres run from the alveolar bone to the cementum in a downward direction (towards the root apex).
   (3) They hold the tooth in its socket and allow it to withstand *masticatory* and *intrusion forces.*
   (4) They are the most powerful fibres of all.
d. *Apical group.*
   (1) They are located at the apex, emerge from the cementum surrounding the apical foramen and are directed towards the deepest part of the alveolus.
   (2) These fibres prevent *lateral* and *extrusion movements* and at the same time dampen *intrusion forces*, with the help of the matrix *ground substance.*
   (3) In this area, the connective tissue is more loose, allowing access for *blood vessels* and *nerves.*
e. *Interradicular group.*
   (1) These fibres are observed exclusively in *multi-rooted teeth.*
   (2) They fan out from the crest of the interradicular septum towards the cementum.
   (3) They prevent *lateral and rotational movements.*

6. Between the principal fibre groups, there are so-called *interstitial spaces*, occupied by *blood vessels*, *lymphatic vessels*, *nerves*, *cells* and *secondary fibres*.
7. The principal fibres run from the alveolar bone to the cementum and cross the entire width of the periodontal ligament.
   a. When viewed in transverse sections of the root, the fibres are arranged like the spokes of a bicycle wheel, as they do not run directly from the bone to the cementum by the shortest route and do not contact the bone or cementum surfaces at right angles.
   b. In addition, they may cross paths with other fibres along the way.
   c. This makes it possible to better resist the rotational forces on the tooth.
8. The principal fibres penetrate the bone at one end and the cementum at the other:
   a. The fibres that penetrate the bone are called *perforating* or *Sharpey's fibres*.
   b. The fibres that penetrate the cementum are called *perforating fibres* and constitute the *extrinsic fibres of the cementum*. These fibres that penetrate the cementum have also been called Sharpey's fibres, like those of bone.
   c. In this book, we will keep a different name for the two types of fibres to avoid confusion:
      (1) When *Sharpey's fibres* are mentioned, this refers to fibres that penetrate the *bone*.
      (2) *Perforating fibres* are those that penetrate the *cementum*.
9. In the human periodontal ligament, there are few *elastic fibres*, which are arranged around the *blood vessels*. In some animals, they are very abundant.
10. *Oxytalan fibres* have been found.
   a. Oxytalan fibres usually connect cementum or bone to blood vessels, and most are arranged parallel to the tooth surface.
   b. They are very abundant in the cervical region, where, on reaching the gingiva, they intertwine with the fibres of the *gingival ligament*.
11. *Elaunin fibres* have also been described in association with blood vessels.
12. *Hyaluronic acid, dermatan sulphate, chondroitin sulphate* and *heparan sulphate* have been identified in the *ground substance* of the periodontal ligament.
13. Two *proteoglycans* have been isolated, one similar to decorin and one similar to biglycan.
14. *Adhesive glycoproteins* typical of connective tissue have also been identified, such as *fibronectin* (evenly distributed throughout the ligament during and after eruption), *tenascin* (concentrated in the vicinity of the alveolar bone and cementum) and *vitronectin* (associated with elastic fibres).

H. *Other structures of the periodontal ligament.*
   1. In addition to cells, the *interstitial spaces contain* blood vessels, lymphatic vessels and nerves.
   2. *Blood vessels*:

a. The periodontal ligament is an exceptionally well-vascularised connective tissue, probably due to the high degree of cellular and molecular turnover of the tissue.
b. The periodontal ligament is supplied by several arteries or arterioles. In most cases, these are vessels that also supply the gingivae and are shown in Fig. 20.5:
   (1) Arteries leading to the dental pulp and penetrating the periodontal ligament through the *periapices.*
   (2) Arteries of the *alveolar bone* that produce branches that leave the alveolar bone and penetrate the ligament. They are called perforating arteries.
   (3) Arteries originating from the *gingiva* and supplying the area closest to the crown.
c. Many blood vessels are oriented in an axial direction (apex-neck), more or less parallel to the tooth surface, with transverse connections, and vascularisation is more abundant in the area close to the bone than in the area close to the cementum.
d. The arterioles form a *capillary plexus* around the tooth.
e. A *postcapillary plexus* has also been described near the alveolar bone, consisting of *postcapillary venules* penetrating the bone.
f. A characteristic feature of the capillaries of the periodontal ligament is that they are usually *fenestrated capillaries*. This may be due to the metabolic requirements produced by the high synthetic activity of fibroblasts.
g. The venous circulation consists mainly of *postcapillary venules*.
   (1) Hardly any veins are seen circulating parallel to the arterioles, as most of the postcapillary venules penetrate the alveolar bone where they join the veins of this tissue.
   (2) There are also venules in the region close to the gingiva that mix with the veins in this area.
h. *Arteriovenous anastomoses* are abundant.
i. Lymphatic vessels usually run parallel to the arterial circulation.

3. *Periodontal ligament nerves*:
   a. Nerves usually accompany blood vessels and have two origins:
      (1) Some originate from the *periapical region* and move up the ligament towards the gingiva.
      (2) Others enter the periodontal ligament from the *alveolar bone* and divide into branches that ascend towards the gingiva and branches that descend towards the apex.
   b. There are myelinated and amyelinated fibres.
      (1) *Myelinated fibres* are all *sensory fibres.*
      (2) *Amyelinated fibres* can be sensory or *autonomic nervous system fibres.* Autonomic nervous system fibres innervate *blood vessels. Sympathetic fibres* have been shown to exist, but not parasympathetic fibres.

c. Sensory fibres are receptors for pain (free endings), touch, pressure, movement and position.
d. Four types of nerve endings have been observed:
   (1) *Free sensory endings*, which are the most frequent. They are arborescent and distributed throughout the periodontal ligament.
   (2) *Ruffini corpuscles*, which are abundant and predominate in the apical region.
   (3) *Coiled endings*, present in the medial region, which have not been studied in detail.
   (4) Structures similar to *Pacinian corpuscle*, which are very rare.

21.5 Alveolar Bone or Alveolar Process

A. The *alveolar bone, alveolar process* or *alveolar ridge* is the part of the upper and lower jaw bones that supports the dental roots.
B. The alveolar process consists of bone that surrounds and contains the *dental alveolus (or tooth socket)*, which is the receptacle that houses the root of the teeth.
C. The alveolar bone is often considered to be composed of two parts: the *alveolar bone proper*, which is the side of the alveolar process that contacts the tooth, and the *supporting alveolar bone*, which is the rest of the alveolar bone.
D. There is no precise anatomical boundary between the alveolar process and the basal bone of the body of the mandible or maxilla.
E. The alveolar bone is *odontodependent*: it forms with the tooth and is destroyed if the tooth is lost.
F. The terms used to describe the components of alveolar bone are varied. To avoid confusion, when reference is made to a structure of the alveolar bone, their alternative names should be given.
G. From an anatomical point of view, the alveolar processes follow the arched contour of the jaws, forming the walls of the dental alveoli. The alveoli can be:
   1. *Simple*, with a cavity to accommodate the *uniradicular teeth.*
   2. *Composites*, with several cavities (*diverticula*) separated by *internal partitions*, to accommodate the roots of *multi-rooted teeth* (two or three cavities for teeth with two or three roots, respectively).
H. Two types of walls can be distinguished in each alveolus (Fig. 21.5):
   1. *Free alveolar walls* (or *free alveolar tables)*. These are the walls of the vestibular (buccal or labial) and lingual or palatal surfaces.
      a. These walls have one side facing the alveolus and a free side facing the vestibule or oral cavity (lingual or palatal).
      b. Each alveolus has two types of free alveolar walls:
         (1) *Vestibular wall.*
         (2) *Lingual wall* (lower jaw) or *palatal wall* (upper jaw).

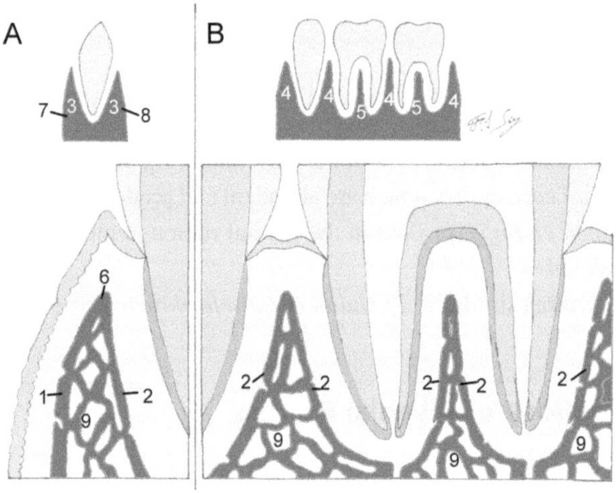

Fig. 21.5 Structure of the alveolar bone in vestibulo-lingual (**A**) and mesiodistal (**B**) sections. A general view is shown above and magnified details below. (1) periosteal plate; (2) periodontal plate; (3) free alveolar wall or tables; (4) interdental septa (septa); (5) interradicular septa or intersepta; (6) alveolar crest; (7) vestibular wall; (8) lingual wall; (9) spongy bone

2. *Alveolar septa*. These bone formations separate the roots of the teeth.
   a. They are best seen in mesiodistal sections.
   b. They can be of two types:
      (1) *Interdental septa, septa* (singular septum, from the Latin saeptum 'hedge, fence, wall') or *interdental bones*. They separate the alveoli of two neighbouring teeth.
      (2) *Interradicular septa, intersepta* (singular interseptum) or *interradicular bones*. They separate the diverticula of a multi-rooted tooth.

I. Organisation of bone tissue in the alveolar process:
   1. When viewed in cross-sections that run from side to side and parallel to the longitudinal axis of the tooth, all of the walls and septa are *triangular in shape*.
      a. The base of the triangle contacts the jaw bone.
      b. The vertex furthest from the jaw bone is called the *alveolar crest (alveolar border* or *alveolar margin)*, which is located near the cervical region of the tooth.
   2. In the free alveolar walls:
      a. The sides of the triangle that emerge from the alveolar crest form the two slopes of the table, which are:
         (1) *Periosteal plate*. This is the *free aspect* (vestibular, lingual or palatal), which is made up of compact bone tissue lined by *periosteum*.
            (a) It has also been called a *cortical plate*.
            (b) Some authors refer to it as '*supporting alveolar bone*', but this nomenclature will not be used here.
         (2) *Periodontal plate*. This is the *alveolar aspect*, which has compact bone lined by the *periodontal ligament* and lacks periosteum.

(a) Some authors call this plate '*alveolar bone proper*' or '*bundle bone*', but this nomenclature will not be used here to avoid confusion.

(b) The periodontal plate is also called *lamina dura* because it is more opaque to X-rays. This opacity does not seem to be due to a higher mineral content, but to the bone being very compact, although it is also because the x-rays must pass through more bone due to the angle at which they pass through.

b. The periodontal plate and the periosteal plate of a free wall contact at the *alveolar crest*, which they reach from converging directions. Some *cancellous* or *spongy bone* is usually found between them.

(1) The alveolar crest and the periosteal table are lined by the gingiva and the dentogingival junction.

(2) The periodontal table, as mentioned above, contacts the periodontal ligament.

3. In the interdental and interradicular septa:
   a. The two sides of the septum are in contact with the periodontal ligament and are formed by *compact bone*, which is called the *periodontal plate*.
   b. Both sides contact at the alveolar crest, and there is usually *spongy bone* between them.
4. *Periodontal plates* are characterised by the fact that the collagen fibres of the periodontal ligament penetrate the bone. These are the so-called *Sharpey's fibres* mentioned above.
   a. Sharpey's fibres are mineralised at the periphery, but not in the interior.
   b. This bone is constantly being remodelled. The newly formed bone lacks Sharpey's fibres, but new fibres are continually forming and penetrating the bone.
   c. The part of the periodontal table that contacts the periodontal ligament and contains Sharpey's fibres is characterised by fewer intrinsic fibres than usual that intertwine with each other. It is therefore not typical lamellar bone, but rather *non-lamellar woven bone* (see Chap. 6).
   d. This woven bone rests on typical lamellar bone, but sometimes all or almost all of the periodontal table is made up of woven bone.
   e. This distribution of lamellar and woven bone is variable, as bone is constantly remodelling, so many different situations can be seen.
   f. *Blood vessels* and *nerves* enter and exit the bone from and to the periodontal ligament through small perforations or foramina, which is why the periodontal plate has also been called *cribriform plate* (from Latin cribrum, 'sieve').
5. *Periosteal plates* consist of superficial bony lamellae overlying bone tissue with Haversian systems.
6. Both in the free walls and in the septa, between the compact plates there is *spongy bone* with *trabeculae*, which, if they are large, may have osteons.
   a. Between the trabeculae, there is *yellow bone marrow*, with many *adipocytes*, although sometimes there may be *red (haemopoietic) bone marrow*.
   b. In anterior teeth, the compact plates may be fused, and therefore the alveolar processes lack spongy bone between them.

Take-Home Message

The attachment periodontium consists of cementum, periodontal ligament and alveolar bone, all of which are formed by the dental sac during root development. Cementum is a mineralised bone-like tissue, synthesised by apposition by cementoblasts and with cementocytes located in lacunae in the matrix, which may contain intrinsic fibres (synthesised by the cementoblasts) and extrinsic fibres (synthesised by the fibroblasts of the periodontal ligament). A distinction can be made between an acellular extrinsic fibre or primary cementum (AEFC) and an intrinsic fibre or secondary cementum, acellular or cellular (CIFC). The periodontal ligament attaches the dental root to the alveolar bone. It has groups of main collagen fibres with different orientations that arise from the alveolar bone and insert into the cementum. The alveolar bone is an odontodependent structure that forms the walls of the dental alveolus where the roots of the tooth are placed. It is formed by free alveolar tables, which are the walls of the vestibular or labial and lingual or palatal sides, and alveolar septa, which separate the contiguous roots of different teeth (interdental septa) or of the same tooth (interradicular septa).

Further Reading

Arzate H, Zeichner-David M (2021) Cementum proteins beyond cementum. Extracell Matrix Biomineraliz Dental Struct 10:157–217

Berkovitz BKB, Holland GR, Moxham BJ (2018a) Alveolar bone. In: Oral anatomy, histology & embryology, 5th edn. Elsevier, pp 237–260

Berkovitz BKB, Holland GR, Moxham BJ (2018b) Cementum. In: Oral anatomy, histology & embryology, 5th edn. Elsevier, pp 195–206

Berkovitz BKB, Holland GR, Moxham BJ (2018c) Development of the root and periodontal ligament. In: Oral anatomy, histology & embryology, 5th edn. Elsevier, pp 402–420

Berkovitz BKB, Holland GR, Moxham BJ (2018d) Periodontal ligament. In: Oral anatomy, histology & embryology, 5th edn. Elsevier, pp 207–236

Bosshardt DD, Selvig KA (2000) Dental cementum: the dynamic tissue covering of the root. Periodontology 13:41–75

Chiego DJ (2023a) Cementum. In: Essentials of Oral histology and embryology, 6th edn. Elsevier, pp 131–138

Chiego DJ (2023b) Periodontium: alveolar process and cementum. In: Essentials of oral histology and embryology, 6th edn. Elsevier, pp 149–157

Chiego DJ (2023c) Periodontium: Periodontal Ligament. In: Essentials of oral histology and embryology, 6th edn. Elsevier, pp 139–148

Davidovitch Z, Nicolay OF, Ngan PW, Shanfeld JL (1988) Neurotransmitters, cytokines, and the control of alveolar bone remodeling in orthodontics. Dent Clin N Am 32:411–435

De Jong T, Bakker AD, Everts V, Smith TH (2017) The intricate anatomy of the periodontal ligament and its development: lessons for periodontal regeneration. J Periodontal Res 52:965–974

Fehrenbach MJ, Popowics T (2020) Periodontium: cementum, alveolar process, and periodontal ligament. In: Illustrated dental embryology, histology, and anatomy, 5th edn. Elsevier, pp 175–196

Foster BL, Nociti FH, Somerman MJ (2022) Development and structure of cementum. In: Naji S, Rendu W, Gourichon L (eds) Dental cementum in anthropology. Cambridge University, pp 46–64

Gómez de Ferraris ME, Campos Muñoz A (2009) Periodonto de inserción: cemento, ligamento periodontal y hueso alveolar. In: Histología, Embriología e Ingeniería Tisular Bucodental, 3rd edn. Médica Panamericana, pp 355–392

Hand AR (2014) Structure and physiology of the periodontium. In: Hand AR, Frank ME (eds) Fundamentals of oral histology and physiology. Wiley Blackwell, pp 115–135

He W, Fu Y, Yao S, Huang L (2023) Programmed cell death of periodontal ligament cells. J Cell Physiol 238:1768–1787

Hirashima S, Kanazawa T, Ohta K, Nakamura K (2020) Three-dimensional ultrastructure imaging and quantitative analysis of the periodontal ligament. Anat Sci Int 95:1–11

Hirashima S, Ohta K, Togo A, Nakamura K (2022) 3D mesoscopic architecture of a heterogeneous cellular network in the cementum-periodontal ligament-alveolar bone complex. Microscopy 71:22–33

Iwayama T, Sakashita H, Takedachi M, Murakami S (2022) Periodontal tissue stem cells and mesenchymal stem cells in the periodontal ligament. Japn Dental Sci Rev 58:172–178

Krstić RV (1991) Digestive apparatus. In: Human microscopic anatomy. An atlas for students of medicine and biology. Springer Verlag, pp 161–255

Martins JFB, Rodrigues CFD, Diogo P et al (2021) Remodelling compartment in root cementum. Folia Morphol (Warsz) 80:972–979

Nanci A (2013) Periodomtium. In: Ten Cate's oral histology, 8th edn. Elsevier, pp 205–232

O'Dowd G, Bell S, Wright S (2023) Oral tissues. In: Wheater's functional histology. Elsevier, pp 241–252

Saffar JL, Lasfargues JJ, Cherruau M (1997) Alveolar bone and the alveolar process: the socket that is never stable

Sokos D, Everts V, de Vries TJ (2015) Role of periodontal ligament fibroblasts in osteoclastogenesis: a review. J Periodontal Res 50:152–159

Takahashi S, Yamamoto T, Takahashi T, Yawaka Y (2023) Incremental lines in human cellular cementum: a histological study. J Oral Biosci 65:55–61

Ten Cate AR (1996) The role of epithelium in the development, structure and function of the tissues of tooth support. Oral Dis 2:55–62

Trujillo AD, DeCasien AR, Janiak MC et al (2022) A comparative genetic analysis of acellular cementum. In: Naji S, Rendu W, Gourichon L (eds) Dental cementum in anthropology. Cambridge University, pp 83–93

Warshawsky H (1988) The teeth. In: Weiss L (ed) Cell and tissue biology—a textbook of histology, 6th edn. Urban & Schwarzenberg, pp 594–640

Wen XY, Pei F, Jin Y, Zhao ZH (2025) Exploring the mechanical and biological interplay in the periodontal ligament. Int J Oral Sci 17:23

Yamamoto T, Hasegawa T, Hongo H, Amizuka N (2019) Alternating lamellar structure in human celular cementum and rat compact bone: its structure and formation. J Oral Biosci 61:105–114

Tooth Eruption

22

What You Will Learn in This Chapter

In this chapter, you will study how primary and permanent teeth erupt, what mechanisms are involved in tooth movements, and what phases of tooth eruption can be distinguished. You will finish by learning how the replacement of primary teeth by permanent teeth takes place, thanks to root resorption (rhizoclasia) and the fall of the primary tooth (exfoliation).

22.1 Essential Notions

A. As indicated in previous chapters, there are two types of dentitions in humans:
   1. *Primary, deciduous* or *milk dentition*.
      a. It appears after birth, during *early childhood*.
      b. It is made up of 20 teeth, arranged by quadrant or hemiarch in 2 *incisors*, 1 *canine* and 2 *molars*.
   2. *Secondary* or *permanent dentition*.
      a. It appears progressively during the *second childhood* and *puberty*.
      b. It is made up of 32 teeth. The hemiarch consists of 2 *incisors*, 1 *canine*, 2 *premolars* and 3 *molars*.
      c. As indicated in Chap. 13, two types of permanent teeth can be considered:
         (1) *Successional teeth*.
            (a) Some permanent teeth replace primary teeth, which fall out due to root loss (*rhizoclasia*).
            (b) Successional teeth are the incisors, canines and premolars, which are formed from the *succession lamina*.

F. J. Sáez, I. Badiola, *Essential Oral Histology*, Learning Materials in Biosciences,
https://doi.org/10.1007/978-3-032-03165-5_22

(2) *Accessional teeth.*
   (a) Other permanent teeth emerge in new positions after the primary teeth, due to the growth of the jaws, and do not replace other teeth.
   (b) The accessional teeth are the molars, which are formed from the *accessional lamina*, which is produced by the backward growth of the general lamina.

B. The existence of two dentitions makes us *diphyodonts* (from the ancient Greek διπηυοσ diphyos 'duplicate').
C. The replacement of primary teeth with permanent teeth can take up to six years.
   1. Each new tooth develops at a different time and each replaced primary tooth is exfoliated at a different time.
   2. During this period, there may be primary teeth coexisting with permanent teeth. This is called *mixed dentition.*
D. The *eruption* process of a primary tooth and a permanent tooth is similar. The difference is that the eruption of the permanent tooth requires prior *exfoliation* of the primary tooth

22.2 Eruption Mechanism

A. During tooth eruption, the tooth migrates from inside the jaw and makes its appearance in the oral cavity. This process involves the movement of the tooth, but also histological changes and the development of certain structures.
B. Eruption is a complex process, and several mechanisms have been proposed to explain it:
   1. *Root growth and development.*
      a. Root growth would raise the tooth into the oral cavity.
      b. Although this mechanism seems obvious, when roots are experimentally removed from growing teeth they are able to erupt.
   2. *Bone remodelling and alveolar bone development.*
      a. During tooth eruption, the bone next to the tooth is modified. Some authors have proposed that the deposit of new bone would push or displace the tooth.
      b. It seems unlikely that this mechanism influences the eruption.
         (1) It has been shown that resorption rather than bone apposition predominates during eruption, so it seems difficult to explain that the bone carries the tooth into the oral cavity.
         (2) During the eruption of the permanent tooth, bone is removed in order to open a channel that allows it to leave the *bony* crypt. In some experiments, it has been shown that this channel opens even if the tooth has previously been removed, so the bone is not resorbed by the pressure of the rising tooth, but independently.

3. *Periodontal connective tissue pressure.*
   a. This pressure would be due to the increase in volume due to vascular supply to the *tissue fluid.*
   b. When an unerupted tooth is removed and replaced with a metal or silicone tooth, it erupts normally.
   c. During eruption, the dental follicle and the periodontium produced from it have *fenestrated capillaries*, which could lead to increase in hydrostatic pressure in the periodontal tissue.
   d. It has also been reported that, during tooth eruption, the periodontium has different compositions of *proteoglycans* and *glycosaminoglycans*, which may therefore influence tissue hydration.
4. *Traction of the periodontal connective tissue.* This traction could be due to two causes:
   a. The formation of new *collagen fibres.* It has been proposed that collagen fibres may have contractile activity, but this has not been demonstrated. If erupting teeth are treated with drugs that prevent the formation of collagen fibres, eruption occurs normally, indicating that traction by collagen fibres does not play an essential role.
   b. The contractile activity of *fibroblasts.* There is not much evidence to support these ideas.
      (1) It is possible that fibroblasts or *myofibroblasts* can contract and transmit this contraction to the collagen fibres.
      (2) It has also been suggested that the migration of these cells performs some kind of traction on the tooth.

C. No definitive evidence has been found to support one or other of the above mechanisms. It does seem clear that:
   1. Eruption is mainly due to the action of the dental follicle and its derivatives (periodontium).
   2. It is a multifactorial process.
   3. Fibroblasts and the hydrostatic pressure of the periodontal ligament are somehow involved.

D. Therefore, we can assume that eruption is a complex process in which all these mechanisms may be involved to a greater or lesser extent

E. It is also not known how tooth eruption is regulated. It is known that it is a process with a complex regulation, because during eruption, tissue destruction and tissue formation have to be coordinated.

F. It has been shown that the eruption process may be regulated by *hormones*, for example:
   1. *Thyroxine*, produced by the *thyroid gland.*
   2. *Hydrocortisone*, produced by the zona fasciculata of the adrenal gland.

G. On the other hand, several molecules have been identified that may be involved in the regulation of the eruption, although the available data are still scarce:
   1. Epidermal growth factor (EGF).

2. Transforming growth factor β1 (TGF-β1).
3. Transforming Growth Factor α (TGF-α).
4. Bone morphogenetic protein 2 (BMP-2).
5. Colony-stimulating factor 1 (CSF-1).
6. Interleukin 1α (IL-1α).
7. Nuclear factor κB (NF κB).
8. C-Fos.
9. Monocyte chemotactic protein 1 (MCP-1).
10. Parathyroid hormone-related protein (PTHrP).

H. The process of tooth eruption involves *bone remodelling*, which requires the recruitment of *monocytes* from the blood to be activated and converted into osteoclasts. It is a complex process, with many molecules involved in its regulation, which in a simple way occurs as follows:
   1. It starts in the *tooth germ* with the production in the stellate reticulum of *IL-1α, TGF-β1* and *PTHrP.*
   2. These molecules induce cells in the *dental sac* to secrete *CSF-1, MCP-1, c-fos* and *NF κB.*
   3. This in turn stimulates the accumulation of *monocytes* in the dental follicle and their fusion to form *osteoclasts.*

22.3 Tooth Movements During Eruption

A. There are several basic movements whose combination produces tooth eruption. These movements are (Fig. 22.1):

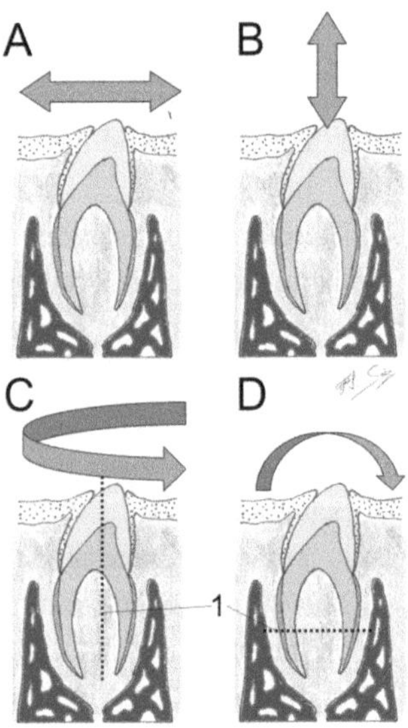

Fig. 22.1 Movements of a tooth. The arrows show the direction of the translational (**A**), axial or vertical (**B**), rotational (**C**) and tilt (**D**) movements. Where there is an axis of rotation, it is indicated by the dashed line (1). In the case of tilt movements (**D**), the axis of rotation is at the fulcrum

1. *Translational movement.* This is a horizontal movement of the tooth.
2. *Axial* or *vertical movement.* Generally towards the occlusal plane.
3. *Rotational movement.* Rotational movements around the vertical axis of the tooth.
4. *Tilt movement.* Movements around the *fulcrum.*

B. These movements can occur before eruption, during eruption and after eruption.

C. The most visible movements occur during eruption, but the tooth can continue to move throughout its life.

22.4 Tooth Eruption

A. To study tooth eruption, it can be divided into three stages (see Fig. 20.4 in Chap. 20):

B. *Preeruptive stage.*

1. When teeth begin to form in the embryo, they are small and have plenty of space available. Until eruption occurs, the tooth germs grow and need more space, but the jaws lengthen and the teeth shift to accommodate the developing bones.
2. The tooth germs are surrounded by the jawbone, which forms a *bony crypt.*
   a. In the primordia of *primary teeth*, the crown is partially surrounded by the bony crypt and separated from the *oral epithelium* by soft tissues (*connective tissue*).
   b. In the primordia of *permanent teeth*, the bony crypts completely surround the crown, except for a hole in the occlusal position called the *gubernacular canal* or *gubernaculum dentis*, which contacts the gingival lamina propria.
3. In permanent teeth, the preeruptive phase extends over many years and during this period the tooth germs undergo many movements (Fig. 22.2).
   (1) *Successional teeth:*
      (a) They are placed in a lingual position with respect to their predecessors and in the same bony crypt.
      (b) From this point onwards, they undergo movement, and there are relative changes in position also due to jawbone growth.
      (c) At the end of this phase:
         1. The *incisors* and *canines* are positioned lingually in relation to the primary teeth, in their own bony cripts.
         2. The *premolars*, which are initially located close to the occlusal area of the primary tooth, are placed under the primary molars, between their roots, which are divergent in shape.
   (2) *Accessional teeth:*
      (a) They are formed from the *accessional lamina*, which develops as the growth of the jaw allows.
      (b) When these teeth begin to form, the lack of space causes them to be tilted.
      (c) Over time, jaw growth allows the tooth to adopt a vertical position.

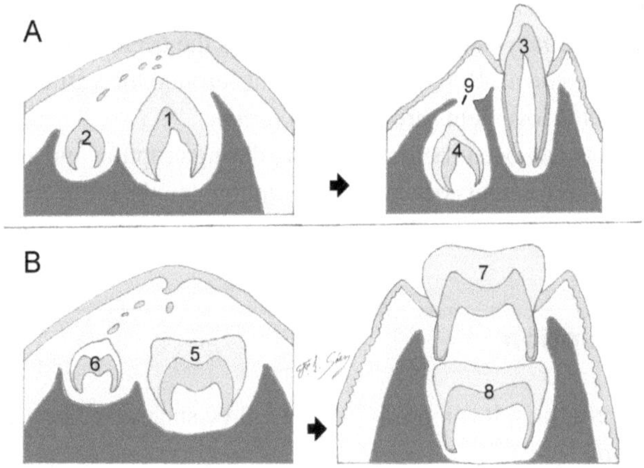

Fig. 22.2 Relative displacement movements between a deciduous incisor and the permanent incisor (**A**) and between a deciduous molar and a permanent premolar (**B**). (1) germ of the deciduous incisor; (2) germ of the permanent incisor; (3) deciduous incisor; (4) permanent incisor prepared for eruption; (5) germ of the deciduous molar; (6) germ of the permanent premolar; (7) deciduous molar with diverging roots; (8) permanent premolar prepared for eruption; (9) *gubernaculum dentis*. The arrows indicate the progress of the developmental process

4. Bony cripts are modified during this phase to adapt to changes in tooth position. It is not known whether bone remodelling occurs as an adaptation to tooth movements, or whether bone remodelling is responsible for tooth movements.

C. *Prefunctional eruptive stage*.
   1. This stage is characterised by:
      a. Root formation.
      b. The destruction of tissues above the tooth (the enamel organ and the bony cript).
      c. The formation of the insertion periodontium (cementum, periodontal ligament and alveolar bone).
   2. Root formation occurs from *Hertwig's epithelial sheath* (the process is discussed in the chapter on odontogenesis).
      a. This sheath has an *inductive function* (it induces the formation of odontoblasts) and a *modelling function*, as it determines the shape of the root.
      b. The growth of the root is accompanied by an increase in the length of the tooth and the consequent displacement of the crown towards the surface of the oral mucosa.
      c. However, although it may seem obvious, it is not clear that root growth causes tooth eruption.
   3. During tooth eruption, the tissues above the tooth are modified.
      a. The connective tissue of the *gobernaculum dentis* has fibres that appear to be directed towards the mucosa. The bone surrounding the canal is resorbed to allow the passage of the dental crown, forming an eruption pathway, in which

macrophages appear and release substances that degrade the connective tissue (*desmolysis*).

b. The enamel organ becomes the *reduced enamel epithelium.*
   (1) Cells of the upper part of the reduced enamel epithelium proliferate towards the connective tissue.
   (2) These cells are thought to secrete enzymes that perform *desmolysis*, allowing the reduced enamel epithelium to move closer to the oral epithelium.

c. The connective *fibroblasts* located between the tooth and the oral epithelium stop synthesising collagen fibres, synthesise *hydrolases* and resorb degraded materials. Finally, their nuclei become pyknotic and die.

d. Eventually, the fusion of the reduced and oral epithelia takes place, forming the *protective periodontium.*
   (1) This is followed by *ischaemic necrosis* in the central area of the fused region. The death of these cells produces a hole through which the tooth begins to erupt.
   (2) The fused zone of the reduced enamel epithelium and oral epithelium, with the underlying lamina propria, forms the *dentogingival junction.*

4. Another process that takes place is the formation of the *insertion periodontium.*
   a. *Cementoblasts*, *fibroblasts* and *osteoblasts* develop in the dental follicle and produce cementum, periodontal ligament and alveolar bone, respectively.
   b. The formation of the periodontal ligament is complex:
      (1) First, a connective tissue with fibrils without a defined orientation is produced, which is called *periodontal membrane.*
      (2) Then the *principal fibre* groups are formed. The first principal fibres appear when the root has not yet begun to develop and form fibres of the cervical area.
      (3) As the root grows, new bundles of principal fibres emerge from the root surface.
      (4) When the tooth makes occlusion, these fibre bundles acquire the definitive organisation of the adult periodontium.
      (5) At the apical end of the root, the so-called *cushion hammock ligament* or *pulp-limiting membrane* is formed by intertwined fibres of the dental papilla, the dental sac and the forming periodontium. This cushion hammock ligament lies in a plane on which the growing root is supported.
      (6) The fibres of the periodontal ligament are constantly being remodelled, which may contribute to tooth traction.

D. *Functional eruptive* or *posteruptive stage.*
   1. It is a very long phase, since it includes all movements from occlusion to tooth loss.
   2. These are slower movements that can have three functions:
      a. *Accommodation of mandibular growth.*
         (1) These movements occur until the end of growth into adulthood.

(2) In order to adapt to the growth of the jaw, the alveolar bones and the tooth apex are remodelled.

(3) They are more frequent in young individuals, aged 14–18 years, and occur earlier in females than in males.

b. *Compensation of occlusal attrition.*

(1) Throughout its life, the tooth wears down on its *occlusal surface* due to friction with the teeth it encounters.

(2) To compensate for the loss of enamel on the occlusal surface, the tooth undergoes *axial movements* that lift it out of its socket.

(3) These movements are accompanied by *cement apposition* at the root to compensate for the loss of tooth material in the occlusal or proximal plane.

c. *Compensation of interproximal attrition.*

(1) Teeth suffer wear at the points of contact with neighbouring teeth on their proximal sides.

(2) Normally, this wear is compensated by the *mesial drift* of the teeth, which causes each tooth to move closer to its anterior neighbour.

(3) These movements are important during orthodontic treatment.

3. Several mechanisms have been proposed that could be responsible for the displacements that cause *mesial drift* of the teeth:
   a. The teeth are usually inclined in the mesial direction, and this causes a force in that direction.
   b. Dental occlusion causes forces directed forward, which would push the teeth together.
   c. The fibres of the *gingival ligament*, especially the *transeptal fibres* that run from one tooth to the neighbouring tooth, exert an attractive force on the teeth.
   d. The tongue and cheeks press and push against the teeth.
   e. Alveolar bone is deposited in the distal part of the tooth sockets, which pushes the tooth in a mesial direction.
4. There are data both supporting and contradicting each of these mechanisms. Mesial drift is likely to be multifactorial in origin.

22.5 Replacement of Primary Dentition

A. The eruption of successional teeth involves the loss and replacement of their corresponding primary teeth. Primary teeth are lost through a process called *exfoliation* (from the Latin exfoliāre 'to defoliate') or *shedding*.

B. For the tooth to fall out, resorption of its root and the periodontium that supports it must first occur. The loss of the root is called *rhizoclasia* (from the Greek terms ῥιζο rizo, 'root' and κλάστης klástēs, 'breaking').

1. Rhizoclasia is caused by pressure from the growing permanent tooth.
2. It is not a continuous process, as there may be alternating periods of activity and rest.

C. Exfoliation occurs almost symmetrically on the right and left sides of the mouth and usually the mandibular teeth are lost earlier than the maxillary teeth, with the exception of the four second molars, which are lost simultaneously.
D. In women, exfoliation tends to occur earlier than in men.
E. There are differences in the exfoliation of different types of teeth:
   1. Permanent *incisors* and *canines* erupt in a lingual position relative to their corresponding deciduous teeth, resulting in root destruction, but much of the pulp chamber of the primary tooth remains intact.
   2. Permanent *premolars* develop between the roots of primary molars, and the process of exfoliation of the roots of primary molars also leads to the destruction of interradicular dentine, part of the dental pulp and sometimes part of the crown dentine and enamel.
F. Rhizoclasia and exfoliation may be stimulated by the pressure exerted by the successional tooth.
   1. In fact, if for some reason a replacement tooth has not developed, rhizoclasia of the deciduous tooth does not occur.
   2. Once rhizoclasia begins and the primary tooth starts to lose attachment, the pressure produced by *mastication* can increase the resorption of the roots.
G. Resorption is carried out by *odontoclasts* or *cementoclasts*.
   1. They are cells apparently identical to osteoclasts and are produced, like them, by the fusion of *monocytes*. They therefore belong to the *mononuclear phagocytic system*.
   2. Odontoclasts are multinucleated, with a highly vacuolated cytoplasm. These cells attach to the surface of cementum and dentine and degrade these materials. On the surface contacting the dental tissues, the plasma membrane shows numerous invaginations forming a *ruffled border*. This structure is typical of osteoclasts.
H. During exfoliation, the periodontium is also disorganised.
   1. The synthesis of collagen and other extracellular matrix molecules is halted.
   2. Fibres embedded in the bone and cementum (*Sharpey's fibres* and *perforating fibres*) are released. Fibre loss occurs quickly.
   3. Fibroblasts die, some probably in response to tissue damage, others by *apoptotic* mechanisms.
I. There is a decrease in the *insertion epithelium*, which causes an increase in the mobility of the tooth and makes it easier for it to fall out.
J. Changes also occur in the *dental pulp*, which atrophies. This process is painless, but it is not well understood how it occurs.
K. The final consequence is that the tooth is loosened by degradation of the root and all periodontal tissues. Thus, as the primary tooth exfoliates, the way is clear for the eruption of the replacement tooth.

Take-Home Message

Humans are diphyodont in that they have a primary dentition which is replaced by a permanent dentition, consisting of successional teeth and accessional teeth. Four mechanisms have been proposed to explain the migration of the erupting tooth: root growth, alveolar bone development, periodontal connective tissue pressure and connective tissue traction. Teeth can have translational, axial, rotational and tilt movements. Throughout the life of the tooth, there are three stages related to eruption: preeruptive, prefunctional eruptive and functional or posteruptive. During the replacement of the primary dentition, rhizoclasia or resorption of the primary tooth root occurs, which allows the tooth to exfoliate or fall out to facilitate the eruption of the permanent tooth.

Further Reading

Berkovitz BKB, Holland GR, Moxham BJ (2018) Development of the dentitions. In: Oral anatomy, histology & embryology, 5th edn. Elsevier, pp 421–441

Chiego DJ (2023) Eruption and shedding of the teeth. In: Essentials of oral histology and embryology, 6th edn. Elsevier, pp 75–87

Fehrenbach MJ, Popowics T (2020) Tooth development and eruption. In: Illustrated dental embryology, histology, and anatomy, 5th edn. Elsevier, pp 52–76

Gómez de Ferraris ME, Campos Muñoz A (2009) Erupción dentaria. In: Histología, Embriología e Ingeniería Tisular Bucodental, 3rd edn. Médica Panamericana, pp 393–410

Hand AR (2014) Tooth eruption and shedding. In: Hand AR, Frank ME (eds) Fundamentals of oral histology and physiology. Wiley Blackwell, pp 136–148

Harakopakis-Hajishengallis E (2007) Physiologic root resortion in primary teeth: molecular and histological events. J Oral Sci 49:1–12

Jacobson A (1983) The physiology of tooth eruption. Birth Deffects-Original Article Series 19:67–82

Kardos TB (1996a) The mechanism of tooth eruption. Br Dental J 181:91–95

Kardos TB (1996b) The mechanism of tooth eruption. Br Dent J 181:91–95

King R (1997) Mechanism of tooth eruption. Br Dent J 182:49–49

Kirilov B (2023) Complex view of the eruption of the permanent teeth. J IMAB 29:5178–5181

Makino Y, Fujikawa K, Matsuki-Fukushima M et al (2021) Role of innate inflammation in the regulation of tissue remodeling during tooth eruption. Dent J 9:7

Marks SC (1987) Tooth eruption—the regulation of a localized, bilaterally symmetrical metabolic event in alveolar bone. Scanning Microsc 1:1125–1133

Marks SC, Schoeder HE (1996) Tooth eruption: theories and facts. Anat Rec 245:374–393

Marks SC, Schroeder HE (1996) Tooth eruption: theories and facts. Anat Rec 245:374–393

Nanci A (2013) Physiologic tooth movement: eruption and shedding. In: Ten Cate's oral histology, 8th edn. Elsevier, pp 233–252

Sasak T, Motegi N, Suzuki H et al (1988) Dentin resorption mediated by odontoclasts in physiological root resorption of human deciduous teeth. Am J Anat 183:303–315

Suda N (2022) Role of reduced enamel epithelium in root resorption. J Oral Biosci 64:43–48

Wang XP (2013) Tooth eruption without roots. J Dent Res 92:212–214

Warshawsky H (1988) The teeth. In: Weiss L (ed) Cell and tissue biology—a textbook of histology, 6th edn. Urban & Schwarzenberg, pp 594–640
Wise GE (1998) The biology of tooth eruption. J Dent Res 77:1576–1579
Wise GE (2009) Cellular and molecular basis of tooth eruption. Orthod Craniofac Res 12:67–73
Wise GE, King GJ (2008) Mechanisms of tooth eruption and orthodontic tooth movement. J Dent Res 87:414–434
Wise GE, Frazier-Browers S, D'Souza RN (2002) Cellular, molecular, and genetic determinants of tooth eruption. Crit Rev Oral Biol Med 13:323–334
Yamaguchi T, Hosomichi K, Shirota T et al (2022) Primary failure of tooth eruption: etiology and management. Jpn Dental Sci Rev 58:258–267
Yao EH, Du JH, Jiang XQ (2024) Tooth root development and homeostasis during eruptive and post-eruptive movement. Chin J Dent Res 29:273–289
Yu Y, Cui C, Guan SY et al (2021) Function of orofacial stem cells in tooth eruption: an evolving perspetiva. Chin J Dental Res 24:143–152

Primary Teeth

23

What You Will Learn in This Chapter
The characteristics of teeth that we have seen in the previous chapters refer mainly to permanent teeth. Although deciduous teeth are similar to permanent teeth, in this chapter, you will study some differential features.

23.1 Human Dentitions

A. Humans are *diphyodont* animals because they have a *primary dentition* and a *permanent dentition*, which replaces the primary dentition.
   1. *Primary, deciduous* or *milk teeth* appear during early childhood, usually from the age of six months to three years.
   2. From the age of six, the primary teeth gradually *exfoliate (tooth shedding)* and are replaced by *secondary* or *permanent teeth.*

B. Primary teeth differ from their permanent counterparts in several respects (Fig. 23.1), which are detailed in the following sections. Some differences influence the decision for dental treatment.

23.2 Size and Shape of Teeth

A. The primary teeth are smaller than their corresponding teeth in the permanent dentition. This is especially evident in the vertical direction (occlusal-apical axis).
   1. *Crown*: The primary teeth have smaller, lower and more rounded crowns. The incisal edges of the incisors have a *fleur-de-lis shape.*
   2. *Root*: In general, the roots are shorter.
      a. The root of uniradicular teeth is ribbon-shaped, whereas in permanent teeth, they are tubular.

F. J. Sáez, I. Badiola, *Essential Oral Histology*, Learning Materials in Biosciences,
https://doi.org/10.1007/978-3-032-03165-5_23

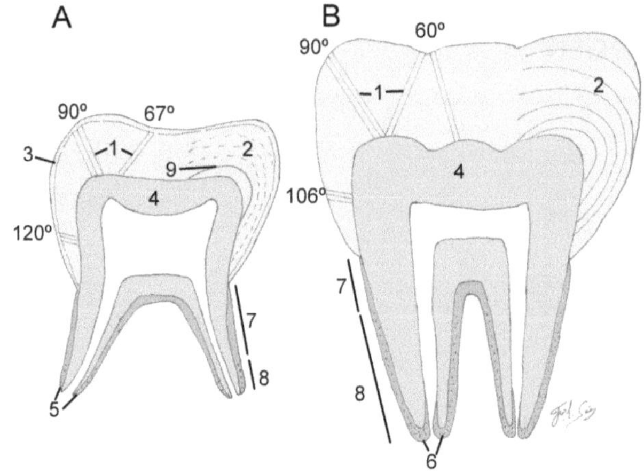

Fig. 23.1 Main differences between a primary tooth (**A**) and a permanent tooth (**B**). (1) enamel rods; (2) Striae of Retzius; (3) aprismatic enamel; (4) dentine; (5) cementum not deposited in the apical foramen; (6) cementum deposited in the apical foramen; (7) acellular cementum; (8) cellular cementum; (9) neonatal enamel line. The angles of the enamel rods (BSUEs) to the surface are indicated

b. The roots of the molars are *divergent*, bifurcate near the cervical region and curve to accommodate the primordia of the permanent premolars. This makes mechanical access to these areas difficult for the dentist.
c. In addition, irregular growth of secondary dentine can occur, forming bridges that make endodontic treatment difficult.

B. Shape of teeth:
   1. Deciduous incisors and canines are similar in shape to the permanent incisors and canines that replace them, respectively, except for the differences already noted.
   2. Deciduous molars differ from the permanent premolars that replace them. Deciduous molars have crowns with a larger *mesiodistal diameter* than those of permanent premolars.

C. The *pulp cavities* are relatively large when compared to those of permanent teeth.
   1. They have well-developed *pulp horns* and wide *root canals.*
   2. *Accessory canals* originate in the bi- or trifurcation zone of the roots, while in permanent teeth, they occur in the apical third.

23.3 Development

A. Primary teeth develop faster than permanent teeth. In addition, their life is shorter and ends with *rhizoclasia* followed by *exfoliation.*

B. The *mineralisation* process begins before birth:
   1. For this reason, enamel and dentine usually have a very marked *neonatal incremental line* that occurs at birth.

2. This implies that primary teeth can be affected by *prenatal systemic disturbances*. In contrast, permanent teeth are more affected by postnatal anomalies.

23.4 Physical and Chemical Properties

A. Most of the *hard tissues* of the tooth, such as enamel and dentine, are thinner in deciduous teeth than in their corresponding permanent teeth.
   1. This is probably why enamel and dentine are more *permeable* in primary teeth than in permanent teeth.
   2. However, in some cases, the dentine is less permeable because it has fewer *dentinal tubules*.
B. Moreover, the composition of the mineral tissues of the tooth is similar to that of the same tissues in permanent teeth.
   1. In general, it must be assumed that the *degree of mineralisation* is somewhat lower.
   2. However, there are contradictory studies, with some detecting lower levels of certain minerals, while other studies report similar levels of the same elements.
   3. There appear to be different levels of *trace elements*. Thus, in the enamel:
      a. There is less Beryllium (Be), Lithium (Li) and Strontium (Sr).
      b. There is more Copper (Cu), Vanadium (V) and Cadmium (Cd).
C. Despite having a similar composition, both enamel and dentine are usually somewhat less hard tissues in deciduous teeth than in permanent teeth. This makes the primary tooth less opaque to *X-rays*.
D. Milk teeth and permanent teeth also differ in *colour*. The primary tooth is bluish-white or greyish-white, while the permanent tooth is usually yellowish-white.

23.5 Enamel

A. The histological organisation of the enamel is the same as that of permanent teeth.
B. However, we can point out the following peculiarities:
   1. There is usually *aprismatic enamel* over the entire crown surface. As a result, the enamel rods never reach the tooth surface.
   2. The *BSUEs* or *enamel rods* have a water drop shape in cross-section, as opposed to the typical old keyhole shape.
   3. The orientation of the BSUEs is different from that of the rods of the permanent teeth in each crown region:
      a. Deep part of grooves and pits of the occlusal surfaces. Acute angles of 67–70^0 are produced here, compared to angles of 60^0 in the BSUEs of permanent teeth.
      b. Cusps. Here, both primary and permanent teeth show BSUEs with right angles (90^0).
      c. Gingival third. In this region, the BSUEs form obtuse angles of 120^0, in contrast to permanent teeth, which have angles of 106^0.

4. There are no prenatal *striae of Retzius*, as the enamel produced in this period is very homogeneous. The large, dark, *neonatal line* separates the prenatal enamel from the postnatal enamel, which does have weak striae of Retzius.
5. Due to the limited development of striae of Retzius and the presence of superficial aprismatic enamel, no perichymata or imbrication lines are observed. *Perikymata* are only visible on the cervical enamel of the second deciduous molar.
6. *Lamellae*. They are abundant and can reach the pulp cavity, facilitating the entry of bacteria.
7. *Enamel spindles* and *penetrating dentinal tubules*. They are found in greater proportion on the cusp side. This does not seem to have an impact on tooth sensitivity, as the development of the nerve plexus is less in primary teeth.
8. There are *Hunter–Schreger bands*. In anterior teeth, they are on the incisal surfaces, in molars, they are in the middle and cervical third.

23.6 Dentin

A. Dentine has a basic histological organisation similar to that of permanent teeth.
B. However, there are some specific differences or characteristics of primary teeth:
   1. *Dentinal tubules*.
      a. Their size and number are similar to that of permanent teeth, but it seems that, at least in molars, there are fewer tubules, with lower density.
      b. The dentinal tubules branch dichotomously at the terminal portion.
      c. There are *giant dentinal tubules* with type I and III collagen fibres, but without odontoblast processes. The type III collagen fibres appear to be derived from von Korff's fibres.
      d. In the cervical region, there may be rectilinear tubules, while in permanent teeth, they are curved.
   2. *Interglobular dentine* or *spaces of Czermack* are scarce or absent.
   3. Structural defects such as fissures or incomplete cavities from the pulp chamber (*recession lines*) may be present in multi-rooted teeth.

23.7 Dental Pulp

A. The dental pulp is relatively abundant, as the pulp cavity is large considering the smaller size of the tooth.
B. Considering that the life of a primary tooth is shorter than that of a permanent tooth, the dental pulp does not reach such a developed histological organisation as it does in permanent teeth.
C. During root development, the pulp is very cellular, but then the number of cells decreases and the concentration of fibres increases until exfoliation.

D. In general, the histological areas observed in the pulp are the same as in a permanent tooth, but they are not always distinguishable:
   1. Odontoblasts form a *pseudostratified* structure due to the varied arrangement of their nuclei. The pseudostratified appearance is greater in the crown, with 5 to 7 strata, than in the root, with only 1 or 2 strata.
   2. *Oligocellular* or *basal zone of Weil* is not well visible.
   3. The *cell-rich zone* is observed only in the coronary zone and does not form a continuous layer. When a tooth forms tertiary dentine, it may eventually disappear.

E. *Arteriovenous anastomoses* are very common in the root pulp.

F. The crown is more vascularised in the middle region, but the rest of the pulp is equally *vascularised*.

G. The degree of *innervation* is less than in a permanent tooth:
   1. The cervical region is quite well innervated (never as in a permanent tooth).
   2. Nerve endings are more abundant in the crown than in the root.
   3. It seems that primary teeth are less sensitive to pain than permanent teeth.

23.8 Cementum

A. Some of the characteristics of cementum in primary teeth are:
   1. The cementum only covers the outside of the root and never enters the apical foramen.
   2. At the *cementum-enamel junction,* overlapping and contact can be observed, but never separation (see types of cementum-enamel junction in the chapter on enamel). Therefore, dentine is never exposed to the surrounding environment.
   3. There is *cellular cementum* and *acellular cementum*. The latter predominates in the upper two-thirds of the root.

23.9 Periodontal Ligament

A. The periodontal ligament of deciduous teeth is wider than that of permanent teeth.

Take-Home Message

Primary teeth have some characteristics that are different from those of permanent teeth. They are generally smaller, often have a marked neonatal incremental line, can be severely affected by prenatal systemic disorders and their molars have divergent roots. Their tissues show a similar degree of mineralisation as permanent teeth. The surface usually shows aprismatic enamel. The enamel rods often show different degrees of inclination than in permanent teeth. The dentine may have giant tubules and little interglobular dentine. The dental pulp does not always show the characteristic zonation of permanent teeth and is less sensitive.

Further Reading

Agematsu H, Imai R, Sakai T et al (1988) A scanning electron microscopic study on secondary dentin in deciduous teeth. Shika Gakuho Dent Sci Rep 88:1333–1345

Angker L, Swain MV, Kilpatrick N (2003) Mirco-mechanical characterization of the properties of primary tooth dentine. J Dent 31:261–267

Degen TC, Lang NP, Bosshardt DD (2002) Immunocytochemical characterization of the dentin-cementum interface in deciduous teeth. J Dent Res 81:A440–A440

Fehrenbach MJ, Popowics T (2020) Primary dentition. In: Illustrated dental embryology, histology, and anatomy, 5th edn. Elsevier, pp 265–274

Gómez de Ferraris ME, Campos Muñoz A (2009) Dientes primarios. In: Histología, Embriología e Ingeniería Tisular Bucodental, 3rd edn. Médica Panamericana, pp 411–425

Hargreave JA, Weatherell JA (1964) Fluoride content of dentin + enamel in human deciduous teeth. J Dent Res 43:959

Harris EF, Lease LR (2005) Mesiodistal tooth crown dimensions of the primary dentition: a worldwide survey. Am J Phys Anthropol 128:593–607

Holbrook IB, Leaver AG (1976) Glycoproteins in dentin of human deciduous teeth. Arch Oral Biol 21:509–512

Isokawa S, Kosakai T, Kajiyama S (1963) Interglobular dentin in deciduous tooth. J Dent Res 42:831

Lynch RJM (2013) The primary and mixed dentition, post-eruptive enamel maturation and dental caries: a review. Int Dent J 63:3–13

Mahoney P (2012) Incremental enamel development in modern human deciduous anterior teeth. Am J Phys Anthropol 147:637–651

Mahoney E, Holt A, Swain M, Kilpatrick N (2000) The hardness and modulus of elasticity of primary molar teeth: an ultra-micro-indentation study. J Dent 28:589–594

Nina S (2012) Enamel of primary teeth—morphological and chemical aspects. Swed Dent J Suppl 222:1–77

Rodd HD, Boissonade FM (2002) Comparative immunohistochemical analysis of the peptidergic innervation of human primary and permanent tooth pulp. Arch Oral Biol 47:375–385

Rodd HD, Boissonade FM (2005) Vascular status in human primary and permanent teeth in health and disease. Eur J Oral Sci 113:128–134

Sabel N (2012) Enamel of primary teeth—morphological and chemical aspects. Swed Dent J 222(Supplement):1–77

Salanitri S, Seow WK (2013) Developmental enamel defects in the primary dentition: aetiology and clinical management. Aust Dent J 58:133–140

Sisca RF, Provenza DV (1972) Initial dentin formation in human deciduous teeth—electron-microscope study. Calcif Tissue Res 9:1–16

Soskolne AW, Bimstein E (1989) Apical migration of the junctional epithelium in the human primary dentition as a multifactorial phenomenon. J Periodont 13:239–242

Sumikawa DA, Strawn SE, Watanabe LG et al (1996) Microstructure of dentin in deciduous teeth. J Dent Res 75:1432–1432

Sumikawa DA, Marshall GW, Gee L, Marshall SJ (1999) Microstructure of primary tooth dentin. Pediatr Dent 21:439–444

Vrbic V, Stupar J, Byrne AR (1987) Trace elements content of primary and permanent tooth enamel. Caries Res 21:37–39

Index

F. J. Sáez, I. Badiola, *Essential Oral Histology*, Learning Materials in Biosciences,
https://doi.org/10.1007/978-3-032-03165-5

C

I

O

U

MIX
Papier aus verantwortungsvollen Quellen
Paper from responsible sources
FSC® C105338

If you have any concerns about our products,
you can contact us on
ProductSafety@springernature.com

In case Publisher is established outside the EU,
the EU authorized representative is:
Springer Nature Customer Service Center GmbH
Europaplatz 3, 69115 Heidelberg, Germany

Printed by Libri Plureos GmbH
in Hamburg, Germany